GALACTIC ASTROPHYSICS AND GAMMA-RAY ASTRONOMY

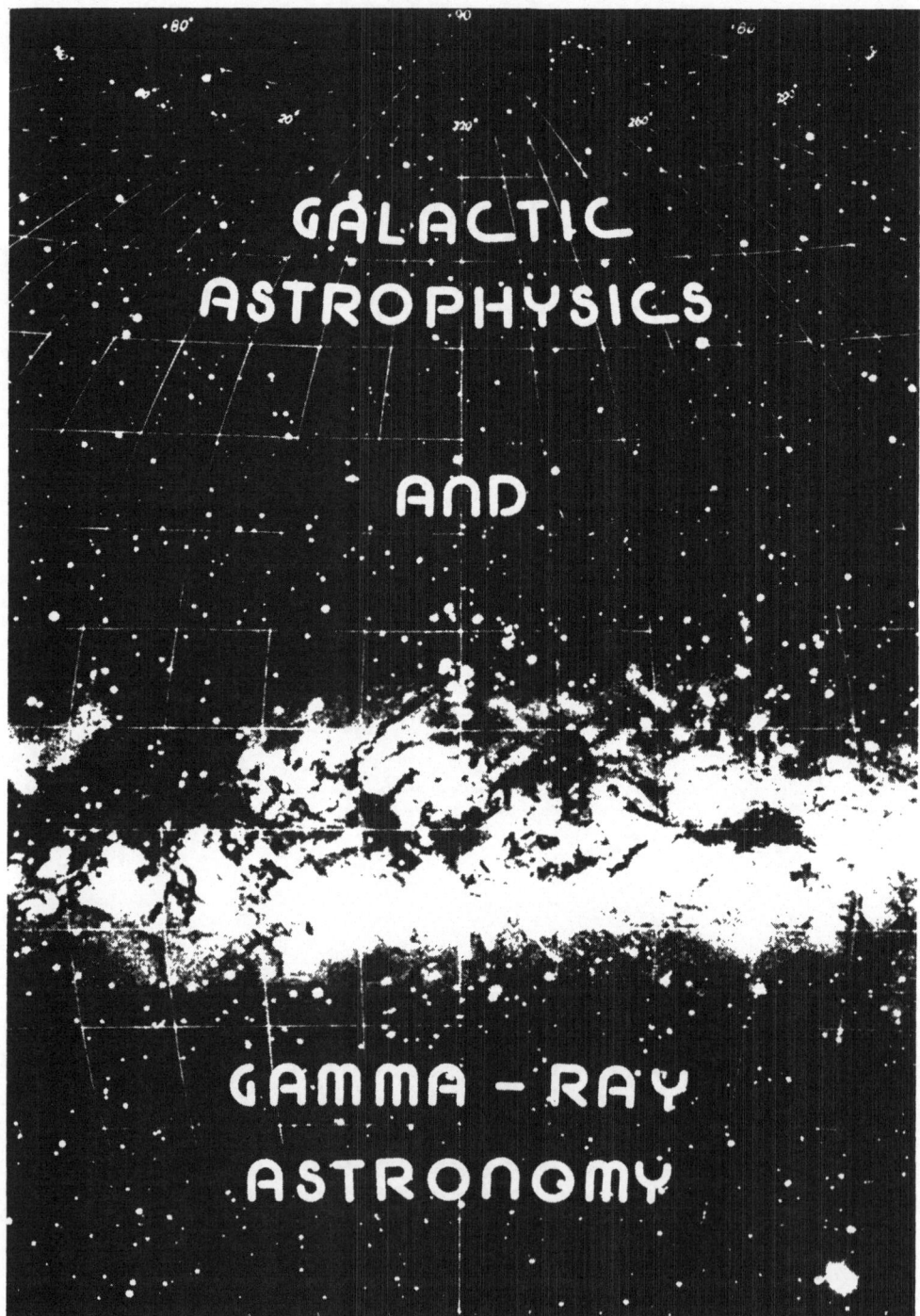

Panorama of the Milky Way.
© 1955 by Astronomical Observatory, Lund, Sweden.
Artwork: Alexandra Waskala.

Galactic Astrophysics
and
Gamma-Ray Astronomy

*Proceedings of a Meeting Organised in the Context of the XVIII General
Assembly of the IAU, held in Patras, Greece, August 19, 1982*

Edited by

G. E. MORFILL

Max-Planck-Institut für Astrophysik, Institut für Extraterrestrische Physik, Garching b. München, F.R.G.

and

R. BUCCHERI

Istituto di Fisica Cosmica e Informatica del Consiglio Nazionale delle Ricerche, Palermo, Italy

Reprinted from

Space Science Reviews, Vol. 36, Nos. 1, 2, and 3

D. Reidel Publishing Company

Dordrecht / Boston

ISBN-13: 978-94-009-7210-0   e-ISBN-13: 978-94-009-7208-7
DOI: 10.1007/ 978-94-009-7208-7

TABLE OF CONTENTS

Galactic Astrophysics and Gamma-Ray Astronomy

PREFACE

This book contains the invited and contributed lectures presented at a meeting organised in the context of the XVIII general assembly of the IAU, held in Patras, August 19, 1982. Roughly one hundred scientists attended this meeting, the discussions were lively – sometimes heated – and the original time span allocated to the meeting was as a result, comfortably exceeded by about 50%.

The aim of this meeting was to determine the role of galactic gamma-ray astronomy within the general concept of galactic astrophysics. The timing, at the end of the COS-B mission, was regarded as opportune, because it gives interested astrophysicists the possibility for interdisciplinary studies using the existing gamma-ray data base (e.g. comparison with infrared, radio, X-ray, etc. astronomies), as well as for theoretical studies. The next generation of gamma-ray detectors will probably not be in operation for another 5 to 10 years, and therefore it is hoped that the proceeding of this meeting can be used (in the intermediate time) as a basis for further studies, as a stimulation for more theoretical work and as an important contribution for defining the aims and operation of future gamma-ray missions. The interrelationship with other branches of astronomy, the astrophysical implications and the study of relevant physical processes using available measurements in the near-Earth environment were important results of the meeting.

Many persons contributed to the success of the meeting, in particular all those who either presented a contribution and/or took part in the discussions.

We wish to thank Profs. Pacini, Scarsi, and Trümper for their support, and the local organisers for their part in arranging this meeting.

We have made a special effort to obtain an up-to-date status of the subject, and wish to thank Prof. de Jager and *Space Science Reviews* for their help in obtaining rapid publication.

G. E. MORFILL
R. BUCCHERI

COSMIC-RAY ACCELERATION AND TRANSPORT, AND DIFFUSE GALACTIC γ-RAY EMISSION*

HEINRICH J. VÖLK

Max-Planck-Institut für Kernphysik, Postfach 10 39 80, 6900 Heidelberg, F.R.G.

Abstract. Cosmic-ray acceleration and transport is considered from the point of view of application to diffuse galactic γ-ray sources. As an introduction we review several source models, in particular supernovae exploding inside or near large interstellar clouds. The complex problem of cosmic ray transport in random electromagnetic fields is reduced to three cases which should be sufficient for practical purposes. As far as diffusive acceleration is concerned, apart from reviewing the basic physical principles, we point out the relation between shock acceleration and 2nd order Fermi acceleration, and the relative importance of the two processes around interstellar shock waves. For γ-ray source models the interaction of cosmic rays with dense clouds assumes great importance. Past discussions had been confined to static interactions of clouds with the ambient medium in the sense that no large scale mass motions in the ambient interstellar medium were considered. The well-known result then is that down to some tens of MeV or less, cosmic-ray nucleons should freely penetrate molecular clouds of typical masses and sizes. The self-exclusion of very low energy nucleons however may affect electron transport with consequences for the Bremsstrahlung γ-luminosity of such clouds.

In this paper we consider also the dynamical interaction of dense clouds with a surrounding hot interstellar medium. Through cloud evaporation and accretion there exist mass flows in the cloud surroundings. We argue that in the case of (small) cloud evaporation the galactic cosmic rays will be essentially excluded from the clouds. The dynamic effects of cosmic rays on the flow should be minor in this case. For the opposite case of gas accretion onto (large) clouds, cosmic-ray effects on the flow will in general be large, limiting the cosmic-ray compression inside the cloud to dynamic pressure equilibrium. This should have a number of interesting and new consequences for γ-ray astronomy. A first, qualitative discussion is given in the last section.

1. Introduction

The observed high energy γ-ray emission ($E_\gamma \gtrsim 30$ MeV) from the Galaxy is due to the interaction of high-energy particles with diffuse interstellar matter or photons, and due to compact as well as extragalactic sources. Two γ-ray pulsars have been identified as compact sources (Bennett *et al.*, 1977); another source is presumably identical with the quasar 3C 273 (Swanenburg *et al.*, 1978).

While the overall contribution of compact and extragalactic sources is not very well known, it is clear that the assumed presence of cosmic rays in the Galaxy *must* lead to γ-ray production (Hayakawa, 1952; Hutchinson, 1952; Morrison, 1958). The intensity, spectrum, and spatial variations can be used as tracers of the interstellar gas density and the energetic particle flux. This ignores inverse Compton interactions but is probably justified for the galatic disk, except perhaps its central regions. Since the galaxy is optically thin, the observable diffuse γ-ray emission is proportional to the line-of-sight-

* Proceedings of the XVIII General Assembly of the IAU: *Galactic Astrophysics and Gamma-Ray Astronomy*, held at Patras, Greece, 19 August 1982.

integral of the product of cosmic-ray intensity and mass density over the instrument's solid angle. Bremsstrahlung quanta come from electrons of comparable and higher energies, whereas π°-decay quanta are due to nucleons of energies above several GeV/nucleon (e.g. Stecker, 1971). Thus, diffuse γ-ray flux variations in certain viewing directions can be due to angular variations in gas density, or in cosmic-ray intensities, or in both, along the line of sight. The observed energy spectrum, although only crudely resolved, appears characterised by π°-decay plus a very sizeable bremsstrahlung contribution (e.g. Lebrun et al., 1982).

To the extent that the cosmic-ray intensity is assumed to be uniform, one can use the observed γ-ray fluxes as a consistency check against the radio-astronomical measurements of the gas density, in particular the density of molecular hydrogen which in dense clouds can only be indirectly determined via the radiation from rotational lines of CO (e.g. Scoville and Solomon, 1975; Gordon and Burton, 1976; Solomon et al., 1979). Apart from learning about compact objects at γ-ray frequencies (e.g. Salvati and Massaro, 1978), the really new and unique prospect of high energy galactic γ-ray astronomy however is the opposite: the determination of the galactic cosmic-ray intensity and its variations in different regions of the Galaxy beyond what is known about electrons from non-thermal radio observations.

Broadly speaking, cosmic-ray intensity variations arise through interactions of energetic particles with low frequency electromagnetic fields. These fields are generated and supported jointly by the interstellar plasma and the cosmic rays themselves. They give rise to particle transport, acceleration and deceleration. Since γ-rays are often produced near regions of strongly enhanced gas density (clouds) where energetic particles lose energy by ionizing or nuclear collisions, and by Bremsstrahlung (electrons), energy loss plays an important role for the relevant cosmic-ray dynamics.

In this paper we shall attempt to review these physical processes. We shall do this in the context of galactic dynamics and shall give examples of how astrophysical phenomena in the interstellar medium affect energetic particles and are in turn modified by their presence. However we do not intend to discuss cosmic rays and their confinement on an overall galactic scale per se (for more recent reviews, see Wentzel, 1974; and in particular Cesarsky, 1980). Instead we consider cosmic-rays as they lead to variations in γ-ray emission. Thus, in the next section we discuss a few examples of possible diffuse γ-ray sources. The relevant physical processes like convective vs diffusive transport and adiabatic energy changes, as well as diffusive shock acceleration are reviewed in Section 3. Cosmic-ray penetration of big molecular clouds and self-confinement are the subject of Section 4. In Section 5 we turn to the question of the dynamical interaction between cold gas clouds and a hot surrounding gas in the presence of cosmic-rays, as it arises in an interstellar medium regulated by supernova explosions. This is a new aspect of cosmic-ray dynamics which has not been discussed up to now and the investigation is not fully completed in all its mathematical detail. Therefore here we shall only discuss the essential physics of this interaction and its qualitative effects on cosmic-rays and the galactic γ-ray emission. Detailed quantitative results will be published elsewhere.

2. Diffuse γ-Ray Sources

In order to differentiate between diffuse γ-ray sources on an astrophysical basis we can introduce the distinction of correlated and uncorrelated sources. For example a sufficiently massive interstellar cloud, simply bathed in the average galactic cosmic-ray intensity, may lead to an uncorrelated diffuse γ-ray source if it is indeed penetrated by the external cosmic rays as we tacitly assume for the average interstellar medium.* At least the γ-ray producing nucleons should penetrate typical interstellar clouds (see Section 4). On this premise already now a group of standard γ-ray sources exists like the Orion molecular cloud that stands out of the background simply because it is massive and close enough to be resolved by the COS-B spark chamber. Many more such sources should be detectable with instruments of higher sensitivity and resolution. If from now on we only consider γ-ray sources with an enhancement over and above the standard intensity, then we eliminate all effects due to the average cosmic-ray intensity assumed to equal that observed near the solar system. Such sources then require a local cosmic-ray enhancement. Again they can be uncorrelated if the cloud happens to be inside or sufficiently near an acceleration region like a supernova remnant (SNR) where the progenitor star has no generic relation to the cloud. Thus, for example, the enhancement of the γ-ray luminosity of the cloud ρ-Oph by a factor of 2 to 3 has been explained by Morfill et al. (1981) as ρ-Oph being located at the border of the SNR North Polar Spur (Loop I) (see Figure 1). Although such an explanation is quite reasonable, it is not unique. First of all the present uncertainty in the mass determination of molecular clouds (Blitz, private communication) makes the argument not compelling. In addition, γ-ray enhancements of this order can occur for large clouds due to mass accretion and cosmic-ray compression by these clouds, without the necessity for an enhanced ambient particle flux (Section 5). So ρ-Oph may either be also a standard uncorrelated, or even a correlated source.

The simplest correlated γ-ray source would be due to a supernova exploding inside a cloud. Indeed, about 16% of all H II regions are radio H II regions (Mezger, 1978), embedded in clouds. This appears to be such a high spatial correlation that the exciting OB-stars are probably still gravitationally bound inside their maternal cloud. In a certain

---

* On average the interstellar medium is probably well penetrated by cosmic rays. First of all, the average age of at least low energy ($\lesssim$ 1 GeV/nucleon) cosmic-ray nucleons is about 2×10^7 yr (Garcia-Muñoz et al., 1977), roughly 10^4 times the straight line escape time from a source in the galactic disk. This implies a storage factor in the Galaxy of the same order. Secondly, studies of meteorites and deep sea sediments suggest that the cosmic-ray intensity near the solar system has not changed by more than a factor of 2 to 3 during various epochs over the last 10^8 to 10^9 yr (e.g. Honda, 1979). This corresponds to a number of galactic rotations with numerous crossings of spiral arms. Finally, the mean gas column density penetrated by cosmic rays during their average lifetime is about 7 g cm^{-2}. The corresponding number density is 0.23 H-atoms cm^{-3}. This is not very much smaller than the interstellar density of about 0.4 cm^{-3} in atomic hydrogen (Burton and Liszt, 1981) plus a roughly equal amount of molecular hydrogen, as averaged over the galactic disk from about 4 to about 15 kpc in radius. The scale height of the cosmic rays perpendicular to the disk should be at least as large as that of the gas. Thus, except for fairly localized regions, limited in overall mass, the cosmic rays appear well mixed with the gas. This degree of mixing allows us to speak of an 'average'-cosmic-ray intensity.

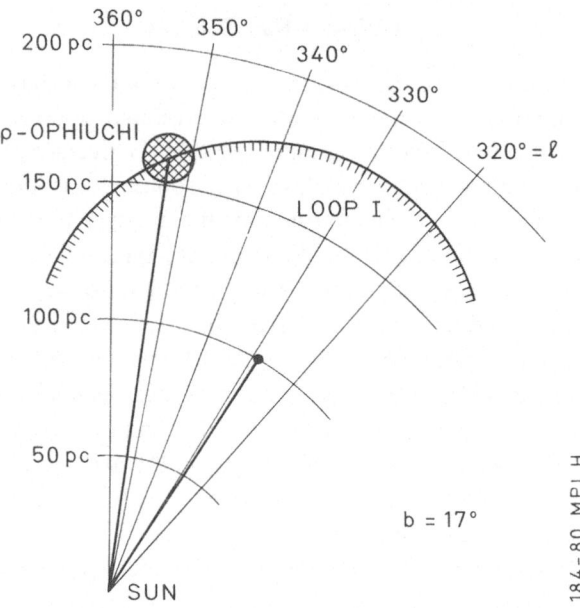

Fig. 1. Possible geometry for the interaction of the accelerated particles from the SNR shock wave of the North Polar Spur (Loop I) with the cloud ρ-Oph (from Morfill *et al.*, 1981).

sense this might lead to a correlated source. However, supernova particle acceleration up to π°-producing nucleon energies inside clouds appears to run into similar problems as stellar wind terminal shock acceleration (Cassé and Paul, 1980; Montmerle and Cesarsky, 1981). It is not very likely to occur inside the stellar wind cavity of the progenitor star because the SNR shock propagates perpendicular to the stellar wind magnetic field direction (Axford, 1981a; Völk and Forman, 1982); outside the stellar wind cavity it is quenched by losses in the dense medium. But SNR will often break out of the cloud, as do H II regions (Tenorio Tagle, 1979) and stellar wind bubbles, in particular if the SN star is not in the very cloud center. With whatever strength it is left with, due to radiation losses (Wheeler *et al.*, 1980; Shull, 1980), the explosion debris will then launch a strong shock into the ambient dilute, hot interstellar medium. This shock will expand laterally, sweep over the field lines intersecting the cloud, and accelerate particles. To the extent that the generated cosmic rays can propagate backward into the cloud against the postshock flow, they will 'illuminate' the cloud and produce (amongst other effects) γ-rays (Völk, 1981; see Figure 2).

To get a rough estimate of the postshock flow speed* we assume the shock to be hemispherical with similar radial flow properties as a Sedov (1959) blast wave, although the shock is a driven one. Taking the total available energy as E_{SN}/η, where E_{SN} is the original supernova energy, and the factor $\eta > 1$ represents radiative energy losses inside

* The following estimate was prompted by a discussion of the author with C. J. Cesarsky.

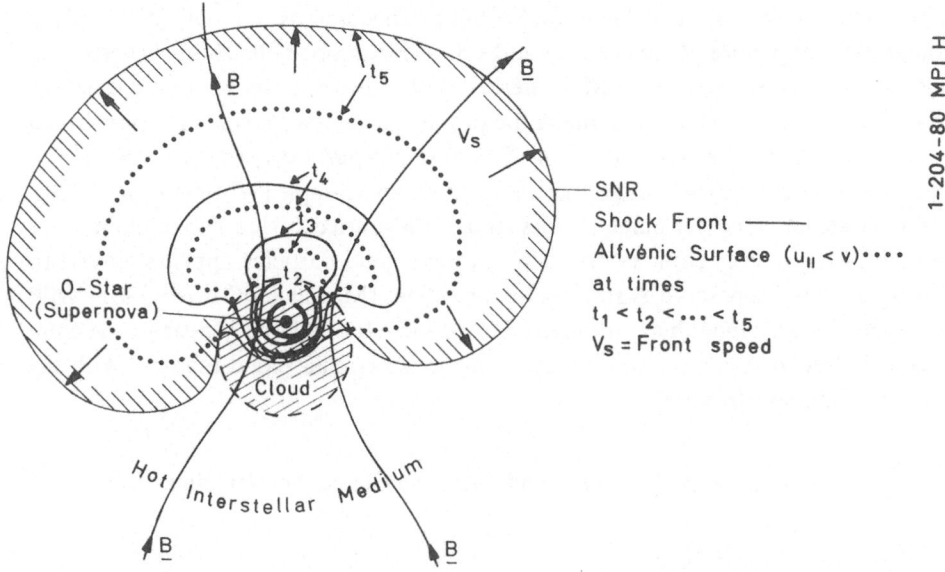

1-204-80 MPI H

SNR
Shock Front ———
Alfvénic Surface ($u_\parallel < v$) ••••
at times
$t_1 < t_2 < \dots < t_5$
V_S = Front speed

Fig. 2. Schematic development of a supernova shock (solid contours) for consecutive times $t_1 < t_2 < t_3 < t_4 < t_5$. An O-star is assumed to explode off-center in a massive cloud, surrounded by a tenous hot interstellar medium. As the shock grows in size it sweeps over the field lines intersecting the cloud. Inside the dotted contours accelerated particles can stream back towards the cloud through the post-shock flow to illuminate it along the field lines (adapted from Völk, 1981).

the cloud, the mean radial flow speed u of the gas inside the SNR shock is roughly given by

$$u(r, t) \simeq \frac{r}{R(t)} u(R) = \frac{r}{R(t)} \frac{\{(\gamma - 1)\dot{R}^2(t)/c_1^2 + 2\}}{(\gamma + 1)\dot{R}^2(t)/c_1^2} \dot{R}(t),$$

where r is the distance from the break-out point, $R(t)$ is the shock front radius with time derivative $\dot{R}(t)$, $u(R)$ is the flow speed right behind the shock, and c_1 is the sound speed in the external medium. For $(\dot{R}/c_1)^2 \gg 1$ and $\gamma = \frac{5}{3}$ we have $u(r, t) \simeq r\dot{R}/(4R)$. Also $R(t) \simeq t^{2/5} \{E_{SN}/(\rho\eta)\}^{1/5}$, where ρ is the upstream mass density. Assuming that cosmic rays in a SNR are so numerous that they cannot stream through the gas faster than with the local Alfvén velocity $v = B(4\pi\rho)^{-1/2}$ due to selfconfinement through excitation of hydromagnetic waves (see Section 3), than $u < v$ some time after the emerging shock has swept over the field line connecting to a given point in the cloud. After shock acceleration and during their backward streaming these particles also lose energy adiabatically in the expanding postshock flow. But most cosmic-ray sources have to cope with this reduction in efficiency in one way or another.

Of course such a model is only the extreme other end of a series of physical models where the particles accelerated by a supernova explosion sweep over a density enhancement to produce a γ-ray source. They can be, but not necessarily are realizations of the phenomenological picture of supernovae and OB associations (SNOB's) suggested on the basis of γ-ray observations by Montmerle (1979). However such a cloud eruption model illustrates on the one hand some of the physical processes involved in non-trivial cases and also shows that the optical or X-ray counterparts of γ-ray sources can be rather unusual. The object η-Carina may be just such an example (Wheeler *et al.*, 1980).

As far as stellar wind terminal shocks and SNR buried inside dense clouds are concerned, their possibly copious production of energetic particles appears limited to MeV energies. They may serve as sources for γ-ray lines (Morfill and Meyer, 1981; Völk and Forman, 1982). Since there are many buried OB-stars relatively near to the Sun, they might indeed be the best candidates to probe interstellar cloud material with the help of γ-ray line spectroscopy.*

3. Cosmic-Ray Transport and Diffusive Shock Acceleration

3.1. Cosmic-ray transport

Discussing the γ-ray sources of the last section we made free use of cosmic-ray transport properties, at least implicitely. In this section we review these properties in somewhat more detail.

As far as 2-body collisions with thermal particles are concerned, cosmic-ray particles are practically collisionless, except at very low energies. Even for a 2 MeV proton the stopping length due to ionisation losses in a gas of density $n = 1\ cm^{-3}$ is $570\ pc$ (Dalgarno and McCray, 1972), larger than the typical size ($\lesssim 100\ pc$) of a large supernova remnant. However, charged energetic particles have a strong interaction with magnetic fields. The average gyroradius of a GeV particle in a 3μG interstellar field is about $10^{-6}\ pc$. Over larger scales, at least perpendicular to the field, the cosmic rays must be considered as a very high temperature *fluid*. Parallel to the mean field direction the situation is considerably more complicated. But in the presence of magnetic irregularities (hydromagnetic waves) with wavelengths on the Larmor radius scale particles are resonantly scattered in pitch angle (Jokipii, 1966). This tends to isotropize the cosmic rays also along the mean field in the rest frame of the waves.

Very interesting reviews of the early ideas relating to the dynamical properties of cosmic rays in the interstellar medium have been given by Parker (1968, 1969). We do not however intend to review the historical development of cosmic-ray transport theory or to go into all its subtleties here. We shall rather consider three different situations which are useful limiting cases in practice.

The first case arises if the resonant irregularities of the field are always of such small amplitude that the resulting pitch angle scattering time is larger than the characteristic

---

* For stellar winds in close binary systems, as opposed to the single stars considered above, the situation may be quite different. The acceleration of particles at the bow shock and in the magnetospheric tail of the companion star should also lead to large intensities of high energy cosmic rays (Dolginov, 1980). This might result in strong γ-ray production in the star's outer atmosphere and make it a high energy γ-ray source.

time variations of the average field which the particle sees along its trajectory. Then cosmic-ray transport must be described on the level of pitch angle diffusion and very roughly the particles can be considered as collisionless. This appears to be often the case in the heliosphere at least in the onset phases of solar flare events, as observed near the Earth.

In the second case there are no waves which exist independently of the cosmic rays. If the energetic particles have a sufficiently strong anisotropy in the frame of the background plasma, for example due to streaming along the field as a result of a cosmic-ray gradient, then hydromagnetic waves are amplified by a resonant instability (Lerche, 1967; Wentzel, 1968; Kulsrud and Pearce, 1969). 'Resonant' means that a given group of particles sees the Doppler shifted wave frequency at its own gyrofrequency. Instability sets in when the mean streaming velocity exceeds the wave phase speed. For smaller streaming no waves are excited. These waves scatter particles and tend to remove the anisotropy in the wave frame. If the damping of the waves is small, as for Alfvén waves in a hot interstellar medium, the waves grow to large amplitudes even for very small anisotropies in the wave frame and the mean cosmic-ray velocity through the plasma is reduced to the hydromagnetic wave speed, typically the Alfvén velocity. If on the other hand the damping is very strong as in dense clouds, then the streaming can become large compared to the Alfvén velocity before being limited by the scattering (Kulsrud and Cesarsky, 1971). In practical applications (see Sections 4 and 5) one may therefore consider very dense, weakly ionized interstellar clouds as scatterfree environments for energetic cosmic rays (e.g. Cesarsky and Völk, 1978), whereas in the fully ionized hot interstellar medium without independent wave sources the cosmic-ray streaming velocity against the direction of their density gradient may come out to lie anywhere between zero and the Alfvén speed in the rest frame of the plasma.

In the third case we assume a sufficiently strong wave field to exist so that the scattering mean free path is small compared to the scales of the average field. The wave field may consist of waves propagating in both directions along the field with different intensities. Through pitch angle scattering waves parallel to $\mathbf{B}$ attempt to create streaming with about the Alfvén speed $\mathbf{v} = \mathbf{B}/(4\pi\rho)^{1/2}$, whereas antiparallel waves attempt to create a streaming speed $-\mathbf{v}$; here $\mathbf{B} \equiv \mathbf{n}B$, with $|\mathbf{n}| = 1$, and ρ denote the average field vector and the background mass density, respectively. In the laboratory frame, assuming Alfvén waves, the resulting mean convection velocity of the cosmic rays is then (Skilling, 1975):

$$\mathbf{U} = \mathbf{u} + \mathbf{v}\left\langle \tfrac{3}{2}(1 - \mu^2)\frac{(P_+ - P_-)}{(P_+ + P_-)}\right\rangle \equiv \mathbf{u} + \bar{\mathbf{v}}. \tag{1}$$

A spatially uniform cosmic ray distribution would be isotropic in zeroth approximation in a frame moving with velocity U. For a spatially nonuniform cosmic-ray population pitch angle scattering will in addition lead to *spatial diffusion* in this frame which amounts to a residual first order anisotropy. In Equation (1) the quantity u denotes the mean mass velocity of the background plasma (neglecting cosmic-ray inertia); the angular brackets denote an average over the cosine of the particle pitch angle μ: $\langle (...) \rangle = (\frac{1}{2})$ $\int_{-1}^{+1} d\mu(...)$; $P_{\pm}(k_{res})$ give the power in waves propagating parallel (+), and antiparallel (−) to **B**, respectively; $k_{res} = m\Omega/\mu p$ denotes the parallel component of the wave number in resonance with particles of rest mass m, rest frame gyrofrequency Ω, and parallel momentum μp. Obviously, for $P_{\pm} = 0$ we have $\tilde{\mathbf{v}} = (\mp 1)\mathbf{v}$. In general $|\tilde{\mathbf{v}}| \leq |\mathbf{v}|$ which means $0 \leq |\tilde{\mathbf{v}}| \leq |\mathbf{v}|$. This result is forced upon the cosmic-ray distribution by the assumed existence of strong wave fields $P_{\pm}$; possible diffusive streaming adds to **U** to give the overall mean velocity of the cosmic-ray component in the laboratory frame. The scattering wave field in this last case will in general have a contribution from the cosmic rays but will predominantly be generated by some other source. This case is therefore applicable to physical situations like the solar wind (and probably stellar winds in general), where the wave field is supposedly produced near the Sun (or the central star).

The above three cases can be described by different transport equations for the average energetic particle distribution function $f(\mathbf{p}, \mathbf{x}, t)$; the average is here over the ensemble of particle fluctuations due to the fluctuations in the fields. We shall assume throughout that the spatial and temporal variations of the average electric and magnetic fields are small enough so that f depends only on $p = |\mathbf{p}|$ and the cosine of the pitch angle μ, but (approximately) not on the particle gyrophase. This is equivalent to conservation of the adiabatic invariants, like the magnetic moment, along the average single particle orbits and appears to be a reasonable approximation even for most oblique collisionless shock transitions (Hudson, 1965; Chen and Armstrong, 1975; Terasawa, 1979). The relevant transport equation for $f(p, \mu, \mathbf{x}, t)$ in case one is a Fokker–Planck equation that describes particle conservation in the phase space $(p, \mu, \mathbf{x})$ along the (adiabatic) average single particle orbits but modified by diffusion terms in μ and p due to pitch angle and momentum scattering in the small scale electromagnetic fields, respectively. Neglecting gradient and curvature drifts perpendicular to the average field (Jokipii and Parker, 1970), this equation has been derived by Skilling (1975). In its general form it is too complicated to be useful. However, an instructive special case is given by a static plasma (**u** = 0) and an energetic particle distribution that does not vary strongly across the average field. Then

$$\frac{\partial f}{\partial t} + w\mu(\mathbf{n} \cdot \nabla)f - \left(\frac{1 - \mu^2}{2}\right) \frac{\partial f}{\partial \mu} w(\mathbf{n} \cdot \nabla) \ln B = \frac{\partial}{\partial \mu} D_\mu \frac{\partial f}{\partial \mu} \qquad (2)$$

if we also neglect momentum scattering. From Equation (2), left to right, f changes as a function of time due to convection with the particle speed μw along **B**, due to pitch angle changes (mirroring) in a spatially varying field strength B, and due to pitch angle diffusion. The pitch angle diffusion coefficient D_μ is proportional to the power spectrum tensor $\mathbf{P}(k)$ of the magnetic fluctuations and, in the simplest case of waves propagating along **B**, given by (Jokipii, 1966):

$$D_\mu = \frac{(1 - \mu^2)}{2} \frac{\pi \Omega^2 m}{\gamma |\mu| p} \frac{P_\perp \left(k = \frac{m\Omega}{|\mu|p}\right)}{B^2}, \qquad (3)$$

where $\gamma = p/(mw)$ is the particle Lorentz factor, and $P_\perp(k) = P_+(k) + P_-(k)$ denotes the magnetic power per unit interval in parallel wave number k of the resonant wave. For more general wave fields, e.g. hydromagnetic waves with an anisotropic distribution of propagation directions and corresponding amplitudes (Lee and Völk, 1975), D_μ is a more complicated expression in terms of the respective wave spectral components. These distinctions are important in the solar wind, where the wave properties are largely determined by the geometry of the system (Völk $et\,al.$, 1974). But for most other astrophysical systems, Equation (3) seems adequate, given all the remaining uncertainties in their configuration.

Thus from Equation (2) we see for instance that an isotropic steady distribution remains spatially uniform and isotropic along all orbits accessible from uniform sources at both ends of the considered magnetic flux tubes (e.g. Parker, 1963), even in the presence of field gradients. On the other hand with unsteady sources the distribution will be neither uniform nor isotropic. For example, in the interplanetary medium, particles from an interior (e.g. solar flare) source will have an exclusively forward anisotropy with $\mu > 0$ further out, before backscattering from beyond the observation point starts to fill in the backward momentum directions $\mu \leq 0$. Finally, a (necessarily anisotropic) distribution of particles, trapped between magnetic mirrors, tends to become isotropized and therefore partly untrapped by pitch angle diffusion. An example would be trapping in quasistatic mirror fields of molecular clouds; this effect may influence the γ-luminosity of such clouds (Cesarsky and Völk, 1978).

In case two, at least if β ($\equiv$ ratio of plasma pressure to magnetic field pressure) exceeds unity and if we do not consider reflection of waves from macroscopic inhomogeneities, all the self-excited waves can, for practical purposes, be assumed to propagate in the same direction (Lee and Völk, 1973). The reason is that, for $\beta > 1$, contrary to the belief of Chin and Wentzel (1972), parallel Alfvén waves do not have any mode coupling that could generate backward propagating waves (Sagdeev and Galeev, 1969).

As long as the mean streaming velocity of the cosmic rays is smaller than $\simeq$ Alfvén velocity or, more generally, as long as the wave damping rate is larger than the growth rate due to streaming, there are no scattering waves of any importance and we come back to case one with $D_\mu = 0$. If the opposite is true then waves are strongly excited. The scattering mean free path λ is approximately given by:

$$\lambda = r_g \frac{B^2}{8\pi} \bigg/ W_w \qquad\qquad (4)$$

where $W_w = (8\pi)^{-1} \int_{1/r_g}^\infty dk\, P_\perp(k)$ is the magnetic energy density in resonant fluctuations for $kr_g > 1$, and $r_g = m\Omega/p$ is the particle gyroradius. Thus $\lambda \gtrsim r_g$ and normally we may assume λ to be small compared to the spatial scales of the background flow. For definitiveness we shall in the sequel confine ourselves to Alfvén waves. In the wave frame, moving with velocity $U = \mathbf{u} \mp \mathbf{v}$, cf. Equation (1) with either $P_+ = 0$ or $P_- = 0$, respectively, the distribution $f(\mu, p, \mathbf{x}, t)$ is to lowest order given by its isotropic part

$\bar{f}(p, \mathbf{x}, t) = \langle f(\mu, p, \mathbf{x}, t) \rangle$ which fulfills the spatial diffusion equation (Skilling, 1971):

$$\frac{\partial \bar{f}}{\partial t} + (\mathbf{u} \pm \mathbf{v}) \cdot \frac{\partial \bar{f}}{\partial \mathbf{x}} - \mathrm{div}\left(\kappa_{\parallel} \, \mathbf{n} \left(\mathbf{n} \cdot \frac{\partial \bar{f}}{\partial \mathbf{x}} \right) \right) -$$

$$-\frac{1}{3} p \frac{\partial \bar{f}}{\partial p} \mathrm{div} \, (\mathbf{u} \pm \mathbf{v}) = \left(\frac{\partial \bar{f}}{\partial t} \right)_s - \left(\frac{\partial \bar{f}}{\partial t} \right)_l. \tag{5}$$

Thus strong pitch angle scattering locks the particles to the waves. It leads to convection along with the waves, and to field aligned spatial diffusion if there is a cosmic-ray density gradient. Since the distribution is held isotropic in the wave frame the particles also change momentum adiabatically if the scattering medium undergoes density variations, i.e. if $\mathrm{div}.(\mathbf{u} \pm \mathbf{v}) \neq 0$. But there is no momentum diffusion in this case. Any particle sources, and energy losses other than the adiabatic ones are formally represented by the r.h.s. of Equation (5). Resonant spatial diffusion perpendicular to **B** is in general negligible (because particles can only random walk by one gyroradius across the field at each pitch angle scattering) although it has been invoked in recent analyses of particle acceleration at the Earth's bow shock (Eichler, 1981; Lee, 1982). The parallel diffusion coefficient $\kappa_{\parallel}$ is given by (Jokipii, 1966; Hasselmann and Wibberenz, 1970)

$$\kappa_{\parallel} = \frac{w^2}{4} \left\langle \frac{(1 - \mu^2)^2}{D_{\mu}} \right\rangle \simeq \frac{1}{3} w \lambda. \tag{6}$$

Thus if there is little scattering ($D_{\mu} \simeq 0$) the spatial diffusion is very large and $\partial f / \partial x \simeq 0$ with, in general, a finite diffusion flux − although of course in this case the use of Equations (5) and (6) is not really justified and should be replaced by a discussion on the level of pitch angle scattering as in case one. The residual field aligned streaming in first order is for particles of given p

$$\langle \mu w f \rangle = \kappa_{\parallel} \left(\mathbf{n} \cdot \frac{\partial \bar{f}}{\partial \mathbf{x}} \right). \tag{7}$$

In the laboratory frame this transforms to the differential current density

$$\mathbf{S} = (\mathbf{u} \mp \mathbf{v}) \cdot f - \mathbf{n} \cdot \kappa_{\parallel} \left(\mathbf{n} \cdot \frac{\partial f}{\partial \mathbf{x}} \right) - \frac{1}{3}(\mathbf{u} \mp \mathbf{v}) \frac{1}{p^2} \frac{\partial}{\partial p} (p^3 f)$$

$$= C_g (\mathbf{u} \mp \mathbf{v}) \cdot f - \mathbf{n} \cdot \kappa_{\parallel} \left(\mathbf{n} \cdot \frac{\partial f}{\partial \mathbf{x}} \right) \tag{8}$$

with the Compton−Getting 'factor'

$$C_g \equiv -\frac{p}{3} \frac{1}{\bar{f}} \frac{\partial}{\partial p}. \tag{9}$$

Integration over all particle momenta results in the spatial current density:

$$\int \mathbf{S} \cdot d^3p = (\mathbf{u} \mp \mathbf{v}) \int f \, d^3p - \mathbf{n} \cdot \int d^3p \, \kappa_\| \left(\mathbf{n} \cdot \frac{\partial f}{\partial \mathbf{x}} \right). \tag{10}$$

Finally, case three is described by the spatial diffusion equation (Skilling, 1975)

$$\frac{\partial \overline{f}}{\partial t} + \frac{\partial \overline{f}}{\partial \mathbf{x}} \cdot \frac{\partial}{\partial (p^3)} (p^3 \mathbf{U}) - \text{div} \left(\kappa_\| \mathbf{n} \left(\mathbf{n} \cdot \frac{\partial \overline{f}}{\partial \mathbf{x}} \right) \right) - p^3 \frac{\partial \overline{f}}{\partial (p^3)} \, \text{div} \, \mathbf{U} -$$

$$- \frac{\partial}{\partial (p^3)} \left(9 p^4 D_p \frac{\partial \overline{f}}{\partial (p^3)} \right) = \left(\frac{\partial \overline{f}}{\partial t} \right)_s - \left(\frac{\partial \overline{f}}{\partial t} \right)_l, \tag{11}$$

where the momentum diffusion (second order Fermi acceleration) is given by the coefficient:

$$D_p \equiv 4 \gamma^2 m^2 v^2 \left\langle \frac{D_{\mu+} D_{\mu-}}{(D_{\mu+} + D_{\mu-})} \right\rangle;$$

$$D_{\mu\pm} \equiv \frac{(1 - \mu^2)}{2} \frac{\pi \Omega^2 m}{\gamma |\mu| p} \frac{P_\pm}{B^2}. \tag{12}$$

The differential current density is

$$\mathbf{S} = \mathbf{U} \cdot \frac{p^3}{f} \cdot \frac{\partial \overline{f}}{\partial (p^3)} - \kappa_\| \mathbf{n} \cdot \left(\mathbf{n} \cdot \frac{\partial \overline{f}}{\partial \mathbf{x}} \right), \tag{13}$$

where $\mathbf{U}$ is given by Equation (1).

Apart from the appearance of the second order Fermi acceleration term, Equations (11) and (13) have a similar interpretation as Equations (5) and (8), respectively, although the former equations are more general since they allow both external and selfexcited wave fields propagating in opposite directions. These characteristics of the wave field can be different for different particle energies.*

3.2. DIFFUSIVE SHOCK ACCELERATION

According to Equation (11) particles can gain energy through momentum diffusion if they scatter in a wave field containing both backwards and forwards propagating waves. This second-order Fermi acceleration (Fermi, 1949) can occur in the general interstellar medium and may be quite efficient under specific circumstances e.g. the sweep-up phase of supernova remnants (Scott and Chevalier, 1975). However, the irreversible first order energy gains due to local compressions div $\mathbf{U} < 0$, combined with spatial diffusion

---

* Both systems of equations do not contain perpendicular drifts. A derivation including drifts, but approximating $\mathbf{U} \simeq \mathbf{u}$ has been given by Jokipii and Parker (1970). These effects are strongly dependent on the geometry of the system and are beyond the scope of the present paper.

($\kappa_\parallel \neq 0$), should be much larger if the flows are super-Alfvénic. This is particularly true in shock waves and we shall turn now to a discussion of diffusive shock acceleration (Krymsky, 1977; Axford *et al.*, 1977; Bell, 1978a, b; Blandford and Ostriker, 1978). The physical picture is given schematically in Figure 3 for a plane, steady, parallel, strong shock. In the shock frame particles are convected towards the shock from upstream ($x > 0$) with velocity $U \equiv V_1 \gg v$. In addition they diffuse spatially, exchanging energy and momentum with waves carried by the background flow. Disregarding the stochastic (second order Fermi) momentum changes $\Delta p/p \simeq (\pm 1)2(v/w)$ per collisions (with waves of different directions ($\mathbf{v} \cdot \mathbf{w}$) $\lessgtr 0$, respectively) *within* either the upstream or the down-stream region, particles gain momentum $\Delta p/p \simeq [2|(V_1 - V_2)|]w$ each time they scatter after crossing the shock. Although all particles are ultimately convected downstream across the shock, there will be particles which cross the shock only once or a few times, whereas others (less in number) will cross many times. This is due to the stochastic nature of the diffusion process and results in a power law spectrum $\sim p^{-q}$, $q > 0$, of the accelerated particles (Bell, 1978a; see also Drury, 1983, for a recent review).

Alternatively, we can calculate this process from the point of view of the transport equations. Although the mean free path λ of the energetic particles is not small compared to the thickness of the collisionfree shock transition in the background plasma (whose scale or at least its subscales are of the order of a *thermal* gyro-radius, or even a Debye length) we shall use Equations (11) and (13) to describe the cosmic-ray population on both sides of the shock (e.g. Toptyghin, 1980; Webb, 1983a). Without second order Fermi acceleration and losses, taking $\mathbf{U} \simeq \mathbf{u}$ for the above plane shock, one can solve Equation (11) for $\overline{f}$ in the uniform media on both sides of the shock. Across the discontinuity not only $\overline{f}$ must be continuous, but also the differential current density S apart from a possible injection current density $Q(p)$ at the shock. These are the 'natural'

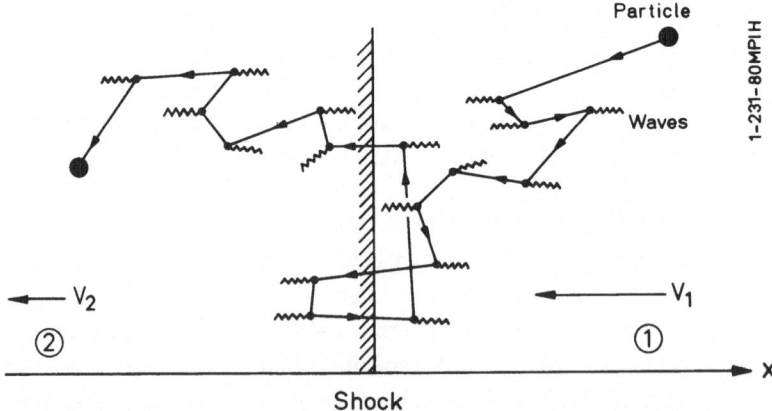

Fig. 3.   Diffusive propagation of an energetic particle through a shock. In the shock frame the medium streams from right to left with velocity V_1 upstream and $|V_2| < |V_1|$ downstream. At each scattering the particle exchanges energy and momentum with waves essentially carried along by the flow. Energy changes within either region 1 or 2 tend to cancel each other (2nd order Fermi effect), whereas at *each* shock crossing the particle gains a large amount of energy (1st order Fermi effect). On average the particle is convected through the shock (adapted from Völk, 1981).

boundary conditions for the transport equation. Denoting upstream (downstream) quantities by a suffix 1(2), the steady state upstream distribution is given by

$$\bar{f}_1(x,p) = \bar{f}_{+\infty}(p) + (\bar{f}_2(p) - \bar{f}_{+\infty}(p)) \exp\left\{-\int\limits_0^x \frac{dx\, |\mathbf{u}_1|}{\kappa_\parallel}\right\}, \tag{14}$$

where $\bar{f}_{+\infty}(p)$ is the asymptotic distribution upstream ($x = +\infty$). The downstream distribution $\bar{f}_2(p)$ on the other hand does not depend on the distance x from the shock

$$\bar{f}_2(p) = qp^{-q} \int\limits_0^p dp'\, p'^{(q-1)} \left\{\bar{f}_{+\infty}(p') + \frac{Q(p')}{|u_1|}\right\}. \tag{15}$$

Here $q = 3r/(r-1)$, and $r = |u_1/u_2|$ is the shock compression ratio. For strong adiabatic shocks as long as the compression is dominated by nonrelativistic particles we have $r < 4$ and therefore $q \geq 4$.

The accelerated spectrum $\bar{f}_2(p)$ tends to be a power law with index q. This can best be seen by considering a mono-energetic injection and/or upstream distribution: $Q/u_1 + \bar{f}_{+\infty} = \delta(p - p_0)\, N(4\pi p_0^2)^{-1}$. Then

$$\bar{f}_2(p) = q(p/p_0)^{-q}\, N(4\pi p_0^3)^{-1} \quad \text{for} \quad \frac{p}{p_0} \geq 1, \tag{16}$$

and zero if p lies below the injection momentum p_0.

If on the other hand the upstream spectrum $\bar{f}_{+\infty}$, convected into the shock, is already a power law $\bar{f}_{+\infty}(p) \sim p^{-\mu}$, then the reaccelerated spectrum is

$$\bar{f}_2(p) \sim \begin{cases} p^{-\mu} & \text{for} \quad \mu < q, \quad \text{(pure amplification)}, \\ p^{-q} & \text{for} \quad \mu > q, \quad \text{(hardening to limiting spectrum)}. \end{cases} \tag{17}$$

The observed galactic spectrum above a few GeV/c has $q \simeq 4.7$, corresponding to $r \simeq 2.8$. The corresponding shock Mach number $M = u_1/c_1$ equals 2.57, where c_1 is the upstream sound speed and $r = (\gamma + 1)M^2\{(\gamma - 1)M^2 + 2\}^{-1}$ with $\gamma = \frac{5}{3}$. However the observed galactic spectrum may be the result of a rigidity dependent galactic escape time $\tau_{\text{esc}} \sim p^{-0.5}$ which implies a harder primary spectrum corresponding to $r \simeq 3.5$ and $M \simeq 4.6$. Thus shock acceleration can produce a power law in momentum as observed or inferred, for quite reasonable shock parameters (Axford, 1981b).

By inspection of Equation (11), the shock acceleration time $\tau_{\text{acc}}^{(s)} = 0(\kappa_\parallel/u_1^2) = (w/u_1)^2\, 0(t_{\text{scatt}})$, where from Equations (3) and (12) $t_{\text{scatt}} = 0((D_{\mu+} + D_{\mu-})^{-1}) \simeq \kappa_\parallel/w^2$ is the mean pitch angle scattering time. A more precise estimate (Axford, 1981b) gives

$$\tau_{\text{acc}}^{(s)} \simeq \frac{3}{|u_1 - u_2|} \int\limits_{p_0}^p \frac{dp}{p} \left[\frac{\kappa_{\parallel 1}}{|u_1|} + \frac{\kappa_{\parallel 2}}{|u_2|}\right] \simeq \frac{6\kappa_\parallel}{|u_1 - u_2|\,|u_1|} \simeq \frac{w^2}{|u_1 - u_2|\,|u_1|}\, t_{\text{scatt}}, \tag{18}$$

where, as above, we take $\kappa_{\parallel}$ as a rough average diffusion coefficient. Since $|u_1|/w = 0(|u_1|/c) \ll 1$ we have $\tau_{\text{acc}}^{(s)} \gg t_{\text{scatt}}$ and the process is quite slow. The reason is of course that only scatterings across the shock lead to a strong energy increase, whereas the vast *majority* of scatterings occur within the upstream or downstream region (Figure 3). The acceleration time $\tau_{\text{acc}}^{(2)}$ for the second order Fermi process due to the latter collisions is from Equation (12) given by

$$\tau_{\text{acc}}^{(2)} \simeq \frac{p^2}{D_p} = \frac{1}{4}\left(\frac{w}{v}\right)^2 \left\langle \frac{D_{\mu+} D_{\mu-}}{(D_{\mu+} + D_{\mu-})} \right\rangle^{-1}. \tag{18}$$

Defining $t_{\text{scatt}}^{(\pm)} = 0(D_{\mu\pm}^{-1})$, so that $(t_{\text{scatt}})^{-1} = (t_{\text{scatt}}^{(+)})^{-1} + (t_{\text{scatt}}^{(-)})^{-1}$, we have

$$\frac{\tau_{\text{acc}}^{(s)}}{\tau_{\text{acc}}^{(2)}} = \frac{v^2}{|u_1 - u_2|\,|u_1|} 0\left(\frac{t_{\text{scatt}}^{(+)} t_{\text{scatt}}^{(-)}}{(t_{\text{scatt}}^{(+)} + t_{\text{scatt}}^{(-)})^2}\right) \tag{19}$$

except for factors of order unity. To the same accuracy we also have $v^2/(|u_1 - u_2|\,|u_1|) \simeq (M^2 - 1)^{-1}(v/c_1)^2$ (e.g. Landau and Lifschitz, 1959). Thus for $t_{\text{scatt}}^{(+)}/t_{\text{scatt}}^{(-)} = 0(1)$ only for strong shocks ($M^2 \gg 1$) is $\tau_{\text{sacc}}^{(s)}/\tau_{\text{acc}}^{(2)} \ll 1$ and second-order Fermi acceleration a priori negligible. However, according to Equation (14) shock acceleration builds up a cosmic-ray gradient which drives a diffusive current upstream against the incoming flow. This will in general excite waves propagating in the same direction. These waves (say waves $P_+(k)$) add to particle scattering (Bell, 1978a) and are in fact expected to dominate the wave field (McKenzie and Völk, 1981, 1982). Therefore $P_+(k) \gg P_-(k)$ at least upstream, and

$$\frac{t_{\text{scatt}}^{(+)} t_{\text{scatt}}^{(-)}}{(t_{\text{scatt}}^{(+)} + t_{\text{scatt}}^{(-)})^2} \simeq \frac{t_{\text{scatt}}^{(+)}}{t_{\text{scatt}}^{(-)}} \ll 1. \tag{20}$$

Thus, even for medium and low Mach number shocks shock acceleration is still expected to dominate since the selfexcited waves tend to depress second order Fermi acceleration. At the shock ($+$) waves may be converted to ($-$) waves (McKenzie and Westphal, 1969) and thus behind the shock second-order Fermi effects may play some role (Vasilyev *et al.*, 1980; Webb, 1983b).

A consequence of the relative slowness of the shock acceleration process is its sensitivity to energy losses and to wave damping (which by decreasing $P_\perp(k)$ increases $\kappa_{\parallel}$ and thus $\tau_{\text{acc}}^{(s)}$). If in the simplest case we describe losses, e.g. due to ionisation, by a momentum dependent spatially uniform loss time $\tau(p)$, then the simple loss-free solutions (14) and (15) will be strongly affected if $\tau_{\text{acc}}^{(s)}/\tau = 0(\kappa_{\parallel}/(u_1^2\tau)) \gtrsim 1$. In Figure 4 the spatial configuration is shown for a plane, steady situation assuming a uniform particle source. Of course the p-dependence of the accelerated spectrum is equally affected. This also plays a role for the injection process at very low energies (Ellison, 1981). The effects of losses on shock acceleration in various environments have been discussed by Völk *et al.* (1981), also for the case when the shock creates its own ionisation precursor in neutral media due to radiation from the hot shocked gas. As a

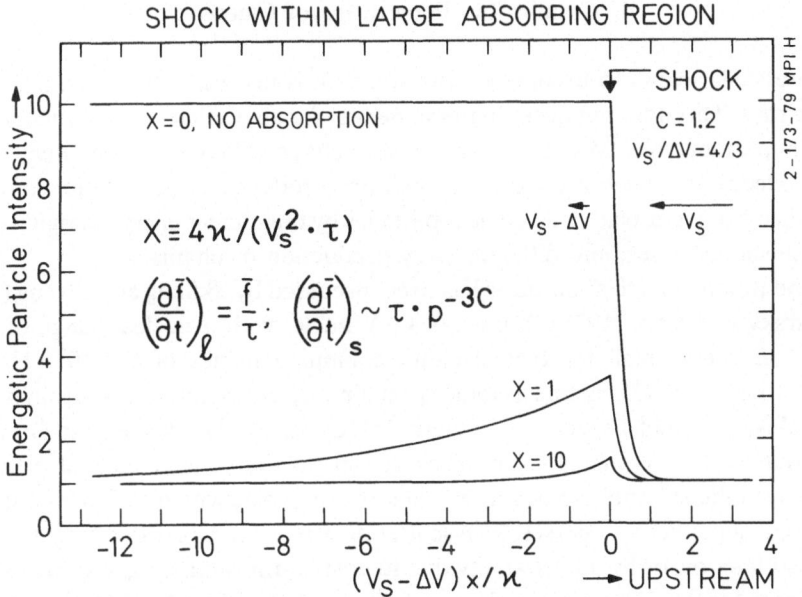

Fig. 4. Spatial structure of the cosmic-ray distribution in a strong plane shock including uniform losses with rate τ^{-1}. The abscissa is spatial distance in units of $\kappa/(V_s - \Delta V)$, where in the text κ is denoted by $\kappa_{\parallel}$ and $V_s - \Delta V$ by u_2. The absorption parameter $X = 4\kappa/(V_s^2\tau)$ determines the decrease of acceleration efficiency from the loss-free case $X = 0$. A uniform source distribution $\bar{f}(p) \sim p^{-3C}$ was assumed (from Völk et al., 1981).

result shock acceleration will generally not be effective in neutral interstellar clouds. The best environment clearly is a hot and dilute, fully ionized medium. To say it somewhat absurdly: cosmic rays are best shock accelerated in a vacuum with a finite Alfvén velocity!

For cosmic-ray pressures small compared to the total momentum flux in the gas, all the average parameters of the gas and field are determined independently of the cosmic rays. However this is not in general the case in the interstellar medium, where gas and cosmic-ray pressures are comparable (e.g. Parker, 1968; see also Section 5 below). It is even less true for diffusive acceleration in strong shocks where a strong backreaction of the accelerated cosmic rays occurs on the thermal flow (Axford et al., 1977; Eichler, 1979; Blandford, 1980; Drury and Völk, 1981; Axford et al., 1982; McKenzie and Völk, 1982). Then, formally, the gradient of the cosmic-ray pressure $p_c \equiv \int \mathrm{d}^3p\, (1/3)wpf$, and the divergence of the cosmic-ray energy flux $\mathbf{F}_c \equiv \int \mathrm{d}^3pE_{\mathrm{kin}} \cdot \mathbf{S}$ have to be added to the overall momentum and energy conservation relations. Together with the cosmic-ray transport equations they determine the combined dynamics of the plasma/cosmic-ray system.*

* A very nice discussion of nonlinear cosmic-ray effects on shock acceleration is contained in the review by Dury (1983).

4. Cosmic-Ray Penetration of Clouds

For an interstellar cloud to become a γ-ray source it is necessary that it be penetrated by cosmic rays. This does not occur in the same way for all particles. Since energy losses play an important role, low energy particles behave differently from high energy particles. Electrons, which are partly secondaries produced in the interior, are to be distinguished from nonrelativistic nuclear particles by their larger energy to rigidity ratio, smaller ionisation losses, and different γ-ray production mechanism.

The penetration of nuclear particles has been discussed by Skilling and Strong (1976), and Cesarsky and Völk (1978). Losses of GeV nuclei with a nuclear range of about $66 \, \text{g cm}^{-2}$ are very small for typical cloud column densities of 5×10^{22} H-atoms $\text{cm}^{-2} = 0.16 \, \text{g cm}^{-2}$. This is true as long as such particles do not diffuse strongly in the cloud. Diffusion would be due to resonant MHD-waves. The waves can be due to various sources. One group of them, discussed in Section 5, is connected with mass accretion from the external rarefied interstellar medium and with cloud evaporation. If we do not consider these processes, then scattering waves can be created by gas motions in the cloud interior, or by an instability due to cosmic-ray streaming into the cloud to offset interior losses. However, at the relevant small wavelengths, MHD waves are strongly damped in a weakly ionised gas due to ion neutral friction (Kulsrud and Pearce, 1969). As shown by Cesarsky and Völk (1978) no conceivable wave source could maintain a sufficiently high level of resonant wave activity in clouds. Thus, without exterior mass motions, energetic nuclei should penetrate clouds freely (see also Lebrun and Paul, 1978); could losses cannot produce significant streaming to excite waves strongly enough.

For nuclei with energies below a few 100 MeV the interior losses tend to become appreciable. Skilling and Strong (1976) estimated that the attendant streaming from the external rarefied medium should excite waves there which are weakly damped only. The waves limit the cosmic-ray streaming velocity and therefore the interior particle population cannot be replenished. This self-shielding of clouds was estimated to start at proton energies below several hundred MeV. However, this does not take into account the compression of the magnetic field by the cloud. If the field is weaker outside, then to the same extent also the cosmic-ray current density is smaller outside than inside along a magnetic flux tube. Assuming a cloud/intercloud field strength ratio of $50 \, \mu\text{G}/3 \, \mu\text{G} \simeq 17$, Cesarsky and Völk (1978) obtained selfshielding only below 50 MeV/nucleon. Such a value is basically consistent with cosmic-ray ionisation rates in clouds. For this estimate the ambient cosmic-ray spectrum was assumed to be given by $\bar{f}(p) \sim p^{-4.5}$, extrapolating the demodulated spectrum of Morfill *et al.* (1976) from about 300 MeV/nucleon to 10 MeV/nucleon. According to Cowsik and Lee (1977) this extrapolation appears to be an upper limit. Therefore also the selfshielding boundary of 50 MeV/nucleon appears to be an upper limit and might be significantly lower. Yet the galactic cosmic-ray energy spectrum well below 200–300 MeV/nucleon is not known observationally and could vary considerably across the Galaxy. As a consequence there is room for speculation.

These effects (and non-effects) on nuclear particles may influence the spectrum of primary and secondary low energy electrons in clouds. Brown and Marscher (1977) and Marscher and Brown (1978) pointed out that secondary electrons could be copiously produced in clouds by penetrating high energy nucleons. Electrons in the hundreds of MeV range would lose energy primarily by Bremsstrahlung and could lead to an increased γ-ray luminosity if they would be confined in the cloud. Cesarsky and Völk (1978) suggested that electrons, in particular those of Brown and Marscher, could be trapped in the cloud by parasitic pitch angle scattering on waves at the boundary, produced by the streaming of low energy nucleons. Indeed, if nucleons below, say, 50 MeV/nucleon excited waves in the ambient intercloud medium, the latter could resonate with electrons up to about 300 MeV and thus trap them. Finally, Morfill (1981, 1982) argued that electrons up to the trapping limit could be accelerated by the confining waves, since they propagate everywhere into the cloud; it follows from Equation (5) for $\mathbf{u} \equiv 0$, but $\text{div}(\pm \mathbf{v}) < 0$. This would lead to an electron enhancement in the cloud as long as the electrons can be treated as test particles, i.e. as long as their energy density is small compared to that of the confining protons at the same rigidity. It should be added here that the confining waves of course also lead to self acceleration of the interior resonant nucleons in a boot-strap manner. It tends to diminish the electron compression.

Whether this electron trapping and acceleration in clouds does importantly contribute to the presumed Bremsstrahlung part in the COS-B γ-ray observations unfortunately depends on the poorly known very low energy nucleon spectrum. It will not be easy to make much progress on this particular question in the near future.

5. Cosmic-Ray and γ-Ray Effects Due to Interactions of Clouds with an Ambient Hot Interstellar Medium

The interaction of a cold cloud with a surrounding hot interstellar medium (HIM) gives rise to mass motions into or away from the cloud: the large heat conduction from the HIM with a temperature $T = O(10^6)$ K (cf. McKee and Ostriker, 1977) into the cold cloud with $T = O(20)$ K causes it to evaporate (Zel'dovich and Pikelner, 1969; Penston and Brown, 1969; Graham and Langer, 1973; Chevalier, 1975; Cowie and McKee, 1977; Balbus and McKee, 1982), or to accrete material (McKee and Cowie, 1977). In detail these are complicated dynamical processes which depend on the heating and cooling effects in the gas, its chemical state, the magnetic field, the cosmic rays, and the overall geometry. Nevertheless, for spherical symmetry, and disregarding for the moment dynamical effects from the cosmic-ray gas, the results appear about as follows: (i) a small cloud with a radius $R \lesssim 10$ pc evaporates supersonically as far as the hot medium is concerned with considerable overpressure (e.g. Cowie and McKee, 1977), (ii) for $R \gtrsim 10$ pc radiative losses outside the cloud are strong enough to offset the conductive heating of its border from the hot environment so that a static equilibrium and presumably also accretion is possible (McKee and Cowie, 1977).

Previously the emphasis has been on cloud evaporation although thermal accretion onto small clouds was included in a refined treatment of the evolution of supernova remnants by Cowie *et al.* (1981). This evolution contains extended low temperature phases where also small clouds accrete mass. It appears however important to also consider thermal accretion flows directly from the $O(10^6)$ K HIM onto big clouds in order to maintain the mass balance between clouds and the HIM. Let us assume that indeed supernova explosions 'regulate' the interstellar dynamics as suggested by McKee and Ostriker (1977). Then the interstellar medium consists of rarefied gas from over-lapping supernova remnants (HIM) together with embedded evaporating and accreting clouds. We have already seen that small clouds tend to evaporate. Big molecular clouds are probably disrupted by the action of massive stars formed in their interior or at their boundaries. McKee and Ostriker (1977) believed that the mass balance of clouds would essentially be maintained by accretion in traversing the cool dense shells formed during the late phases of supernova remnants. This accretion may sometimes occur even though shell formation turns out to be a rare process (e.g. Cowie *et al.*, 1981). Also coagulation of clouds may play a role before the collision partners have evaporated. However it is not clear that these processes are sufficiently frequent to achieve the mass balance. Thus, if accretion, directly from the HIM onto large clouds, is dynamically possible and efficient then this could be a simple solution for the mass balance problem. Roughly one can estimate this as follows. Neglecting gravitational effects, the accretion is basically a standing rarefaction wave leading to a subsonic flow in the HIM. In fact we shall later argue that cosmic-ray compression will keep the flow even sub-Alfvénic. Assuming spherical flow the mass accretion rate is given by

$$\dot{M} \equiv 4\pi r^2 \rho(r) v(r) \simeq 4\pi R^2 \rho_\infty v_\infty. \tag{21}$$

Here $\rho_\infty = 5 \times 10^{-27}$ g cm^{-3} and $v_\infty = 3 \times 10^6$ cm s^{-1} are assumed HIM values for the mass density and the Alfvén speed, respectively. Taking $R = 10$ pc we get from Equation (21) that $\dot{M} \simeq 3 \times 10^{-3} M_\odot$ yr^{-1}. This means that the mass accreted during a cloud lifetime of 10^6 to 10^7 yr can be of the same order as the cloud mass. Equation (21) may even underestimate $\dot{M}$; therefore direct accretion from the HIM may well be able to offset evaporation losses from small clouds.

Thus, let us assume the above picture of accreting big clouds and evaporating (occasionally in dense neutral shells also accreting) small clouds in dynamical equilibrium with a HIM. What are the dynamical effects of the cosmic rays whose pressure is about equal to the gas pressure? Conversely, how are the cosmic rays affected by these interactions and what can we expect for the γ-ray emission of such a system?

In the case of cloud evaporation, the internal overpressure (over the external gas pressure) will in most cases also overpower the external cosmic-ray pressure. Assuming $\beta > 1$, i.e. the thermal gas pressure to be greater than the magnetic field pressure, a supersonic outflow is also super-Alfvénic. Since the evaporation speeds are $O(100)$ km s^{-1} like for the solar wind, but over scales $O(10)$ pc $\simeq 10^4$ times the size of the solar cavity, we expect by analogy a very strong modulation by the outflow, i.e. an exclusion of practically all cosmic rays of interest from evaporating small clouds. Small clouds accreting dense, neutral material should leave the cosmic rays largely indifferent, because of the strong damping of all resonantly scattering waves (Section 4).

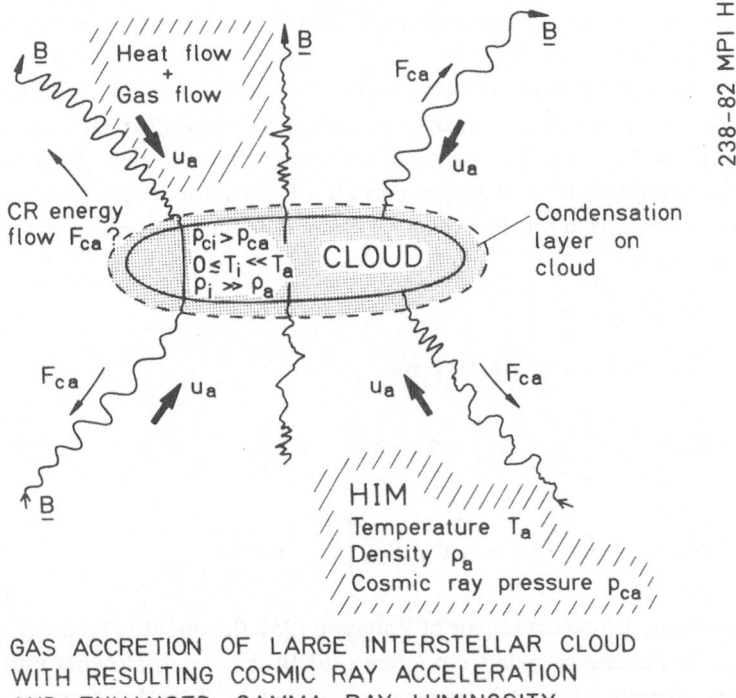

GAS ACCRETION OF LARGE INTERSTELLAR CLOUD
WITH RESULTING COSMIC RAY ACCELERATION
AND ENHANCED GAMMA RAY LUMINOSITY

Fig. 5.   Accretion flow from a hot interstellar medium (HIM) onto a dense, cold cloud. The heat flow is accompanied by a gas flow which tends to convect cosmic-ray (CR) particles into the cloud and to accelerate them by compression. Due to scattering on the irregular magnetic field **B** the maximum speed of relative flow between gas and cosmic rays is the Alfvén velocity, to which u_a should reduce in a steady state. Possibly the cloud becomes a source of cosmic-ray energy with current density F_c.

Accreting big clouds can do so only subsonically; the driving force is the gas pressure gradient between the cold and hot gases. Therefore the cosmic-ray pressure can influence the flow strongly. If the flow is super-Alfvénic it tries to convect the cosmic rays into the cloud (Figure 5). Since they do not cool efficiently except at very low energies this leads to a compression of the cosmic-ray component in the cloud. The resulting adverse cosmic-ray pressure gradient will decelerate the incoming flow until it is sub-Alfvénic to allow a steady state by outward cosmic-ray streaming through the accretion flow. Dynamic equilibrium in the steady state gives

$$p_{ci} - p_{ca} \simeq p_{ga} - p_{gi} + \rho u_\infty^2 . \tag{22}$$

Here the suffices c and g denote cosmic rays and gas, respectively, inside (i) the cloud, and in the ambient (a) medium. p_{gi} may effectively be zero or even negative due to self-gravity of the cloud; $\rho u_\infty^2 \simeq B^2/4\pi$ due to Alfvénic streaming. Therefore Equation (22) yields

$$p_{ci} - p_{ca} \lesssim p_{ga} \left(1 + \frac{1}{\beta} \right) \lesssim p_{ga} . \tag{23}$$

Taking $p_{ca} = p_{ga}$, the interior overpressure of cosmic rays can be about equal or twice the exterior gas pressure.

The theoretical description of these processes is as follows: the cosmic-ray distribution function is given by the transport equations (11)–(13). However we cannot prescribe the background flow independently of the cosmic rays. In other words, the cosmic rays cannot be treated as test particles. Rather we postulate mass, momentum and energy conservation for the system of gas and cosmic rays as a whole. This implies in the steady state:

$$\operatorname{div}(\rho \cdot \mathbf{u}) = 0, \tag{24}$$

$$\operatorname{div}\{\rho \mathbf{u} : \mathbf{u} + (p_g + p_c) \cdot \mathbf{I}\} = 0, \tag{25}$$

$$\operatorname{div}\left\{\rho \cdot \mathbf{u} \cdot \left[\frac{u^2}{2} + W\right] + \mathbf{F}_c - \chi(T) \cdot \nabla T\right\} + \frac{\rho^2}{m^2} \Lambda = 0 \tag{26}$$

$$\operatorname{div} \mathbf{F}_c = (\hat{\mathbf{U}} \cdot \nabla)p_c + \int \mathrm{d}^3 p \, E_{\mathrm{kin}}\left\{\left(\frac{\partial \overline{f}}{\partial t}\right)_s - \left(\frac{\partial \overline{f}}{\partial t}\right)_l\right\}. \tag{27}$$

In the momentum transport tensor of Equation (25), the quantity $\mathbf{I}$ denotes the diagonal unit tensor. In the energy balance relation (26) W is the enthalpy per unit mass of the thermal gas whereas the cosmic-ray energy flux density $\mathbf{F}_c$ and the cosmic-ray pressure p_c are given by

$$\mathbf{F}_c \equiv \int \mathrm{d}^3 p \, E_{\mathrm{kin}} \cdot \mathbf{S} \simeq \frac{1}{(\gamma_c - 1)} \left\{-\mathbf{n}\kappa_{\parallel}\left(\mathbf{n} \cdot \frac{\partial p_c}{\partial \mathbf{x}}\right) + \gamma_c \hat{\mathbf{U}} p_c\right\}, \tag{28}$$

$$p_c \equiv \int \mathrm{d}^3 p \, \frac{1}{3} \, wp\overline{f}, \tag{29}$$

using relation (13).

The parameter γ_c is the adiabatic index of the cosmic-ray gas ($\gamma_c \simeq \frac{4}{3}$) and $\hat{\mathbf{U}} = \hat{\mathbf{u}} + \hat{\tilde{\mathbf{v}}}$ (see Equation (1)). The parameters $\hat{\tilde{\mathbf{v}}}$ and $\hat{\kappa}_{\parallel}$ are suitable averages of $\tilde{\mathbf{v}}$ and $\kappa_{\parallel}$ over the momentum distribution $\overline{f}(p)$. The heat conductivity is $\chi(T) \sim T^{5/2}$ for a collision dominated plasma (e.g. Balbus and McKee, 1982). Finally, Λ is the radiative cooling function, and m the mean molecular weight of the gas.

In contrast to a pure gas dynamic system, now the quantities p_c and F_c appear in the overall momentum and energy balance. Equation (27) can be considered as the cosmic-ray 'energy-equation' in this cosmic-ray 'hydrodynamics'. It follows by taking the E_{kin}-moment of Equation (11). In the two distinct approximations: $\tilde{\mathbf{v}} = 0$, and $|\tilde{\mathbf{v}}| = |\mathbf{v}|$, these equations were used earlier by Axford et al. (1977), and McKenzie and Völk (1982), respectively.

The solution of Equations (24)–(28) is clearly beyond the scope of this paper and will be given elsewhere. However it is evident that these equations admit static solutions with

p_c = const, and div $\mathbf{F}_c = 0$, for $(\partial \overline{f}/\partial t)_s = (\partial \overline{f}/\partial t)_l = 0$, if the gas by itself is in static equilibrium. Therefore the dividing line between evaporation and accretion (McKee and Cowie, 1977), if it indeed exists, remains the same in the presence of a strong cosmic-ray population.

The consequences for the high energy γ-ray emission from individual clouds and for the integrated emission from different galactic directions depend of course on the quantitative size of the effects. But *qualitatively* we can conclude the following:

– Galactic regions of predominantly molecular hydrogen should be primarily represented by big accreting clouds. Their γ-ray emission should be *increased* by a factor up to 2 or even 3 over the standard value.

– Regions of atomic hydrogen (represented mainly by small evaporating clouds) have on average a *depressed* γ-ray emission, even though some of these clouds should accrete cool gas and thus be γ-astronomically neutral.

– The compression of roughly all cosmic-ray particles tends, at small energies, to offset the losses in the cloud.

– Very high energy particles may be less compressed in clouds than lower energy particles which carry the bulk of the cosmic-ray pressure and therefore determine the dynamics. This would tend to decrease the general galactic ratio of secondary to primary cosmic rays with increasing energy.

– The cosmic-ray compression in the galactic regions of large clouds corresponds to an enhancement of the cosmic-ray intensity I_{CR} with gas column density ρ. This gives some *physical* justification for phenomenological models (Cesarsky *et al.*, 1977; Hartmann *et al.*, 1979) which assume $\partial I_{CR}/\partial \rho > 0$.

– The local compression of cosmic rays might alone explain the γ-ray enhancement observed in the cloud ρ-Oph.

– Assuming the decrease of molecular hydrogen with galactocentric distance to be due to a decrease in the number of accreting large clouds and correspondingly assuming the uniformity of the surface density of atomic hydrogen to be due to a roughly uniform surface density of small evaporating clouds may possibly account for part or all of the apparent cosmic-ray intensity decrease towards the galactic anticenter (Dodds *et al.*, 1975).

Acknowledgements

The author is indebted to Drs Cesarsky, Dolginov, Dorfi, Drury, Kanbach, Mayer-Hasselwander, and Webb for discussions about different aspects of this paper. He also enjoyed a conversation with Drs Balbus and McKee on the question of nucleon modulation in evaporating clouds at the Space and Astrophysical Plasmas Workshop, Santa Barbara, USA. Finally, he would like to thank Drs Drury and McKee for valuable comments on the manuscript.

The editors thank M. Forman for assistance in evaluating this paper.

References

Axford, W. I.: 1981a, in G. Setti, G. Spada, and A. W. Wolfendale (eds.), 'Origin of Cosmic Rays', *IAU Symp.* **94**, 339.
Axford, W. I.: 1981b, *Proc. 17th Int. Cosmic Ray Conf. (Paris)* **12**, 155.
Axford, W. I., Leer, E., and Skadron, G.: 1977, *Proc. 15th Int. Cosmic Ray Conf. (Plovdiv)* **11**, 132.
Axford, W. I., Leer, E., and McKenzie, J. F.: 1982, *Astron. Astrophys.* **111**, 317.
Balbus, S. A. and McKee, C. F.: 1982, *Astrophys. J.* **252**, 529.
Bell, A. R.: 1978a, *Monthly Notices Roy. Astron. Soc.* **182**, 147.
Bell, A. R.: 1978b, *Monthly Notices Roy. Astron. Soc.* **182**, 443.
Bennett, K., Bignami, G. F., Boella, G., Buccheri, R., Hermsen, W., Kanbach, G., Lichti, G. G., Masnou, J. L., Mayer–Hasselwander, H. A., Paul, J. A., Scarsi, L., Swanenburg, B. N., Taylor, B. G., and Will, R. D.: 1977, *Astron. Astrophys.* **61**, 279.
Blandford, R. D.: 1980, *Astrophys. J.* **238**, 410.
Blandford, R. D. and Ostriker, J. P.: 1978, *Astrophys. J. Letters* **221**, L29.
Brown, R. L. and Marscher, A. P.: 1977, *Astrophys. J.* **212**, 659.
Burton, W. B. and Liszt, H. S.: 1981, in G. Setti, G. Spada, and A. W. Wolfendale (eds.), 'Origin of Cosmic Rays', *IAU Symp.* **94**, 227.
Cassé, M. and Paul, J. A.: 1980, *Astrophys. J.* **237**, 236.
Cesarsky, C. J.: 1980, *Ann. Rev. Astron. Astrophys.* **18**, 289.
Cesarsky, C. J. and Völk, H. J.: 1978, *Astron. Astrophys.* **70**, 367.
Cesarsky, C. J., Cassé, M., and Paul, J. A.: 1977, *Astron. Astrophys.* **60**, 139.
Chen, G. and Armstrong, T. P.: 1975, *Proc. 14th Int. Conf. Cosmic Rays (Munich)* **5**, 1814.
Chevalier, R. A.: 1975, *Astrophys. J.* **200**, 698.
Chin, Y. C. and Wentzel, D. G.: 1972, *Astrophys. Space Sci.* **16**, 465.
Cowie, L. L. and McKee, C. F.: 1977, *Astrophys. J.* **211**, 135.
Cowie, L. L., McKee, C. F., and Ostriker, J. P.: 1981, *Astrophys. J.* **247**, 908.
Dalgarno, A. and McCray, R.: 1972, *Ann. Rev. Astron. Astrophys.* **10**, 375.
Dodds, D., Strong, A. W., and Wolfendale, A. W.: 1975, *Monthly Notices Roy. Astron. Soc.* **171**, 569.
Dolginov, A. Z.: 1980, *Soviet Astron. Letters* **6**(2), 132.
Drury, L.O'C.: 1983, *Rep. Progr. Phys.*, in press.
Drury, L.O'C. and Völk, H. J.: 1981, *Astrophys. J.* **248**, 344.
Eichler, D.: 1981, *Astrophys. J.* **244**, 711.
Ellison, D.: 1981, Ph.D. Thesis, The Catholic University of America, Washington, D.C.
Fermi, E.: 1949, *Phys. Rev.* **75**, 1169.
Garcia-Munoz, M., Mason, G. M., and Simpson, J. A.: 1977, *Astrophys. J.* **217**, 859.
Gordon, M. A. and Burton, W. B.: 1976, *Astrophys. J.* **208**, 346.
Graham, R. and Langer, W. D.: 1973, *Astrophys. J.* **179**, 469.
Hartmann, R. C., Kniffen, D. A., Thompson, D. J., Fichtel, C. E., Ögelman, H. B., Tümer, T., and Ozel, M. E.: 1979, *Astrophys. J.* **230**, 597.
Hasselmann, K. and Wibberenz G.: 1970, *Astrophys. J.* **162**, 1049.
Hayakawa, S.: 1952, *Prog. Theor. Phys.* **8**, 571.
Honda, M.: 1979, *Proc. 16th Int. Cosmic Ray Conf. (Kyoto)* **14**, 159.
Hudson, P. D.: 1965, *Monthly Notices Roy. Astron. Soc.* **131**, 23.
Hutchinson, G. W.: 1952, *Phil. Mag.* **43**, 847.
Jokipii, J. R.: 1966, *Astrophys. J.* **146**, 480.
Jokipii, J. R. and Parker, E. N.: 1970, *Astrophys. J.* **160**, 735.
Krymsky, G. F.: 1977, *Dokl. Akad. Nauk SSR* **234**, 1306.
Kulsrud, R. M. and Cesarsky, C. J.: 1971, *Astrophys. Letters* **8**, 189.
Kulsrud, R. M. and Pearce, W.: 1969, *Astrophys. J.* **156**, 445.
Landau, L. D. and Lifshitz, E. M.: 1959, *Fluid Mechanics*, Pergamon Press, p. 329ff.
Lebrun, F. and Paul, J. A.: 1978, *Astron. Astrophys.* **65**, 187.
Lebrun, F., Bignami, G. F., Buccheri, R., Caraveo, P. A., Hermsen, W., Kanbach, G., Mayer-Hasselwander, H. A., Paul, J. A., Strong, A. W., and Wills, R. D.: 1982, *Astron. Astrophys.* **107**, 390.
Lee, M. A.: 1982, *J. Geophys. Res.* **87**, 5063.
Lee, M. A. and Völk, H. J.: 1973, *Astrophys. Space Sci.* **24**, 31.

Lee, M. A. and Völk, H. J.: 1975, *Astrophys. J.* **198**, 485.

Lerche, I.: 1967, *Astrophys. J.* **147**, 689.

Marscher, A. P. and Brown. R. L.: 1978, *Astrophys. J.* **221**, 583.

McKee, C. F. and Cowie, L. L.: 1977, *Astrophys. J.* **215**, 213.

McKee, C. F. and Ostriker, J. P.: 1977, *Astrophys. J.* **218**, 148.

McKenzie, J. F. and Völk, H. J.: 1981, *Proc. 17th Int. Cosmic Ray Conf. (Paris)* **9**, 242.

McKenzie, J. F. and Völk, H. J.: 1982, *Astron. Astrophys.* **116**, 191.

McKenzie, J. F. and Westphal, K. O.: 1969, *Planet. Space Sci.* **17**, 1029.

Mezger, P. G.: 1978, *Astron. Astrophys.* **70**, 565.

Montmerle, T.: 1979, *Astrophys. J.* **231**, 95.

Montmerle, T. and Cesarsky, C. J.: 1981, *Proc. 17th Int. Cosmic Ray Conf. (Paris)* **1**, 173.

Morfill, G. E.: 1982, *Monthly Notices Roy. Astron. Soc.* **198**, 583.

Morfill, G. E.: 1982, *Astrophys. J.* **262**, 749.

Morfill, G. E. and Meyer, P.: 1981, *Proc. 17th Int. Cosmic Ray Conf. (Paris)* **9**, 56.

Morfill, G. E, Völk, H. J., Drury, L., Forman, M., Bignami, G. F., and Caraveo, P. A.: 1981, *Astrophys. J.* **246**, 810.

Morrison, P.: 1958, *Nuovo Cimento* **VII**, Nr. 6, p. 858.

Parker, E. N.: 1963, *Interplanetary Dynamic Processes*, Interscience Publ., New York, London, p. 165.

Parker, E. N.: 1968, in B. M. Middlehurst and L. A. Aller (eds.), *Stars and Stellar Systems*, Vol. VII: 'Nebulae and Interstellar Matter', The University of Chicago Press, p. 707ff.

Parker, E. N.: 1969, *Space Sci. Rev.* **9**, 651.

Penston, M. V. and Brown, F. E.: 1970, *Monthly Notices Roy. Astron. Soc.* **150**, 373.

Sagdeev, R. Z. and Galeev, A. A.: 1969, *Nonlinear Plasma Theory*, W. A. Benjamin, Inc., New York.

Salvati, M. and Massaro, E.: 1978, *Astron. Astrophys.* **67**, 55.

Scott, J. S. and Chevalier, R. A.: 1975, *Astrophys. J.* **197**, L5.

Scoville, N. Z. and Solomon, P. M.: 1975, *Astrophys. J.* **199**, L105.

Sedov, L. I.: 1959, *Similarity and Dimensional Methods in Mechanics*, Academic Press, New York.

Shull, J. M.: 1980, *Astrophys. J.* **237**, 769.

Skilling, J., 1971, *Astrophys. J.* **170**, 265.

Skilling, J.: 1975, *Monthly Notices Roy. Astron. Soc.* **172**, 557.

Skilling, J. and Strong, A. W.: 1976, *Astron. Astrophys.* **53**, 253.

Solomon, P. M., Scoville, N. Z., and Sanders, D. B.: 1979, *Astrophys. J. Letters* **232**, L89.

Stecker, F. W.: 1971, *Cosmic Gamma Rays*, NASA SP-249, p. 106ff.

Swanenburg, B. N., Bennett, K., Bignami, G. F., Caraveo, P., Hermsen, W., Kanbach, G., Masnou, J. L., Mayer–Hasselwander, H. A., Paul, J. A., Sacco, B., Scarsi, L., and R. D. Wills: 1978, *Nature* **275**, 298.

Tenorio–Tagle, G.: 1979, *Astron. Astrophys.* **71**, 59.

Terawasa, T.: 1979, *Planet. Space Sci.* **27**, 193.

Toptyghin, I. N.: 1980, *Space Sci. Rev.* **26**, 157.

Vasilyev, V. N., Toptyghin, I. N., and Chirkov, A. G.: 1980, *Kosmissled.* **18**, 556 (English translation in *Cosm. Res.* **18**, 401).

Völk, H. J.: 1981, *Izv. AN SSSR, Ser. fiz.* **45**, No. 7, (in Russian), p. 1122 (English translation: *Bull. Acad. Sci. USSR, Phys. Ser.* **45**, No. 7, p. 1, Allerton Press Inc.).

Völk, H. J. and Forman, M. A.: 1982, *Astrophys. J.* **253**, 188.

Völk, H. J., Morfill, G. E., Alpers, W., and Lee, M. A.: 1974, *Astrophys. Space Sci.* **26**, 403.

Völk, H. J., Forman, M. A., and Morfill, G. E.: 1981, *Astrophys. J.* **249**, 161.

Webb, G. M.: 1983a, to appear in *Astron. Astrophys.*

Webb, G. M.: 1983b, to appear in *Astrophys. J.*

Wentzel, D. G.: 1968, *Astrophys. J.* **152**, 987.

Wentzel, D. G.: 1974, *Ann. Rev. Astron. Astrophys.* **12**, 71.

Wheeler, J. C., Mazurek, T. J., and Sivaramakrishnan, A.: 1980, *Astrophys. J.* **237**, 781.

Zel'dovich, Ya. B. and Pikel'ner, S. B.: 1969, *J. Eksp. Theor. Phys.* **29**, 170.

ACCELERATION AND TRANSPORT PROCESSES: VERIFICATION AND OBSERVATIONS*

J. R. JOKIPII

Dept. of Planetary Sciences, Univ. of Arizona, Tucson, AZ 85721, U.S.A.

Abstract. The general problem of diffusive transport and acceleration of energetic charged particles is considered. The transport of solar-flare particles, solar modulation of galactic cosmic rays and shock acceleration processes on the solar wind are examined and observational tests are summarized. It is concluded that the basic diffusive transport equation is a useful approximation in situations like the solar wind, where turbulent scattering by magnetic irregularities is sufficient to maintain near isotropy. The application of this equation to the interstellar medium and other, more distant astrophysical regimes is then discussed and implications for gamma-ray astrophysics are outlined. Finally the evidence for interstellar turbulence is reviewed and its consequences briefly discussed.

1. Introduction

Essential to an understanding of most areas of high-energy astrophysics is an understanding of the physics of the acceleration and transport of energetic charged particles or cosmic rays. The *in situ* study of these processes of acceleration and transport in interplanetary space has in many cases resulted in substantial deepening of our conceptual understanding and in an increased confidence in the applicability of the theories. For, one of the most severe tests of an astrophysical theory is to subject it to close scrutiny in a regime where the parameters and boundary conditions are as well determined as they can be in interplanetary space.

The transport and acceleration of charged particles in interplanetary space are quite similar to that expected in the interstellar medium. In both instances we confront the problem of charged particle motion in a collisionless, tenuous background plasma consisting mostly of protons and electrons. Perhaps the most important difference between the two is that in the interstellar medium, the energy density of the energetic particles is comparable with that of convective motions and the magnetic field, whereas in the solar wind the energetic-particle energy density is small. This can result in some differences between the two situations but, nontheless, they are expected to be quite similar to each other.

2. Basic Theory

The acceleration and transport of charged particles in astrophysics may be described quite succinctly by the equation of motion

$$\dot{\mathbf{p}} = q\left[\mathbf{E} + \frac{\mathbf{w} \times \mathbf{B}}{c}\right], \tag{1}$$

* Proceedings of the XVIII General Assembly of the IAU: *Galactic Astrophysics and Gamma-Ray Astronomy*, held at Patras, Greece, 19 August 1982.

where the electromagnetic fields $\mathbf{E}$, $\mathbf{B}$ are determined by the state of the background plasma (which contributes the mass and momentum). The basic problem is to call upon our knowledge of the structure and dynamics of the plasma to determine the electric and magnetic fields to be used in Equation (1) and to determine the initial and boundary conditions for the accelerated particles. Since the equation is impossible to solve in general, a number of approximations have been utilized.

The most important and generally-used approximation is the diffusion approximation. This is based chiefly on the observation that energetic particles in many important cases are observed to have a very nearly isotropic pitch angle distribution relative to the local plasma. This isotropy is a consequence of the 'scattering' of the particles by turbulent fluctuations in the ambient magnetic field, and may be derived from a consideration of the equation of motion (for reviews see Jokipii, 1971; Volk, 1975; or Fisk, 1978). Space does not permit a discussion of this theory here. The theory, in the near-isotropic limit results in a diffusion convection with energy change, which may be written with the various physical effects noted next to the corresponding terms:

$$\frac{\partial f}{\partial t} = \frac{\partial}{\partial X_i}\left[K_{ij}\frac{\partial f}{\partial X_j}\right] \qquad \text{(diffusion)}$$

$$-V_i\frac{\partial f}{\partial X_i} \qquad \text{(convection)}$$

$$+\frac{1}{3}\frac{\partial V_i}{\partial X_i}\frac{\partial f}{\partial \ln p} \qquad \text{(adiabatic energy change)} \tag{2}$$

$$+\frac{1}{2}\frac{\partial}{\partial p}\left[D_{pp}\frac{\partial f}{\partial p}\right]\text{(2nd-order Fermi accel.)}$$

where $f(\mathbf{r}, p, t)$ is the omnidirectional particle distribution as a function of position r, momentum magnitude p, and time t. $\mathbf{V}$ is the background convection velocity, K_{ij} is the diffusion tensor which may be expressed in terms of the spectrum of magnetic irregularities in the turbulence, D_{pp} is the rate of acceleration due to the random motion of the magnetic irregularities, and Q is the local source function. Equation (2) contains, as a special case, nearly all of the cosmic-ray acceleration and transport mechanisms discussed in recent years and will be the basis of the present discussion. Of course, in many cases, some of the terms in Equation (2) may be omitted to simplify the analysis. In addition, it must be kept in mind that under certain conditions the scattering is week and the distribution may be highly anisotropic, in which case Equation (2) may not be sufficiently accurate.

Another way of viewing Equation (2) which shows that it should be in fact a reasonable approximation in widely varied circumstances, is to note that Liouville's theorem states that a homogeneous, isotropic distribution in an arbitrary static magnetic field is in a steady state. Equation (2) essentially describes the first-order consequences of gradients in the distribution function and plasma flow velocity (which, among other things produces the first-order $\mathbf{v} \times \mathbf{B}$ magnetic field). Hence, even if we do not know some of the transport coefficients accurately, the general form of Equation (2) may be adequate for many purposes.

Nearly all models of cosmic-ray transport in astrophysics make use of Equation (2), or some simplification of it. For example, the diffusion of cosmic rays in the interstellar gas or in supernova remnants follows Equation (2). The basic theory of acceleration at hydromagnetic shocks or the venerable second-order Fermi acceleration also use Equation (2).

In view of the wide-reaching importance of this equation, the fact that it applies also to transport in the solar wind, which may be observed in much more detail than more-distant regions, makes it possible to subject it to more stringent tests than might otherwise be possible.

3. Applications in the Interplanetary Medium

A. SOLAR FLARE EVENTS

The first quantitative application of diffusive transport to explain observations was to the time-intensity profile of relativistic particles during the February 23, 1956, solar particle event by Meyer *et al.* (1956). For these particles the diffusive term is the only important one on the rhs of Equation (2), and it was established that a diffusive profile with a reasonable value for the diffusion coefficient provided a compelling interpretation of the event. A more-recent comparison of theory and observation is shown in Figure 1. The agreement is clearly quite good. Palmer (1982) has recently reviewed the interpretation of solar events in terms of Equation (2). His compilation of the diffusion mean-free path as a function of energy as determined by a number of authors is shown in Figure 2 together with theoretical determinations of the mean free path determined from a straightforward application of quasi-linear theory. Clearly, we may conclude that although the basic diffusion appears to be valid, the value of the mean free path is significantly larger than predicted at low energies. There is as yet no generally accepted explanation for the discrepancy. Perhaps the most promising approach is that of Morfill and Volk (1978) and Scholer and Morfill (1982) who suggested that medium scale fluctuations in the magnetic field result in a larger mean free path.

It should also be mentioned that a number of highly-anisotropic events have been observed at very low energies and these of course cannot be interpreted in terms of Equation (2). An approach to these events in terms of focussing by the diverging interplanetary magnetic field and a low scattering rate has been developed principally by Earl (see, e.g. Earl, 1972) and applied successfully to these events. Earl's equations may be regarded as an extension of Equation (2) to situations where the physics of particle motion force a large anisotropy. However, in order to accomplish this extension, some effects such as perpendicular transport and energy change were neglected.

In summary, the transport of solar cosmic rays, particularly at high energies, appears

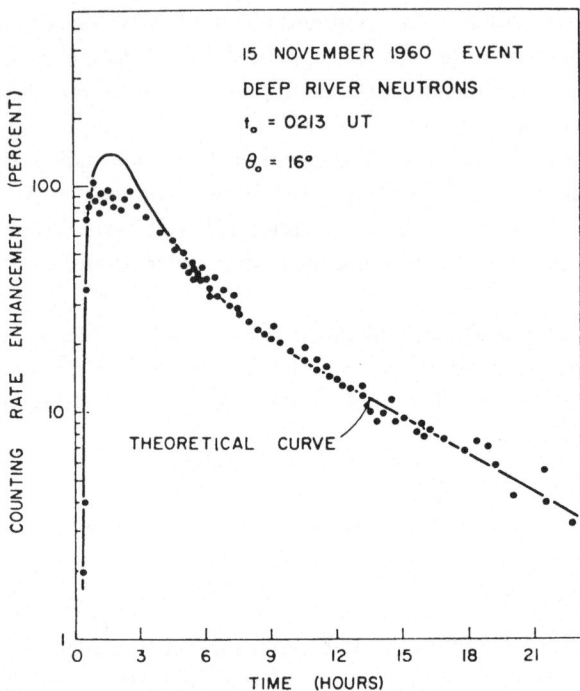

Fig. 1. A theoretical fit, using the diffusion part of Equation (2), to the Deep River neutron monitor data for the November 15, 1960, solar particle event. θ_0 is the angle between the flare and the average interplanetary magnetic field line passing through the point of observation (Burlaga, 1967).

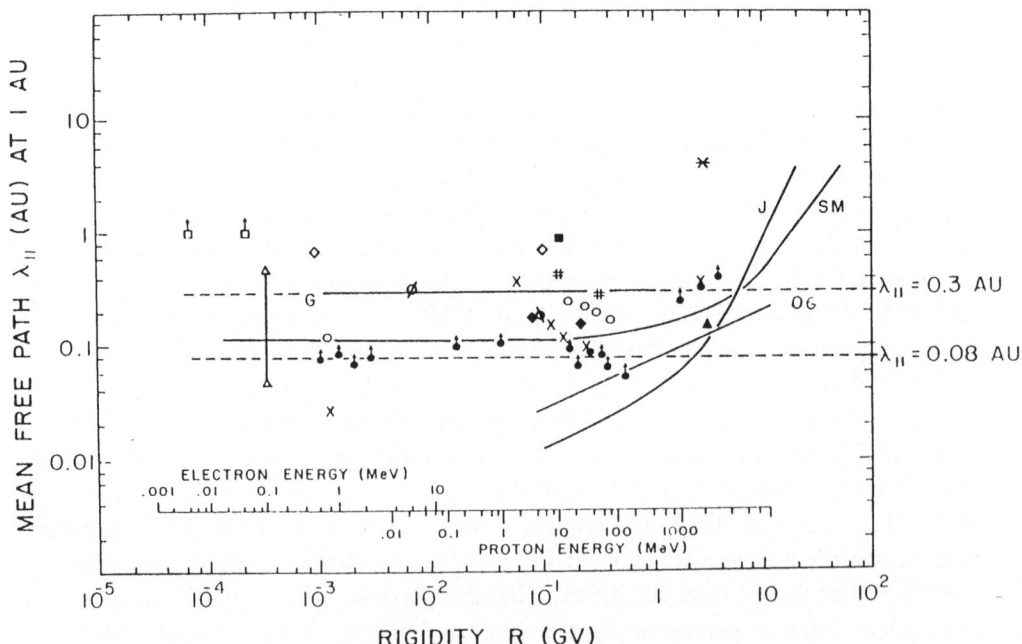

Fig. 2.   Composite plot of diffusion mean free path parallel to the magnetic field vs particle rigidity for both electrons and protons (Palmer, 1982).

to be reasonably well explained by Equation (2). The most important term in this case is the spatial diffusion term and so we may say that solar cosmic rays have tested the general validity of spatial diffusive transport and verified that it is a good approximation, at least at high energies.

B. SOLAR MODULATION OF GALACTIC COSMIC RAYS

Possibly the most complicated and detailed application of Equation (2) has been to the modulation of galactic cosmic rays by the Sun. Here the solar system is regarded as residing in a constant, isotropic bath of galactic cosmic rays. The outflowing solar wind acts to partially exclude these particles from the inner solar system, and the resulting quasistationary balance results in the modulated cosmic-ray intensity (see Figure 3). In this case, all the terms in the equation except for the D_{pp} term play important roles.

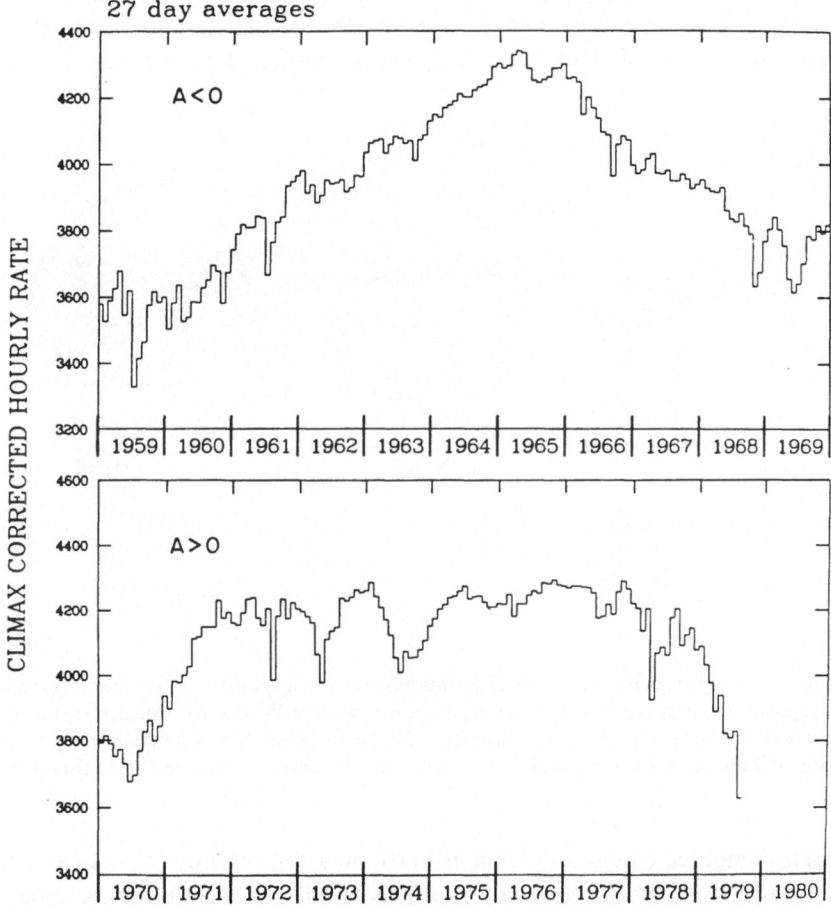

Fig. 3.   Illustration of the solar-cycle modulation of cosmic rays by the solar wind, as shown in the Climax neutron monitor counting rate, which responds to ~ 2 GeV protons. The two sunspot cycles correspond to different solar magnetic field field directions, and appear to have different shapes. Current theory produces similar shape differences as a natural consequence of the magnetic field reversal. (Data from J. A. Simpson, private communication, 1980.)

Indeed, the full form of Equation (2) (without the D_{pp} term) was first written down in response to the challenges presented by solar modulation (e.g., Parker, 1965; Axford, 1965; Jokipii and Parker, 1969). It appears that, at least for particle energies greater than approximately 1 MeV, the D_{pp} term is small and may be safely neglected. Hence the modulation may be regarded as a balance between the inward random walk or diffusion, the outward convection by the solar wind, gradient and curvature drifts caused by the large-scale structure of the interplanetary magnetic field, and the adiabatic cooling due to the radial expansion of the solar wind.

Sophisticated fully three-dimensional numerical solutions have recently been obtained, utilizing the presently accepted picture of the solar wind and its entrained magnetic field illustrated schematically in Figure 4. The model is quite accurate in the equatorial regions, where there are many direct observations, but the values of the parameters such as ambient magnetic field and flow velocity in the polar regions are quite uncertain. This uncertainty notwithstanding, the calculated properties of the model are in quite good agreement with the observed properties of cosmic rays.

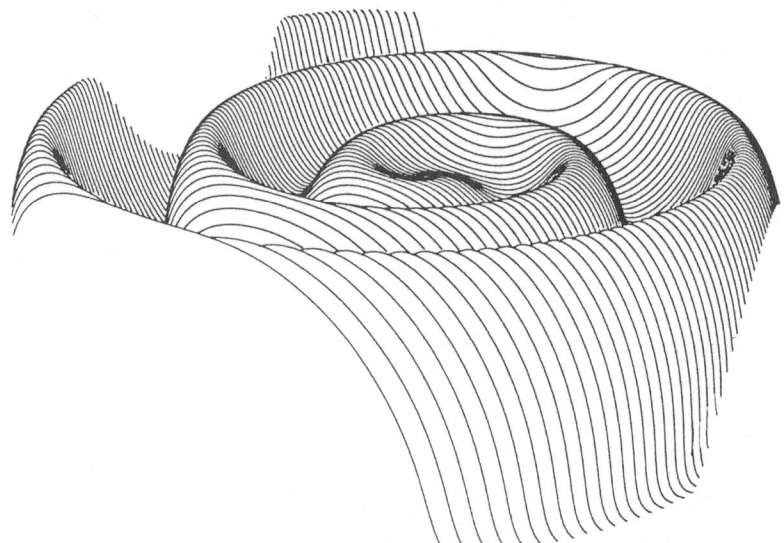

Fig. 4.   Illustration of the general structure of the interplanetary magnetic field as inferred from spacecraft data. The surface is the (thin) current sheet separating the oppositely directed Archimedean spiral magnetic fields in the northern and southern solar hemispheres. The shape of the sheet is determined by the rotation of the Sun and the outward solar wind flow. In this view the observer is some 75 AU from the Sun.

A sample computed energy spectrum at Earth obtained recently by Kota and Jokipii (1982) is shown in Figure 5. They used a fully three-dimensional numerical code which was a finite-difference approximation to the full Equation (2). The only restriction in the model was that the modulation be time-independent in a frame co-rotating with the Sun. The parameters used correspond to a diffusion coefficient with a scattering mean free path of several particle gyro-radii, a solar wind which is independent of latitude, and

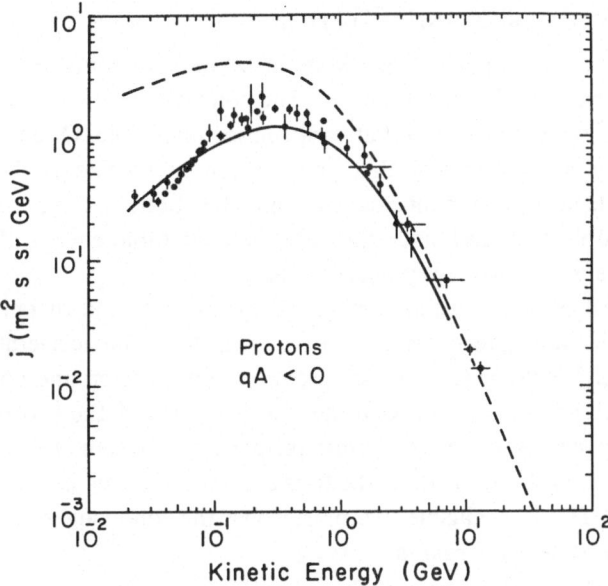

Fig. 5.   Computed energy spectrum of galactic cosmic-ray protons during the 1965 solar minimum period compared with observations. The dashed line is the assumed interstellar spectrum and the solid line gives the theoretical calculation. Computations carried in collaboration with J. Kota.

a magnetic field which is a Parker archimedean spiral which is either inward or outward in the northern hemisphere and in the opposite direction in the southern hemisphere. The surface at which the field changes direction abruptly is the so-called heliospheric current which near solar sunspot lies near the heliospheric equator. These assumptions represent a conservative extrapolation of observations near the heliospheric equator to high heliographic latitudes.

A remarkable consequence of the model is that the particle gradient and curvature drifts due to the large scale spiral magnetic field have a dominant effect on the solutions over a wide range of parameters. The model predicts that the full cycle of cosmic-ray modulation is 22 years, and that alternate solar sunspot cycles will exhibit somewhat different modulation. This is in agreement with observations. Of course, detailed comparison with observations awaits observations at high heliographic latitudes. One may conclude that if the field at high latitudes is a reasonable extrapolation of that observed, drifts must be a significant effect in modulation. In addition, local phenomena such as shocks generated by solar flares and regions of fast and slow flow in the solar wind also produce effects such as Forbush decreases and it is likely that combinations of these effects may be an important factor in the modulation of galactic cosmic rays (see, e.g. McDonald *et al.*, 1981).

In summary, the modulation of cosmic rays by the solar wind provides a reasonably complete verification of the basic transport equation, with the exception of the Fermi acceleration term.

C. Acceleration at Bow and Interplanetary Shocks

In the past several years, particle acceleration by the electromagnetic fields at shock fronts has received considerable attention. The concept has a long theoretical history, but only recently has detailed observational support available. One may now say with a reasonable degree of certainty that acceleration at shocks is the only generally applicable particle acceleration mechanism which has been observed. Since shocks are common in astrophysics, and since they have considerable energy, they may well be most important site of cosmic-ray acceleration.

Again, the most popular current version of this acceleration theory is described by Equation (2) with the D_{pp} term set equal to zero. Indeed, the acceleration comes from the adiabatic energy-change term, which provides compression and consequent energy increase at the shock front. To reveal the basic physics of the process most clearly, consider a locally plane shock with uniform properties upstream and downstream of the shock (see Figure; 6). If we work in the frame at rest with respect to the shock front, and assume that D_{pp} is negligible, we must solve the following equation in both the upstream and downstream regions

$$\frac{\partial f}{\partial t} = \frac{\partial}{\partial X}\left[K_{Xj}\,\frac{\partial f}{\partial X_j}\right] - V_i\,\frac{\partial f}{\partial X_i}\,, \tag{3}$$

where, in many cases, the steady state solution is desired and the time derivative may be set equal to zero. The solutions so obtained must be matched across the shock. The matching conditions contain the acceleration which corresponds physically to the particles being statistically compressed in the region of flow compression at the shock.

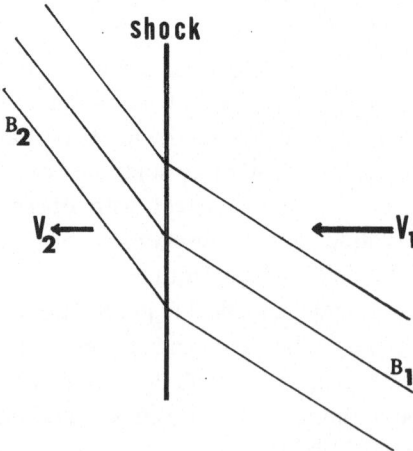

Fig. 6.   Schematic illustration of the parameters relevant to diffusive shock acceleration. The x axis is along the flow direction and the y direction is normal to the plane containing the upstream magnetic field and flow velocity.

Although it is not completely obvious, a sufficiently accurate jump condition may be obtained by integrating Equation (2) across the shock (for a discussion of this point, see Toptygin, 1979). If we denote the upstream and downstream regions by 1 and 2, we obtain the general jump condition

$$\left[K_{Xj} \frac{\partial f}{\partial X_j} + \frac{V_X}{3} \frac{\partial f}{\partial \ln p} \right]_1^2 = Q_* ,\tag{4}$$

where Q_* is a possible source function at the shock.

The first solutions to these equations in the steady state and for a no transverse structure were published by Blandford and Ostriker (1978), Bell (1978), and Axford *et al.* (1978). Solutions which incorporated more general magnetic fields, the resultant drifts and transverse structure were published by Jokipii (1982). In the simplest case of a steady state with no transverse structure gives rise to the solution

$$f(p) \propto p^{-q} ,\tag{5}$$

where $q = 3V_{1X}/(V_{1X} - V_{2X})$. As has been pointed out elsewhere in this symposium, this solution has many attractive features in the context of galactic cosmic rays. The waves and turbulence which scatter the particles may in some cases be partially generated by the particles themselves through instabilities.

Although the simple infinite, plane, steady state solution (5) is physically revealing and interesting, more complex solutions are required for shocks in the solar system. The most thoroughly studied shock is, of course, the Earth's bow shock. This differs from the simple case discussed above chiefly in that it is not infinite and plane. Nontheless, energetic particles are frequently observed upstream of the shock, and it appears quite likely that they are accelerated in a time-dependent and spatially dependent version of the above. It also appears in this case that the scattering of the particles upstream is enhanced by particle-driven instabilities. An example of an event observed upstream of the Earth's bow shock is shown in Figure 7.

In addition, particles are clearly observed to be accelerated in the vicinity of co-rotating interaction regions, regions associated with shock waves generated by long-lived fast solar streams overtaking solwer streams as a consequence of solar rotation (see, e.g. Tsurutani *et al.*, 1982). It seems likely that these particles are accelerated at the shock fronts, much as discussed above.

It appears, then, that the application of Equation (2) to the general problem of acceleration has been successful. As does the modulation of cosmic rays by the solar wind, this application tests the full equation with the exception of the D_{pp} term.

D. OTHER APPLICATIONS IN THE SOLAR WIND

The basic transport Equation (2) has been applied in other contexts in the solar wind, and I will only mention them briefly here. The transport of energetic electrons from Jupiter has been successfully described by diffusive transport (see, e.g. Pyle, 1979, for a recent survey). Transient decreases in the cosmic rays caused by propagating solar-

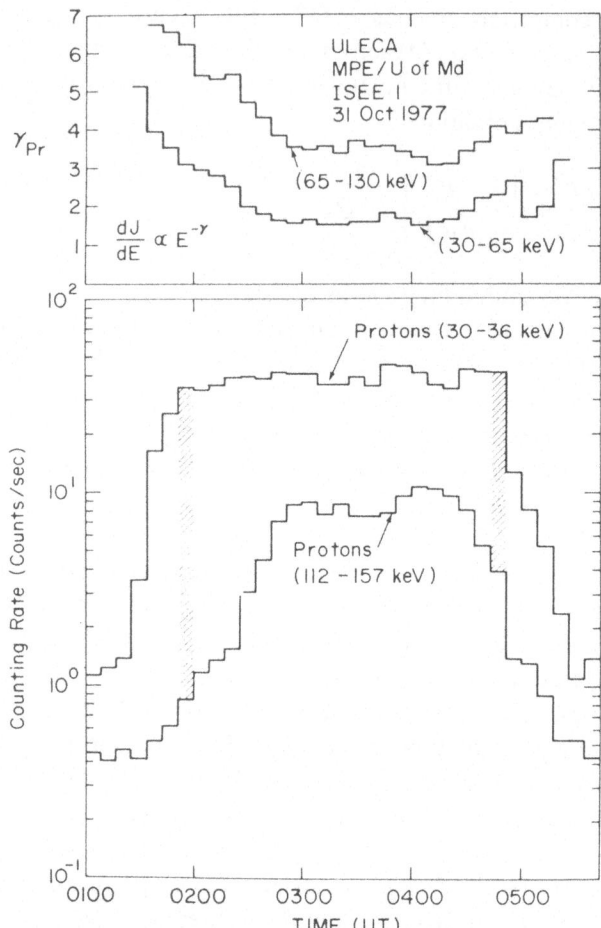

Fig. 7. Particles accelerated at the Earth's bow shock and observed upstream of the shock (Ipavich *et al.*, 1979).

wind shock waves, called Forbush decreases, also canbe understood in terms of the basic equation.

E. GENERAL CONCLUSIONS FROM THE INTERPLANETARY MEDIUM

From the above examples we can conclude that the general approach to cosmic-ray transport embodied in Equation (2) is an accurate approximation in cases where the particle pitch-angle distribution is reasonably close to isotropy. The various physical processes appearing in the equation, and enumerated above have been tested, with the exception of the second-order Fermi acceleration term. Although there is no evidence up to now that this latter term is important, there is also no evidence against it.

Since isotropy is more-readily attained in regions having large scale, and which are far from condensed objects, it appears that the constraint of near isotropy may be even more readily satisfied in the interstellar medium. Hence, provided that there is turbulence or magnetic fluctuations, one should expect these equations to be valid in interstellar space. This question of interstellar turbulence is addressed in the next section.

4. Is the Interstellar Medium Turbulent?

As was pointed out above, application of the basic transport Equation (2) to the transport of cosmic rays in the interstellar gas requires that there be enough turbulent fluctuations to keep the cosmic-ray distribution function nearly isotropic. Indirect evidence that this indeed is the case comes from the fact that the observed anisotropy of galactic cosmic rays at high energies (where the interplanetary magnetic field does not prevent the galactic anisotropy from being seen at Earth) is very small. Observations at energies $> 10^{11}$ eV give an anisotropy of less than 10^{-4} (see, e.g. Cutler *et al.*, 1981; Elliot, 1974). For this reason it is likely that some form of Equation (2) is a good approximation in the interstellar gas.

Over the past three decades, in fact, special cases of Equation (2) have been used in many theoretical treatments of the origin and transport of cosmic rays in the interstellar medium. See, e.g. Ginzburg and Syrovatsky (1962) and Axford (1981) for reviews spanning the early ideas to the most recent. It is my purpose in this section to review briefly the observational evidence for interstellar turbulence. As we will see, there is suggestive, but not definitive evidence that the interstellar medium is indeed turbulent, with a standard Kolmogorov-like* power spectrum. This lends further support to the use of Equation (2) in this regime.

Observations of the interstellar medium are necessarily indirect and our knowledge of its structure is correspondingly incomplete. Nonetheless there is a growing body of observational evidence that can be used to support the view that the interstellar gas is turbulent. The interstellar parameter most relevant to cosmic-ray transport is the interstellar magnetic field.

There are four basic types of observations which give rise, more or less directly, to estimates of the magnitude and/or direction of the interstellar magnetic field. They are: (a) Faraday rotation of linearly polarized radiation: (b) polarization of starlight by oriented dust grains; (c) synchrotron radiation; and (b) Zeeman effect.

The best observations of the local structures of the interstellar magnetic field involve utilization of Faraday rotation of waves from extragalactic sources and pulsars.

If the frequency of the observed wave ω is much larger than either the electron cyclotron frequency or plasma frequency the rotation θ is given by the integral along the line of sight s:

$$\theta(s, \lambda) = \theta_0(\lambda) + \lambda^2 C_R \int_0^s N_e \, \mathbf{B} \cdot d\mathbf{s} = \theta_0(\lambda) + \lambda^2 R(s), \tag{6}$$

* By this I mean a power spectrum which is of the general form of inertial-subrange turbulence in ordinary fluid mechanics, $P(k) \propto k^{-\alpha}$, where k is wavenumber and the power-law index α is 3.5 to 3.9.

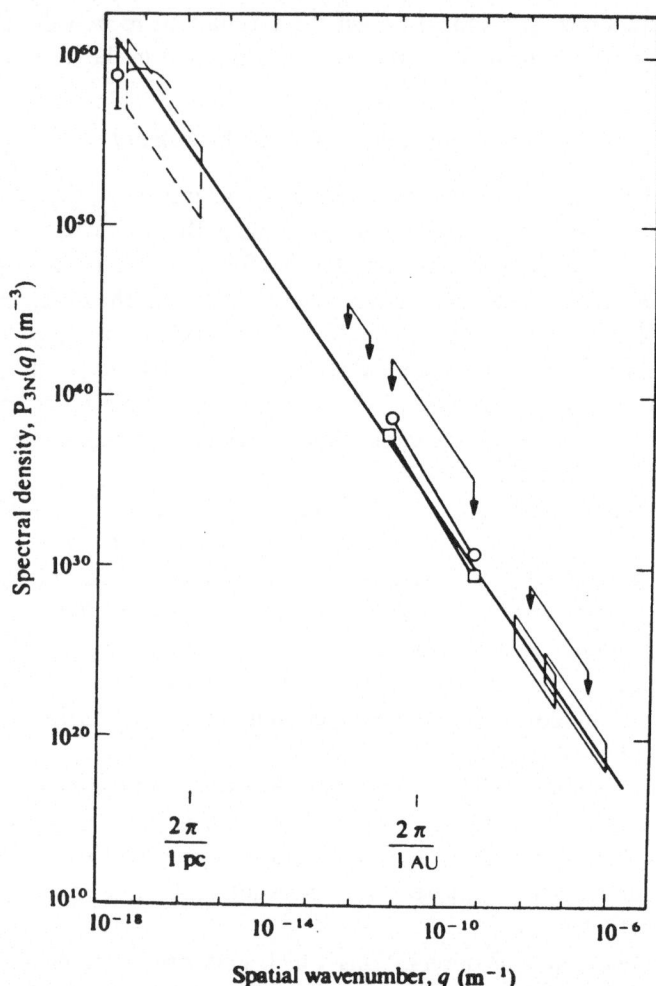

Fig. 8.   Estimates or upper limits on the power spectral density of density fluctuations in the interstellar medium, obtained by Armstrong *et al.* (1980) from various observation techniques. The straight line is the Kolmogorov spectrum.

where θ_0 is the initial angle of polarization at the source, C_R is a known constant, and R is defined as the rotation measure.

The appearance of the unknown N_e in Equation (6) is bothersome and observations of pulsars have been used to attempt to correct for this by making use of signal dispersion. In addition to Faraday rotation a signal suffers a time delay

$$\tau(s, \lambda) = \tau_0 + \lambda^2 C_D \int_0^s N_e \, ds = \tau_0(\lambda) + \lambda^2 D(s). \qquad (7)$$

Observation of both $D(s)$ and $R(s)$ are available for some pulsars (e.g. Manchester, 1972) and one can hope to eliminate N_e. One finds for the *average* magnetic field near the solar system a uniform field of magnitude 3×10^{-6} G effectively parallel to the plane of the disk. It is, however, also evident that there are real, large fluctuations about this uniform field which, at least in part, must be regarded as random, or turbulent. For example, Jokipii and Lerche (1969) and Wright (1973) found evindence for random fluctuations in the magnetic field with a coherence scale of the order of 100 pc.

The same general structure for *velocity* fluctuations was reported by Kaplan (1966) who obtained the correlation function of the radial (w.r.t. the observer) velocity in clouds as a function of the transverse coordinate. The velocities are measured using Doppler shifts of resonance lines. Again, the data suggest random, or turbulent fluctuations with a coherence scale of about 100 pc and a Kolmogorov-like spectrum down to scales ~ 1 pc. More recent work by Larson (1979), based on stellar kinematics, also suggests a turbulent velocity spectrum.

The structure of interstellar fluctuations at small scales is more difficult to determine. The most convincing data come from observations of the interstellar scintillations of pulsar signals. The scintillations are caused by electron-density fluctuations with scales 10^{11} cm. The theory relating the fluctuations to the scintillations is complicated (see, e.g. Lee and Jokipii, 1974, for a modern treatment). The most comprehensive recent review of both observation and theory is that of Rickett (1978). As pointed out by Lee and Jokipii and by Rickett, the data are consistent with a spatially continuous distribution of fluctuations along the line of sight. It is remarkable that the data also suggest a Kolmogorov spectrum which extrapolates smoothly to that inferred above at scales of the order of parsecs.

A recent, important and interesting paper by Armstrong *et al.* (1981) considers observations of many types which cover the whole range of scales from tens of pc. down to 10^{10} cm. Figure 8 shows the compendium of data presented by Armstrong *et al.* They conclude that the data a strikingly suggestive of a generally turbulent medium with a Kolmogorov spectrum.

In summary, available data suggest the possibility of a generally turbulent interstellar medium, with a coherence scale of the order of 100 pc and possibly a Kolmogorov turbulence spectrum. If this were to be close to the actual situation, the consequences for cosmic-ray transport would be substantial. It would provide a natural mechanism for the confinement of cosmic rays to the galaxy and assist in the acceleration of cosmic rays either by shocks as discussed above, or by a second-order Fermi process in interstellar space.

The existence of such a turbulence spectrum is not without theoretical difficulties. It has been pointed out (e.g. Cesarsky, 1971) that the damping of the fluctuations may be large enough to make the problem of maintaining the turbulence non-trivial. Similarly, the question of the origin of the turbulence must be addressed (see, e.g. Fleck, 1980) for one recent approach to this problem. Finally, it is clear that the interstellar medium is not statistically homogeneous, and the relation of any turbulence to the various phases of the interstellar gas is yet to be addressed in any overall model.

Acknowledgements

This work was supported, in part, by the National Aeronautics and Space Administration under Grant NSG–7101 and by the National Science Foundation under Grant ATM–220–18.

The editors thank M. Scholer for assistance in evaluating this paper.

References

Armstrong, J. W., Cordes, J. M., and Rickett, B. J.: 1981, *Nature* **291**, 561.
Axford, W. I.: 1965, *Planet. Space Sci.* **13**, 115.
Axford, W. I.: 1981, 'Acceleration of Cosmic Rays by Shock Waves', Preprint, MPAE.
Axford, W. I., Leer, E., and Skadron, G.: 1977, *Proc. 15th Int. Cosmic Ray Conf., Plovdiv.* **11**, 132.
Bell, A. R.: 1978a, *Monthly Notices Roy. Astron. Soc.* **182**, 147.
Bell, A. R.: 1978b, *Monthly Notices Roy. Astron. Soc.* **182**, 443.
Blandford, R. D. and Ostriker, J. P.: 1978, *Astrophys. J.* **221**, L29.
Burlaga, L. F.: 1967, *J. Geophys. Res.* **72**, 4449.
Cesarsky, C.: 1972, Ph.D. Thesis, Harvard University.
Cutler, D. J., Bergeson, H. E., Davis, J. F., and Groom, D. E.: 1981, *Astrophys. J.* **248**, 1166.
Earl, J. A.: 1976, *Astrophys. J.* **205**, 900.
Elliot, H.: 1974, *Phil. Trans. Roy. Soc. London* **277**, 381.
Fisk, L. A.: 1978, in Kennel, Lanzerotti, and Parker (eds.), *Space Plasma Physics*.
Fleck, R. C.: 1980, Astrophys. J. **242**, 1019.
Ginzburg, V. L. and Syrovatsky, S. I.: 1964, *The Origin of Cosmic Rays*, Pergamon Press, New York.
Ipavich, F. M., Scholer, M., and Gloeckler, G.: 1981, *J. Geophys. Res.* **86**, 11153.
Jokipii, J. R.: 1971, *Rev. Geophys. Space Phys.* **9**, 27.
Jokipii, J. R.: 1982, *Astrophys. J.* **255**, 716.
Jokipii, J. R. and Lerche, I.: 1969, *Astrophys. J.* **157**, 1137.
Jokipii, J. R. and Parker, F. N.: 1967, *Planet. Space Sci.* **15**, 1375.
Kaplan, S. A.: 1966, *Interstellar Gas Dynamics*, Pergamon Press, New York.
Kota, J. and Jokipii, J. R.: 1982, *Astrophys. J.*, in press.
Larson, R. B.: 1979, *Monthly Notices Roy. Astron. Soc.* **186**, 479.
Lee, L. C. and Jokipii, J. R.: 1975, *Astrophys. J.* **206**, 735.
Manchester, R. N.: 1972, *Astrophys. J.* **172**, 43.
McDonald, F. B., Lal, N., Trainor, J. H., Van Hollebeke, M. A. I., and Webber, William, R.: 1981, *Proc. 17th Int. Cosmic Ray Conf., Paris* **00**, 000.
Meyer, P., Parker, E. N., and Simpson, J. A.: 1956, *Phys. Rev.* **104**, 761.
Morfill, G. and Volk, H.: 1978, *J. Geophys. Res.* **83**, 5531.
Palmer, I. D.: 1982, *Rev. Geophys. Space Phys.* **20**, 335.
Parker, E. N.: 1965, *Planet. Space Sci.* **13**, 9.
Pyle, K. R.: 1959, *Rev. Geophys. Space Phys.* **17**, 587.
Rickett, B. J. and Barney, J.: 1977, *Ann. Rev. Astron. Astrophys.* **15**, 470.
Scholer, M. and Morfill, G.: 1982, *Astrophys. J.*, in press.
Toptygin, I. N.: 1980, *Space Sci. Rev.* **26**, 157.
Tsurutani, B. T., Smith, E. J., Pyle, K. R., and Simpson, J. A.: 1982, 'Energetic Protons Accelerated at Corotating Shocks: Pioneer 10 and 11 Observations from 1 to 6 AU', preprint, JPL.
Volk, H. J.: 1976, *Rev. Geophys. Space Phys.* **13**, 547.
Wright, W.: 1973, Ph.D. Thesis, Calif. Inst. Tech.

THE RELEVANCE OF MOLECULAR OBSERVATIONS TO GAMMA-RAY ASTRONOMY*

T. W. HARTQUIST**

Department of Physics and Astronomy, University College London, London WC1E 6BT, England

Abstract. The interaction of cosmic rays with interstellar clouds may produce some of the observed gamma-ray sources. The use of molecular observations to estimate the cloud masses, which are used to derive cosmic-ray fluxes, is reviewed. Molecular diagnostics of high cosmic-ray ionization rates are discussed, and a detailed application of those diagnostics is summarised and presented as evidence that second-order Fermi acceleration is important in old supernova remnants and can produce cosmic rays of too low energy to induce gamma-ray emission.

1. Introduction

Stecker (1969) pointed out that the interaction of cosmic-ray protons of energies greater than 300 MeV with gas in dense interstellar molecular clouds can lead to the production of π mesons, which decay into gamma rays. Black and Fazio (1973) initially proposed that cloud masses could be estimated from the strength of the gamma-ray sources if the cosmic-ray fluxes were known. More recently, the observed intensities of gamma-ray sources and cloud masses estimated from moleclar observations have been used to derive the fluxes of cosmic rays of several hundred MeV (Issa and Wolfendale, 1981; Morfill *et al.*, 1981). The derived cosmic-ray fluxes vary considerably.

The cosmic-ray flux below 300 MeV is of major relevance to the interstellar medium and in particular to interstellar molecular chemistry. Cosmic rays with energies below several hundred MeV are important sources for the heating and the ionization of cool instellar gas. In the earliest two phases models (cf. Field, 1975) the gross thermal and ionization structures of the interstellar medium were explained by assuming a cosmic-ray ionization rate of about 10^{-15} s^{-1}. A more realistic estimate of the total cosmic-ray ionization rate is 7×10^{-18} s^{-1} (Spitzer and Tomasko, 1968). A rate of this order would occur if the flux of cosmic rays below about 100 MeV is small but it is still sufficient as will be discussed more fully below, to create enough ionization to drive the ion-molecule reaction sequences which occur in dense interstellar clouds.

The emphasis of this paper will be on whether the masses of molecular clouds are well enough known to be used for inferring cosmic-ray fluxes from gamma-ray observations and on the molecular diagnostics of high cosmic-ray fluxes. In Section 2, a review of the observations and methods used to determine cloud masses is given. Section 3 contains a discussion of the molecular signatures of high-ionization rates, while Section 4 cautions against assuming that a high-ionization rate implies an enhancement

* Proceedings of the XVIII General Assembly of the IAU: *Galactic Astrophysics and Gamma-Ray Astronomy*, held at Patras, Greece, 19 August 1982.
** Royal Society Jaffé Donation Fellow.

in the flux of several hundred MeV cosmic rays and presents molecular evidence that second-order Fermi acceleration of cosmic rays occurs.

2. Estimating the Masses of Molecular Clouds

The galaxy contains a variety of different types of interstellar clouds. Many have masses of roughly several hundred $M_\odot$ and extend a few parsecs. The most massive clouds are really cloud complexes containing about 10^4–$10^6 M_\odot$ and have dimensions of 10–100 parsecs. It is this class of giant molecular clouds or cloud complexes which can be gamma-ray sources.

The masses of molecular clouds are generally estimated by mapping the antenna temperature of ^{13}CO emission, converting the antenna temperature to a ^{13}CO column density by assuming that the level populations are in LTE and that the line is optically thin, and multiplying the ^{13}CO column density by an assumed H_2 to CO density ratio to obtain the H_2 column density. Dickman (1978) proposed a value of 4.0×10^5 after obtaining the ^{13}CO column densities towards a number of clouds in the way described above and comparing them to the clouds' visual extensions, which are proportional to the total column densities of hydrogen. However, the optical properties of dust in giant molecular clouds are likely to differ from those of dust in the diffuse clouds which have been observed to establish the relationship between extinction and total hydrogen column density. The number of grains and the extinction will decrease in giant molecular clouds if grains agglomerate (cf. Whittet, 1981) and if grain material can be converted into gas by sputtering in the shocks (cf. Aannestad and Purcell, 1973) which occur in giant molecular clouds. Blitz and Shu (1980) argued that this H_2 to ^{13}CO conversion factor is also suggested by comparing ^{13}CO (Solomon *et al.*, 1980) and infrared (Becklin and Neugebauer, 1969; Spinrad *et al.*, 1971) and ^{13}CO and soft X-ray absorption (Cruddace *et al.*, 1978) data towards the galactic center. The infrared arguments again are based on the assumption that dark cloud grains are like those in less dense regions, and the soft X-ray absorption results are influenced by the forms taken for the background sources' intrinsic spectra.

Much larger values of the H_2 to ^{13}CO ratio have also been used (cf. Solomon and Sanders, 1980). The larger values seem to be appropriate for at least that class of giant molecular cloud in which central condensation occurs. Stenholm *et al.* (1981) have made a thorough study of the structure of these clouds. For the centrally condensed clouds, it was possible to define the cloud center as the point of greatest extinction and to define a radius for an antenna temperature or line width contour as the mean of the distances of all points on that contour from the cloud center. Hence, spherical models of the clouds were produced with each cloud being assigned a cloud radius, R_0, taken as that radius at which that cloud's CO brightness temperature, obtained by extrapolating the brightness temperature radius-relationship, equalled zero. For a number of clouds the brightness temperature and line widths of spectral features of ^{12}CO, ^{13}CO, HCO^+, H_2CO, and CS were given as functions of R/R_0, where R is the radius. The function for each given feature was then averaged over all clouds to produce a 'mean' model.

Performing a complete non-LTE level population calculation and using a technique which properly treated the radiative transfer even when the line widths had equal contributions from systemic and turbulent motions, a model spherical cloud, which reproduced the mean cloud characteristics, was constructed. Over much of the model the density fell as R^{-2} and the clouds appeared to be in collapse. The model mass was $6.2 \times 10^4 M_\odot$ and the ratio of H_2 to ^{13}CO was a bit less than 10^7 in the regions where most of the cloud mass is contained; it is interesting to note that the ^{12}CO to H_2 ratio dropped substantially from its central value as the outer radius was approached.

In addition to the uncertainty that incomplete knowledge of the H_2 to ^{13}CO ratio introduces into the estimated cloud masses, the assumption that the ^{13}CO level populations are in LTE introduces a major complication. In gas with hydrogen number densities below about 10^3 cm^{-3}, the LTE assumption leads to what can be a substantial underestimate of the population in the ground rotational state of ^{13}CO. Especially those clouds which are not centrally condensed are known to have very irregular clumpy structures. It is conceivable that the clouds consist of high density clumps which are detected in ^{13}CO emission embedded in a far lower density interclump medium which is difficult to detect but which contributes substantially to the cloud mass. The ^{13}CO will also be underestimated if its line is optically thick.

Stark and Blitz (1978) have proposed an ingenious means of setting lower limits on cloud masses which show that the ^{13}CO counting technique underestimates cloud masses by a factor of at least two or three. The clouds, because they have great linear extent, suffer strong tidal distortion due to the differential rotation of the galactic disk. By requiring that a cloud's self-gravity is strong enough to hold it together in the presence of the tidal forces, a lower limit to the cloud's mass can be set.

Blitz (1980) has suggested that a correction factor of 2 or 3 be used to multiply the masses derived in the optical thin, LTE approximation when the H_2 to ^{13}CO ratio is taken as 4×10^5. In their work on molecular cloud gamma ray sources Issa and Wolfendale take the 'Blitz factor' equal to 2. Multiplications of the LTE masses by factors of this order give masses roughly equal to the lower bounds.

Stark and Blitz also proposed a means of setting an upper bound on the root-mean-square giant molecular cloud mass. Spitzer and Schwarzschild (1953) argued that kinetic energy is transferred from clouds to stars in distant encounters. The observed velocity dispersion of stars limits the root-mean-square mass to being about $5 \times 10^5 M_\odot$ or less. If the cloud-star encounters are responsible for the stars having the observed velocity dispersion, then the masses of the clouds are about 2.5 times greater than the 'Blitz factor' corrected masses derived from ^{13}CO observations.

Hartquist (1980) has suggested another means of probing the mass distribution in molecular clouds. As mentioned above clouds are clumpy, and the role that magnetic fields have in the stability and evolution of clouds has been discussed frequently (cf. Mouschovias, 1974). In a magnetically supported cloud, a clump will sink at a rate such that the tension in the magnetic field lines, which are distorted by the clump's motion, will be equal to the weight of the clump. The tension will cause the field and the ions attached to them to stream through the clump's neutral material. Friction between the streaming ions and the neutrals will heat the clump at a rate independent of the field strength and proportional to the square of the local gravitational field of the cloud. Observations of clump temperatures can set upper bounds on the gravitational field.

The discussion above raises a number of the problems faced when determinating cloud masses. The uncorrected optically thin, LTE method discussed by Blitz and Shu would give the mass of one of the centrally condensed clouds considered by Stenholm *et al.* as about an order of magnitude lower than the one given by the cloud modelling procedure and a typical cloud mass a factor of about five lower than that derived from stellar dynamics. It is probably dangerous to make strong conclusions about the cosmic-ray flux from 'known' cloud masses and gamma-ray source strengths.

3. Molecular Signatures of High Ionization Rates in Giant Molecular Clouds

While potential barriers prevent the occurrence of many neutral-neutral reactions, many ion-molecule reactions can proceed rapidly at low temperatures, and in recent years ion-molecule reaction schemes (cf. Dalgarno and Black, 1977; Watson, 1978) have been used to explain the abundances of a number of interstellar species. In dark clouds which are too optically thick for ultraviolet photons to penetrate, cosmic rays with energies exceeding about 100 MeV can provide the dominant source of ionization. In these schemes cosmic rays ionize H_2

$$CR + H_2 \rightarrow H_2^+ + e + CR$$

$$\rightarrow H^+ + e + CR$$

H_2^+ reacts rapidly with H_2

$$H_2^+ + H_2 \rightarrow H_3^+ + H$$

and H_3^+ is formed. H^+ and H_3^+ can react with any of a number of species to initiate a series of hydrogen abstraction reactions. For instance,

$$O \quad + H^+ \rightarrow O^+ \quad + H$$

$$O \quad + H_3^+ \rightarrow OH^+ + H_2$$

$$O^+ \quad + H_2 \quad \rightarrow OH^+ + H$$

$$OH^+ + H_2 \quad \rightarrow OH_2^+ + H$$

$$OH_2^+ + H_2 \quad \rightarrow OH_3^+ + H .$$

Such a series produces trace amounts of the intermediate ions like OH^+ and OH_2^+ and large abundances of an ion like OH_3^+, terminating the sequence. The terminating ion recombines with electrons to form neutral species like OH and H_2O

$$OH_3^+ + e \rightarrow H_2O + H$$

$$\rightarrow OH + H_2 .$$

The neutral species are removed in reactions with H_3^+, He^+, C^+, and other ions, or in reactions with other neutral species when the reaction barrier is small or nonexistent. Examples of these removal reactions are

$$OH + C^+ \rightarrow CO + H^+$$

$$H_3^+ + CO \rightarrow HCO^+ + H_2$$

$$OH + O \rightarrow O_2 + H.$$

Reactions amongst the molecules produced by simple ionization, hydrogen abstraction, recombination, and removal sequences can build even larger molecules.

As the chemical sequences are initiated by simple ionization events, and many species are removed by ions or in recombination with electrons, the abundances of some of the species depend sensitively on the ionization rate. If the ionization is caused by cosmic rays with energies greater than 100 MeV, one would expect regions with molecular abundances reflecting high ionization rates to be prime candidates for gamma-ray sources.

Oppenheimer and Dalgarno (1974) calculated the electron abundance in dense interstellar clouds as a function of ionization rate, ζ. Previously the electrons were thought to be removed primarily by dissociative recombination reactions like

$$HCO^+ + e \rightarrow H + CO$$

but Oppenheimer and Dalgarno (1974) argued that charge transfer with metals such as magnesium should occur.

$$HCO^+ + Mg \rightarrow HCO + Mg^+ .$$

The metals would be the most abundant ions. Because they are removed by radiative recombination reactions, which are much slower than the dissociative recombination reactions which destroy molecular ions, the presence of metallic ions would result in a much higher electron abundance than would occur if molecular ions contained most of the positive charge. Oppenheimer and Dalgarno found that the electron abundance should be proportional to $\zeta^{1/2}$ if molecular ions were more abundant than metallic ions; otherwise the dependence should be $\zeta^{1/3}$.

The higher electron abundances which would occur if metallic ions were prevalent, would lead to rapid removal of the molecular ions such as HCO^+. In fact, the HCO^+ abundances are comparable to the upper bounds on electron abundances derived from observations of the ratio of the abundances of DCO^+ and HCO^+ (Guelin *et al.*, 1977). This ratio is proportional to that of the abundances of H_2D^+ and H_3. H_2D^+ is formed by the reaction

$$H_3^+ + HD \rightarrow H_2D^+ + H_2 + \Delta E$$

and is removed by

$$H_2D^+ + e \rightarrow H_2 + D$$

$$\rightarrow HD + H .$$

Other removal mechanisms also operate but by taking dissociative recombination to be dominant one we can calculate an upper bound to the electron density. By equating the formation and destruction rates of H_2D^+ one finds

$$n(HD)n(H_3^+) \propto n(H_2D^+)n(e),$$

where the n's denote number densities. By observing the ratio of protonated and deuterated forms of the same species, Guelin *et al.* found the fractional ionization to be less than about 10^{-8} in regions with number densities of about 10^5 cm^{-3} to 10^6 cm^{-3}. The similarity between the electron densities and the HCO^+ abundances showed that HCO^+ is the dominant positive ion. Elmegreen (1979) has suggested that this result can be understood even if there is no depletion of metals if recombination occurs through positive ion capture onto negatively charged grains.

If such recombination occurs on grain surfaces the abundance of HCO^+ is proportional to $\zeta^{1/2}$. Searches for areas with particularly high abundances of HCO^+ then may reveal likely sources of gamma rays. HCO^+ has been found to have enhanced abundances in several molecular regions in which shock heating occurs (Dickinson *et al.*, 1980). In such regions reactions occurring in the hot postshock gas actually should remove HCO^+ and reduce its abundance (Iglesias and Silk, 1978) and the enhanced abundances must reflect a substantially increased HCO^+ formation rate due to a high ionization rate. Such regions should be considered in detail as potential gamma-ray sources. As discussed later, stochastic acceleration of cosmic ray (cf. Morfill and Scholer, 1979) in supernova remnants can be a source of cosmic rays capable of providing the additional ionization.

Li *et al.* (1982) have noticed an interesting correlation between the abundance ratio $(H_2CO)/(CO)$ and the number density of supernova remnants. In a large theoretical study, incorporating a large network of ion-molecular reactions, of the dependence of molecular abundances on ionization rate, Krolik and Kallman (1982a, b) have shown that the abundance ratios of many complicated carbon molecules to CO should be sensitive to ζ. Their results do not include grain surface recombination which would diminish somewhat the abundance dependences on ζ. They argue that the abundance of species like HC_3N, HC_5N, HC_7N, and HC_9N may be particularly sensitive to ζ.

However, Krolik and Kallman were concerned primarily with a source of ionization other than cosmic rays. They argued that X-rays from early type stars can provide a sorce of ionization which is high locally but which diminishes away from the stars as absorption occurs. Cosmic rays of more than 100 MeV would be expected to provide a uniform ionization rate in a cloud. Their X-ray ionization models appear to explain the high abundance of CN seen in a number of clouds (Churchill, 1980) while the abundances of C_2H (Wooten *et al.*, 1980) exceed by factors of 10^5 and 40 those expected in a uniform low-level cosmic ray ionization model and the attenuated X-ray model they give. The X-rays from supernova remnants may often be of too low energy to penetrate far enough into a cloud to lead to any substantial increase in ionization and the HCO^+ enhanced regions observed by Dickinson *et al.* (1980) are probably not affected by soft X-rays.

4. Molecular Evidence for the Second Order Fermi Production of Low Energy Cosmic Rays

The Per OB2 region appears to be a shell which has formed behind a shock driven at a speed of 5 km s^{-1} by the expansion of an old supernova remnant (Sancisi, 1974). From observations of H and the populations of J levels in the H$_2$ ground vibrational state it was possible to construct models of the shell material along several lines of sight. Those models and the abundances of OH and HD, which form following the ionization of H by low energy cosmic rays, were used by Hartquist *et al.* (1978a, b) to deduce the average cosmic ray ionization rates along those lines of sight. They found that the ionization rate toward o Per is anomalously high.

Comparison of observations of CO radio emission and CO ultraviolet absorption against o Per indicate that the star lies on the near side of a density enhancement (Snow, 1976) while comparison of OH emission and absorption in the same direction show that little OH lies beyond the star. Hence, Hartquist and Morfill (1983) have argued that the ionization rate is low beyond it. However, the HD column density seen in absorption against o Per is so high that the cosmic-ray density must be high in most of the material between us and the star. Hartquist and Morfill suggested that the enhanced ionization rate towards o Per implies that the supernova remnant which lies between us and the shell must be a source of low energy cosmic rays which propagate directly into the shell in the direction of o Per as magnetic field lines will tend to concentrate in this dense region and will direct cosmic rays into it. They pointed out that to explain the OH and HD observations the cosmic rays must suffer significant ionization losses over a column density comparable to that towards o Per requiring that their energy must be about a few MeV.

The lower ionization rates in the other directions result from the magnetic field entering the shell at very small angles as a consequence of the shock enhancing the component of the field perpendicular to the shell. Thus, low energy cosmic rays suffer large ionization losses in a thin region at the shell's inner edge where very little material is molecular and OH cannot be formed efficiently.

The analysis of the molecular data for the Per OB2 association shows that the ionization rate can be high due to an enhanced density of cosmic rays far too low in energy to induce gamma-ray emission. Gamma-ray observations of regions having high ionization rates would provide a valuable means of constraining the cosmic-ray spectrum in them.

In light of Jokipii's (1983) remarks in this volume that solar system observations have provided no evidence that second order Fermi acceleration occurs, the conclusion of Hartquist and Morfill (1983) that stochastic acceleration of cosmic rays in the supernova remnant explains the ionization structure in the shell seems particularly important.

Wave turbulence in the ionized region of the supernova remnant can be produced by shock accelerated particles during the phase in the remnant evolution when it is expanding at speeds greater than the Alfvén speed in the ambient interstellar medium. Hartquist and Morfill argued that the shortest frequency waves generated in this manner will have frequency $f_0 \sim 10^{-9}$ Hz. The wave amplitudes at higher frequencies were taken to be related to that at f_0 by the Kolmogoroov spectrum, and the total wave energy was taken to be $\frac{1}{2}\alpha^2 B^2$ with B being the large-scale field strength. α was the only free parameter in the model.

The wave spectrum was used to calculate the spatial diffusion coefficient κ_r, applying the quasilinear wave-particule interaction theory leg (cf. Hasselmann and Wibberenz, 1968; Jokipii, 1971). The acceleration time is given by

$$\tau_F = \kappa_r / v_A^2 \, ,$$

where v_A is the local Alfvén speed. The loss time due to spatial diffusion is

$$\frac{R^2}{\pi^2 \kappa_r}$$

where R is the radius of the remnant. Adiabatic losses were found to be less important. The acceleration time was found to be greater than the loss time for cosmic rays of several MeV when $\alpha^2 = 0.2$ but losses dominated at energies above this. Hartquist and Morfill showed that $\alpha^2 \approx 0.2$ was to be expected in a remnant of this age. Hence, second-order Fermi acceleration in an evolved supernova remnant produces cosmic rays of precisely the correct energy to understand a wide range of molecular observations.

If most regions with high cosmic-ray ionization are near old supernova remnants, they may not be unusually strong gamma-ray sources. However, good upper bounds on gamma-ray emissivity from such regions will give insight into second order Fermi acceleration.

Acknowledgement

The editors thank B. Elmegreen for assistance in evaluating this paper.

References

Aannestad, P. and Purcell, E. M.: 1973, *Ann. Rev. Astron. Astrophys.* **11**, 309.
Becklin, E. E. and Neugebauer, G.: 1979, *Astrophys. J. Letters* **157**, L31.
Black, J. H. and Fazio, G. G.: 1973, *Astrophys. J.* **185**, L7.
Blitz, L.: 1980, in P. M. Solomon and M. G. Edmunds (eds.), *Giant Molecular Clouds in the Galaxy*, Pergamon Press, Oxford.
Blitz, L. and Shu, F. H.: 1980, *Astrophys. J.* **238**, 148.
Churchill, E.: 1980, *Astrophys. J.* **240**, 811.
Cruddace, R. G., Fritz, G., Shulman, S., Friedman, H., McKee, J., and Johnson, M.: 1978, *Astrophys. J. Letters* **222**, L95.
Dalgarno, A. and Black, J. H.: 1977, *Rept. Prog. Phys.* **39**, 573.
Dickinson, D. F., Rodriguez Kuiper, E. N., St. Clair Dinger, A., and Kuiper, T. B. H.: 1980, *Astrophys. J. Letters* **237**, L213.
Dickman, R. L.: 1978, *Astrophys. J. Suppl.* **37**, 407.
Elmegreen, B. G.: 1979, *Astrophys. J.* **232**, 729.

Field, G. B.: 1975, in R. Balion, P. Encrenaz, and J. Lequeux (eds.), *Atomic and Molecular Physics and the Interstellar Matter: Les Houches Session XXVI*, American Elsevier, New York.

Guélin, M., Langer, W. D., Snell, R. L., and Wooten, H. A.: 1977, *Astrophys. J. Letters* **217**, L165.

Hartquist, T. W.: 1980, *Monthly Notices Roy. Astron. Soc.* **191**, 49.

Hartquist, T. W. and Morfill, G. E.: 1983, *Astrophys. J.* **266**, 271.

Hartquist, T. W., Black, J. H., and Dalgarno, A.: 1978a, *Monthly Notices Roy. Astron. Soc.* **185**, 643.

Hartquist, T. W., Doyle, H. T., and Dalgarno, A.: 1978b, *Astron. Astrophys.* **68**, 65.

Hasselmann, K. and Wibberenz, G.: 1968, *Z. Geophys.* **34**, 353.

Iglesias, E. R. and Silk, J. K.: 1978, *Astrophys. J.* **226**, 851.

Issa, M. R. and Wolfendale, A. W.: 1981, *Nature* **292**, 430.

Jokipii, J. R.: 1972, *Rev. Geophys. Space Phys.* **9**, 27.

Jokipii, J. R.: 1983, *Space Sci. Rev.* **36**, 27 (this volume).

Krolik, J. H. and Kallman, T. R.: 1982a, in R. S. Roger and P. E. Dewdney (eds.), *Regions of Recent Star Formation*, D. Reidel Publ. Co., Dordrecht, Holland, p. 000.

Krolik, J. H. and Kallman, T. R.: 1982b, *Astrophy. J.*, in press.

Li, T., Riley, P. A., and Wolfendale, A. W.: 1982, *J. Phys.* **G8**, 1141.

Morfill, G. E. and Scholer, M.: 1979, *Astrophys. J.* **232**, 473.

Morfill, G. E., Völk, H. J., Drury, L., Forman, M., Bignami, G., and Caraveo, P. A.: 1981, *Astrophys. J.* **246**, 810.

Mouschovias, T. Ch.: 1974, *Astrophys. J.* **192**, 37.

Oppenheimer, M. and Dalgarno, A.: 1974, *Astrophys. J.* **187**, 231.

Sancisi, R.: 1974, in F. J. Kerr and S. C. Simonson (eds.), 'Galactic Radio Astronomy', *IAU Symp.* **60**, 115.

Snow, T. P.: 1976, *Astrophys. J.* **204**, 759.

Solomon, P. M. and Saunders, D. B.: 1980, in P. M. Solomon and G. M. Edmunds (eds.), *Giant Molecular Clouds in the Galaxy*, Pergamon Press, Oxford.

Solomon, P. M., Scoville, N. Z., and Saunders, D. B.: 1979, *Astrophys. J. Letters* **232**, L89.

Spinrad, H., Liebert, J., Smith, H. E., Schweizer, F., and Kuhi, L. V.: 1971, *Astrophys. J.* **165**, 17.

Spitzer, L. and Schwarzschild, M.: 1953, *Astrophys. J.* **118**, 106.

Spitzer, L. and Tomasko, M. G.: 1968, *Astrophys. J.* **152**, 971.

Stark, A. A. and Blitz, L.: 1978, *Astrophys. J. Letters* **225**, L15.

Stecker, F. W.: 1969, *Nature* **222**, 865.

Stenholm, L. G., Hartquist, T. W., and Morfill, G. E.: 1981, *Astrophys. J.* **249**, 152.

Watson, W. D.: 1978, *Ann. Rev. Astron. Astrophys.* **16**, 585.

Whittet, D. C. B.: 1981, *Quart. J. Roy. Astron. Soc.* **22**, 3.

Wooten, A., Bozyan, E. P., Garrett, D. B., Loren, R. B., and Snell, R. L.: 1980, *Astrophys. J.* **239**, 844.

GAMMA RAYS AND NEUTRINOS AS COMPLEMENTARY PROBES IN ASTROPHYSICS*

M. M. SHAPIRO

University of Iowa and University of Bonn

and

R. SILBERBERG

Naval Research Laboratory, Washington, DC 20375, U.S.A.

Abstract. Electrons are more susceptible to energy losses in magnetic fields and photon fields than protons. Hence, photons at various wavelengths, including gamma rays, bring more readily information on high-energy electrons than on protons. Neutrinos provide a unique tracer for protons. Furthermore, at high energies the neutrino flux can considerably exceed the gamma-ray flux, as gamma rays above ~ 1 MeV are degraded by γ-γ interactions in compact high-intensity sources. Active galactic nuclei (AGN) with outputs $> 10^{45}$ ergs s^{-1} and dimensions $\sim 10^{14}$ cm would constitute such sources. If the AGN are powered by ultra-massive black holes, then these numerical conditions are satisfied, and at high energies the flux $J_v > J_y$. Berezinsky and Ginzburg have pointed out that the photon intensity around spinars is not sufficient to cause gamma-ray degradation. These authors have demonstrated that the measurement of neutrino flux, combined with the measurement (or upper limit) of gamma-ray flux would show whether the active galactic nuclei are powered by massive black holes or spinars. We estimate that gamma rays would be degraded at spinars, too, at energies > 1 GeV.

1. Introduction

Neutrino astrophysics is complementary to gamma-ray astrophysics for learning about regions where particles are accelerated to very high energies: cosmic gamma rays (and X-rays) yield information on highly relativistic electrons and their energy loss processes in magnetic fields and intense photon fields. Because electrons are much more susceptible to these energy loss processes than protons, their contribution to gamma rays appears to dominate over that of protons in the 'point sources' observed. Neutrinos, on the other hand, provide a probe for acceleration of protons, the intensity of energetic protons and the density of matter at (or near) the regions of acceleration.

Furthermore, while gamma-ray detectors most commonly are used to explore the energy drain near 10^8 eV, a neutrino detector like the proposed DUMAND array would explore neutrinos near 10^{12} eV, and yield information on acceleration of protons and nuclei to very high energies ($E > 10^{13}$ eV).

At such high energies, gamma rays are likely to be absorbed near their sources, when these are sufficiently powerful and compact. Such sources are pulsars and active galactic nuclei.

Berezinsky and Ginzburg (1981) showed that a combination of gamma-ray observations (the energy at which the spectrum steepens, due to photon-photon interactions), with neutrino observations permits a discrimination between black hole and spinar

---

* Proceedings of the XVIII General Assembly of the IAU: *Galactic Astrophysics and Gamma-Ray Astronomy*, held at Patras, Greece, 19 August 1982.

Space Science Reviews **36** (1983) 51–56. 0038–6308/83/0361–0051$00.90.

models of active galactic nuclei. At high energies, the neutrino luminosity exceeds considerably that of gamma rays in case of the black hole model.

Eichler (1979) showed that gamma rays can be degraded in compact sources by absorption in matter. He pointed out (Eichler, 1980) that the suggested model of SS 433 – accreting neutron star in a binary system – is consistent with a high ratio of neutrino to gamma-ray fluxes.

2. Types and Models of Active Galactic Nuclei

Recent observations with X-ray and γ-ray telescopes indicate that many active galactic nuclei (AGN), in addition to being powerful sources of radio and infrared radiation, are also powerful emitters of X-rays and γ-rays. The power output in electromagnetic radiation of several AGN peaks at energies near 1 MeV (the quasar 3C 273, and type 1 Seyferts NGC 4151 and MCG8–11–11). The energy output rates of AGN are 10^{42} to 10^{48} ergs s^{-1}, i.e., 10^8 to 10^{14} times the solar luminosity.

The AGN are subdivided into several classes. Radiogalaxies are elliptical galaxies; they are characterized by having pairs of extensive radio lobes. Seyfert galaxies (classes 1 and 2) are spiral galaxies, which have fast moving clouds. In the case of class 2, the nucleus is obscured by dust. Another class of compact nuclei are the BL Lac objects, probably in large elliptical galaxies. These objects are relatively rapidly variable, and are characterized by absence of strong emission lines. The most powerful class of AGN are the quasars, which require up to tens of solar masses per yera to be converted into energy.

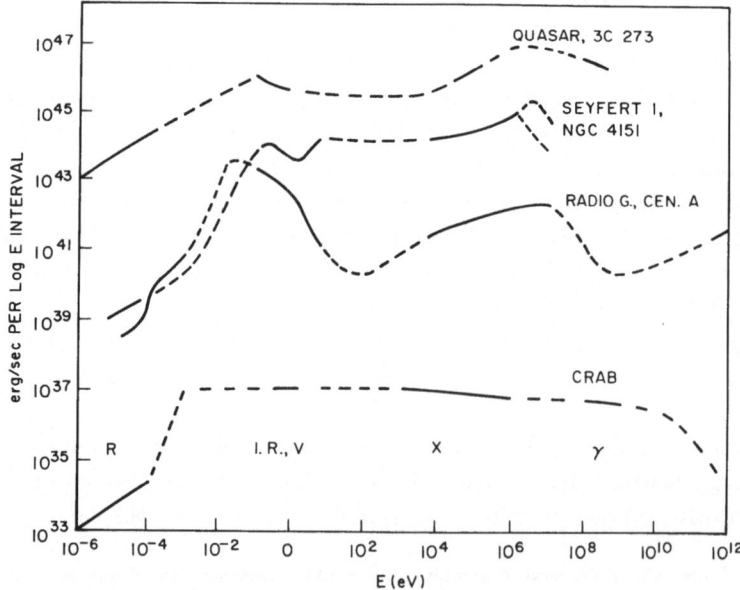

Fig. 1.   Rate of electromagnetic energy output (ergs s^{-1}) per log$E$ interval for several active galaxies: 3C 273, NGC 4151, and Cen A. For comparison, the output of the Crab nebula (with its pulsar) is also shown.

The central power source of an AGN is highly compact; variability on a time scale of hours has been reported (Delvaille *et al.*, 1978; Tananbaum *et al.*, 1978). Such a time scale implies dimensions of the order of 10^{15} cm.

The energy spectra of photons from various types of AGN are shown in Figure 1. These are expressed in the power per $\log E$ interval. For comparison, the spectrum of a galactic non-thermal source, the supernova remnant Crab nebula and pulsar, is also shown. Its luminosity is $\sim 10^{-9}$ times that of 3C 273.

Various models have been proposed to explain the properties of the AGN. These compact, supermassive objects could be magnetized, rotating plasma masses, e.g., of the 'spinar' or 'magnetoid' type (Ginzburg and Ozernoy, 1977), or black holes accreting matter; in the latter case the accretion disk itself might contain magnetic fields, or it could set up large-scale fields as it rotates. The nature of the central source might be determined by high energy neutrino measurements (Silberberg and Shapiro, 1979; Berezinsky and Ginzburg, 1981). Supermassive black holes seem to provide a more natural acceleration model than spinars, since they are as compact as possible, and more efficient in converting accretion energy into radiation. We concentrate in the next section on particle acceleration mechanisms in the vicinity of black holes.

3. Probing Acceleration Processes in Active Galactic Nuclei

Compared to protons, electrons readily suffer energy losses in magnetic fields and intense photon fields, and increasingly so at high energies. The lifetime against radiative synchrotron loss of electrons at 1 GeV in a magnetic field of 10^3 G (as for black-hole ergosphere models) is 0.4 s. Hence it is not at all surprising that photons at various wavelengths tend to bring us information on electrons rather than protons. Furthermore, due to their energy losses electrons cannot be accelerated to as high energies as protons. While some gamma rays are produced in nucleon-nucleon collisions via π° decay, it should be noted that the Compton process, and particularly the synchrotron self-Compton process, contribute importantly to cosmic γ-radiation, as does the bremsstrahlung process.

We shall discuss electrons here rather briefly, noting, however, that they are especially important for gamma-ray production processes. The acceleration of electrons by a dynamo and magnetic-flare mechanism has been studied by Sturrock and Barnes (1972), Harrison (1976), Takahara (1979), and Lovelace *et al.* (1979). The production of electron pairs by breakdown of vacuum and acceleration in magnetic fields has been explored by Blandford and Znajek (1977).

The acceleration of protons in the accretion disk of an ultra-massive black hole in the center of active galaxies by the Fermi mechanism has been explored by Kafatos *et al.* (1981), and by a dynamo mechanism with acceleration to ultra-high energies by Lovelace (1976).

Gamma-ray studies of extragalactic sources have barely begun: only four such sources (three of them shown in Figure 1) have been observed. The mechanisms considered for the γ-ray production in these sources are synchrotron self-Compton

(Grindlay, 1975; Mushotzky *et al.*, 1978) and Penrose–Compton or Penrose pair production (Leiter and Kafatos, 1978; Kafatos and Leiter, 1979; and Kafatos, 1980). Hence the gamma-ray fluxes have been interpreted as due to high-energy electrons, or in terms of energy-gain processes of photons near a black hole.

Neutrinos, on the other hand, would provide a unique tracer for acceleration of protons, and of the intensity of energetic protons and the matter density in active galactic nuclei. Since protons form the dominant component of galactic cosmic rays, and also of energetic solar flare particles, we expect them to be present also near active galactic nuclei.

4. Suppression of High-Energy Gamma Rays from Sources Surrounded by Intense Photon Fields

A large fraction of the total energy emitted from some AGN is observed in the form of gamma rays. Time variations in X-ray emission, as short as a couple of hours, have been observed for Cen A (Delvaille *et al.*, 1978), and possibly even 10 min for NGC 4151 (Tananbaum *et al.*, 1978). Hence these sources must be extremely compact. We have pointed out that for $r < 10^{14}$ cm and an energy output $> 10^{45}$ ergs s^{-1}, gamma rays can be degraded by $\gamma + \gamma \to e^+ + e^-$ reactions (Silberberg and Shapiro, 1979).

Consider a region of radius 3×10^{13} cm (10^3 light-seconds). Inside this volume of $\sim 10^{41}$ cm^3, about 10^{45} ergs are generated *per second*, so that the energy content of the region approaches 10^{48} ergs. (We shall take 3×10^{47} ergs.) If some 10% of this energy goes into ~ 100 keV X-rays, then the supply of X-ray photons is $\sim 6 \times 10^{53}$, and the photon density is 6×10^{12} cm^{-3}. This is enough to degrade the γ-ray photons.

Gould and Schreder (1967) have shown that the γ-ray absorption probability per unit path length for $\gamma\gamma$ reactions is

$$\frac{dP_{abs}}{dx} = \pi r_0^2 C \left(\frac{E}{m_0^2 c^4} \right)^{\alpha - 1} F_\alpha . \tag{1}$$

Here $\pi r_0^2 = 2.4 \times 10^{-25}$ cm^2, E is the energy of the γ-ray photons, and m_0 is the mass of the electron. $C = N(\varepsilon)\varepsilon^\alpha$, $N(\varepsilon)$ is the photon density per unit energy interval at energy ε. For $\alpha = 2$ and $E < (m^2 c^4)/\varepsilon_0$, F_α is 0.33. (Here ε_0 is the threshold energy above which the exponent α is applicable.) With $\alpha = 2$, $C \simeq 3 \times 10^5$ ergs cm^{-3}. Consider the probability of γ-ray absorption at energies of 10 MeV, i.e., when $E \simeq 1.6 \times 10^{-5}$ erg. Using Equation (1), the probability is 6×10^{-13} per cm. Hence the γ-rays will be absorbed long before traversing the source dimension of $\sim 3 \times 10^{13}$ cm.

The expression of Gould and Schreder (1967) for the probability of γ-ray absorption in collisions with softer (e.g., X-ray) photons reduces to $P \approx (E_\gamma/m_0 c^2) L_{45}/R_{15}$ (Cavaliere, 1979). Here m_0 is the electron mass, R_{15} is the distance traversed by a γ-ray in units if 10^{15} cm, and L_{45} is the luminosity per $\log E$ interval in units of 10^{45} ergs s^{-1}. (We have assumed that the differential energy spectrum of the photons has an exponent -2.0.) It is possible that the 'shelf' and steepening of the diffuse extragalactic gamma-ray spectrum and of NGC 4151 just above the pair production threshold could be due to the γ-γ process. On the other hand, no such steepening would be expected in the neutrino spectrum, until about 10^{13} eV, when p-γ interactions could start to steepen the proton spectrum at compact, ultra-energetic sources (Blumenthal, 1970).

The degradation of the high-energy gamma rays, moreover, destroys information on acceleration of particles to energies $E \gg 10^9$ eV, hence also on the nature of the acceleration processes in AGN. Neutrinos are not thus affected, at least up to 10^{13} eV, as described above.

5. Neutrino and Gamma-Ray Measurements for Discriminating Between Black Hole and Spinar Models

We have pointed out (Silberberg and Shapiro, 1979) that neutrino astronomy along with X-ray and γ-ray astronomy can probe the nature of active galactic nuclei: are they super-massive black holes, or giant spinars or supernova clusters? This problem was explored in great depth by Berezinsky and Ginzburg (1981). They showed that gamma rays (at modest energies) would not be degraded in the spinar or magnetoid model. In this model, with radius $\sim 10^{16}$ cm, the photon intensity is not high enough to degrade the γ-rays. We find, however, that even for a spinar, the γ-rays at higher energies, > 1 GeV, *are* suppressed. For acceleration near a black hole, on the other hand, the X-ray density is sufficiently high to degrade γ-rays at energies as low as ~ 1 MeV.

6. Conclusions

Gamma rays are good tracers of electron acceleration processes in AGN, while neutrinos are unambiguous tracers of energetic protons, i.e., of cosmic rays. At high energies ($E > 1$ GeV), gamma rays from the more powerful AGN are likely to be degraded, and neutrinos provide a tracer of particle acceleration to higher energies. Thus, a dearth of high-energy gamma rays need not lead to an expectation of a low neutrino flux. A combination of gamma ray observations ($1-10^3$ MeV) and neutrino observations (~ 1 TeV) is likely to permit discrimination between black-hole and spinar models.

Acknowledgements

One of us (MMS) gratefully acknowledges the support and encouragement of Prof. James A. Van Allen and the Office of Naval Research. He thanks the Alexander von Humboldt Foundation for the grant of a Senior U.S. Scientist Award, and Prof. Wolfgang Priester of the Institute for Astrophysics in Bonn for his gracious hospitality. This paper was presented in a special symposium at the IAU General Assembly in Patras, Greece, at the kind invitation of Dr G. E. Morfill and Prof. L. Scarsi.

The editors thank T. Hartquist for assistance in evaluating this paper.

References

Berezinsky, V. S. and Ginzburg, V. L.: 1981, *Monthly Notices Roy. Astron. Soc.* **194**, 3.

Blandford, R. D. and Znajek, R. L.: 1977, *Monthly Notices Roy. Astron. Soc.* **179**, 433.

Blumenthal, G. R.: 1970, *Phys. Rev.* **D1**, 1596.

Cavaliere, A.: 1979, *16th Internat. Cosmic Ray Conf., Kyoto* **14**, 91.

Delvaille, J. P., Epstein, A., and Schnopper, H. W.: 1978, *Astrophys. J.* **219**, L81.

Eichler, D.: 1979, *Astrophys. J.* **232**, 106.

Eichler, D.: 1980, *Proc. 1980 DUMAND Symposium, Hawaii* **2**, 266.

Ginzburg, V. L. and Ozernoy, L. M.: 1977, *Astrophys. Space Sci.* **50**, 23.

Gould, R. J. and Schreder, G. P.: 1967, *Phys. Rev.* **155**, 1404.

Grindlay, J. E.: 1975, *Astrophys. J.* **199**, 49.

Harrison, E. R.: 1976, *Nature* **264**, 525.

Kafatos, M.: 1980, *Astrophys. J.* **236**, 99.

Kafatos, M. and Leiter, D.: 1979, *Astrophys. J.* **229**, 46.

Kafatos, M., Shapiro, M. M., and Silberberg, R.: 1981, *Comm. Astrophys.* **9**, 179.

Leiter, D. and Kafatos, M.: 1978, *Astrophys. J.* **226**, 33.

Lovelace, R. V. E.: 1976, *Nature* **262**, 649.

Lovelace, R. V. E., McAuslan, J., and Burns, M.: 1979, in J. Arons, C. McKee, and C. Max (eds.), *Particle Acceleration Mechanisms in Astrophysics*, p. 399, publ. by Amer. Inst. Physics.

Mushotzky, R. F., Boldt, E. A., Holt, S. S., Pravdo, S. H., Serlemitsos, P. J., Swank, J. H., and Rothschild, R. H.: 1978, *Astrophys. J.* **226**, L65.

Silberberg, R. and Shapiro, M. M.: 1979, *16th International Cosmic Ray Conf., Kyoto* **10**, 357.

Sturrock, P. A. and Barnes, C.: 1972, *Astrophys. J.* **176**, 31.

Takahara, F.: 1979, *Progr. Theoret. Phys.* **62**, 629.

Tananbaum, H., Peters, G., Forman, W., Giacconi, R., Jones, C., and Avni, Y.: 1978, *Astrophys. J.* **223**, 74.

ON PARTICLE ACCELERATION IN SUPERNOVA REMNANTS*

L. DRURY

Max-Planck Institut für Kernphysik, 6900 Heidelberg, F.R.G.

Abstract. The applicability of first and second order Fermi acceleration to electrons in supernova remnants is briefly examined.

In most models of gamma-ray sources the photons are produced by the interaction of energetic particles with ambient matter; thus current ideas on the acceleration of particles in supernova remnants (SNR's) are of some interest to gamma-ray astronomy (despite the rather surprising fact that no SNR has been detected as a gamma-ray source (cf. Blandford and Cowie, 1982)). Excluding acceleration by pulsars (Kulrud and Gunn, 1972) and perhaps by other compact objects (e.g. binary neutron stars, Kundt, 1980) there are essentially two theories for acceleration in the remnants themselves. The older is second order Fermi acceleration associated with turbulence inside the remnant (Scott and Chevalier, 1975; Chevalier *et al.*, 1976, 1978; Cowsik, 1979; Morfill and Scholer, 1979), the more recent is diffusive shock acceleration (also a Fermi acceleration mechanism, but of first order: Krymsky, 1977; Axford *et al.*, 1977; Bell, 1978a, b; Blandford and Ostriker, 1978) at the outer boundary of the remnant or at internal shocks associated with cloud crushing (Blandford and Cowie, 1982).

Almost all information about energetic particles in SNR's is obtained by interpreting their radio emission as the synchrotron radiation of relativistic electrons. The radio luminosity of SNR's is usually summarized as a relationship between surface brightness (Σ) and linear diameter (D). Clark and Caswell (1976) obtain

$$\Sigma \propto D^{-3}. \tag{1}$$

Caswell and Lerche (1979) include a dependence on height (z) above the Galactic plane and obtain

$$\Sigma \propto D^{-3} \exp(-|z|/175 \text{ pc}) \tag{2}$$

while Göbel *et al.* (1981) argue for a dependence on the radio spectral index (α) and obtain

$$\Sigma \propto D^{-2(1+\alpha)}. \tag{3}$$

---

* Proceedings of the XVIII General Assembly of the IAU: *Galactic Astrophysics and Gamma-Ray Astronomy*, held at Patras, Greece, 19 August 1982.

Space Science Reviews **36** (1983) 57–60.   0038–6308/83/0361–0057$00.60.

The relationship is poorly determined, due both to intrinsic scatter in the data and to uncertainty in the estimates of D. However as pointed out by Shklovsky (1960) if there were no continuing acceleration in SNR's and they expanded adiabatically one would expect the much steeper relationship

$$\Sigma \propto D^{-4(1+\alpha)} . \tag{4}$$

This is evidence for continuing particle acceleration (at least of the electrons) in the observed SNR's, but the relationship is too uncertain to say much more. The usual 'minimum energy' argument suggests a total energy in relativistic electrons of 10^{47}–10^{49} erg; these data have been collected by Palumbo and Vitollini (1981) and Cavallo (1982). The required acceleration efficiencies are high, but seem possible for both Fermi mechanisms.

A more stringent test is provided by the spectral information. Remnants are generally observed to have power law spectra, $S_\nu \propto \nu^{-\alpha}$, from which it is inferred that the electron distribution function, f, is also a power-law in momentum, p, of the form

$$f \propto p^{-(3+2\alpha)} . \tag{5}$$

Here the simple theory of diffusive shock acceleration (i.e. ignoring reaction effects) makes a clear prediction: at a shock of compression ratio r the accelerated particles should have a spectrum

$$f \propto p^{-3r/(r-1)} \tag{6}$$

corresponding to a radio spectral index

$$\alpha = \frac{3}{2(r-1)} . \tag{7}$$

For strong shocks we expect $r = 4$ and $\alpha = 0.5$. However (see Figure 6 of Clark and Caswell, 1976) for the galactic SNR's α has an apparently random distribution (roughly a gaussian of mean 0.45 and standard deviation 0.15) and shows no correlation with the age or diameter of the remnant.

While the mean agrees well with the predicted 0.5 the dispersion is surprising. However let us consider in more detail the compression ratio of a strong shock. Denoting up- and down-stream values by the subscripts 1 and 2, density by ρ, velocity by U, pressure by P and internal energy density by E the conservation laws, applied in the shock frame, give

$$\rho_1 U_1 = \rho_2 U_2 = A , \tag{8}$$

$$A U_1 = A U_2 + P_2 , \tag{9}$$

$$\tfrac{1}{2} A U_1^2 = \tfrac{1}{2} A U_2^2 + U_2(P_2 + E_2) + \dot{Q} , \tag{10}$$

where $\dot{Q}$ represents any additional energy flux out of the shock and in a strong shock we neglect P_1 and E_1. These equations at once imply that the compression ratio

$$r = 1 + 2 \left(\frac{E_2}{P_2} + \frac{\dot{Q}}{U_2 P_2} \right). \tag{11}$$

The ratio of internal energy density to pressure has a minimum value of $\frac{3}{2}$ for a monatomic non-relativistic gas with no internal degrees of freedom; in a real gas disassociation and excitation lead to larger values. Thus in the absence of thermal effects ($\dot{Q} = 0$) four is the *minimum* compression ratio of a strong shock. Cooling in the shock front ($\dot{Q} > 0$) further increases the compression ratio. We conclude that strong shocks can only have compression ratios less than four if $\dot{Q}$ is negative, i.e. if there is an additional flux of energy *into* the shock. The only reasonable source for such an energy flux is thermal conduction from the hot interior of the remnant and indeed on assuming that conduction inside the remnant is so efficient that the interior is effectively isothermal Solinger *et al.* (1975) find that the compression ratio is reduced to 2.4. However as pointed out by Lerche (1980) the thermal conductivity will not in general be this high and depending on its exact value we expect a range of compression ratios from 2.4 upwards (corresponding to radio spectral indices from 1.1 downwards). This important point (which I did not fully appreciate until after the oral presentation in Patras) offers an explanation for the observed range in spectral indices (although the neglect of reaction effects is serious, cf. Drury and Völk, 1981).

Models based on second order Fermi acceleration can also explain the obseved dispersion; their problem is rather to explain why it is not greater. The transport equation in the diffusion approximation for the isotropic part of the phase space density f is

$$\frac{\partial f}{\partial t} + \mathbf{U} \cdot \nabla f = \nabla(\kappa \nabla f) + \tfrac{1}{3} \nabla \cdot \mathbf{U} p \frac{\partial f}{\partial p} + \frac{1}{p^2} \frac{\partial}{\partial p} \left(p^2 D \frac{\partial f}{\partial p} \right), \tag{12}$$

where $\mathbf{U}$ is the velocity field, κ the spatial diffusion tensor, and D the momentum pace diffusion coefficient. For simplicity assume uniform expansion of the remnant with a time scale τ_{exp} (i.e. $\nabla \cdot \mathbf{U} = 3/\tau_{\text{exp}}$), change the independent variable to $s = \ln(p/p_0)$ and use $n(s)$, the distribution function of particles in the remnant wrt s

$$n(s) = 4\pi p^3 \int_{\text{remnant}} f \, \mathrm{d}^3 x \tag{13}$$

as the dependent variable. Then

$$\frac{\partial n}{\partial t} = \frac{1}{\tau_{\text{exp}}} \frac{\partial n}{\partial s} + \frac{\partial}{\partial s} \left[\frac{1}{\tau_2} \left(\frac{\partial n}{\partial s} - 3n \right) \right], \tag{14}$$

where $\tau_2 = p^2/D$ is the acceleration time scale (in general τ_2 depends on s, but is always assumed to be independent because this leads to power-law spectra). By suitably choosing the time scales the observed spectral indices can then be obtained, but there is no obvious reason why they should be related in such a way that the observed distribution is reproduced. However the optical and radio morphology of some remnants is perhaps more suggestive of acceleration in a turbulent shell than at a spherical shock and Cowsik's second order Fermi model for Cas-A (1979) does give a remarkably good fit both to the spectrum and its time variation. Also various estimates (e.g. Morfill and Scholer, 1977) suggest that significant second order Fermi acceleration should occur in SNR's.

In conclusion the observational evidence does not exclude one or the other theory, rather it suggests that while both processes can occur in the observed remnants, diffusive shock acceleration is probably the more important.

References

Axford, W. I., Leer, E., and Skadron, G.: 1977, *Proc. 15th ICRC (Plovdiv)* **11**, 132.
Bell, A. R.: 1978a, *Monthly Notices Roy. Astron. Soc.* **182**, 147.
Bell, A. R.: 1978b, *Monthly Notices Roy. Astron. Soc.* **182**, 443.
Blandford, R. D. and Cowie, L. L.: 1982, *Astrophys. J.* **260**, 625.
Blandford, R. D. and Ostriker, J. P.: 1978, *Astrophys. J.* **221**, L29.
Caswell, J. L. and Lerche, I.: 1979, *Monthly Notices Roy. Astron. Soc.* **187**, 201.
Cavallo, G.: 1982, *Astron. Astrophys.* **111**, 368.
Chevalier, R. A., Robertson, J. W., and Scott, J. S.: 1976, *Astrophys. J.* **207**, 450.
Chevalier, R. A., Oegerle, W. R., and Scott, J. S.: 1978, *Astrophys. J.* **222**, 527.
Clark, D. H. and Caswell, J. L.: 1976, *Montly Notices Roy. Astron. Soc.* **174**, 267.
Cowsik, R.: 1979, *Astrophys. J.* **227**, 856.
Drury, L. O'C. and Völk, H. J.: 1981, *Astrophys. J.* **248**, 344.
Göbel, W., Hirth, W., and Fürst, E.: 1981, *Astron. Astrophys.* **93**, 43.
Krymsky, G. F.: 1977, *Soviet Phys. Dokl.* **23**, 327.
Kulsrud, R. M. and Gunn, J. E.: 1972, *Phys. Rev. Letters* **28**, 636.
Kundt, W.: 1980, *Ann. NY Acad. Sci.* **336**, 429.
Lerche, I.: 1980, *Astron. Astrophys.* **85**, 141.
Morfill, G. E. and Scholer, M.: 1979, *Astrophys. J.* **232**, 473.
Palumbo, G. G. C. and Vettolini, G.: 1981, *Proc. 17th ICRC (Paris)* **2**, 299.
Scott, J. S. and Chevalier, R. A.: 1975, *Astrophys. J.* **197**, L5.
Shklovsky, I. S.: 1960, *Soviet Astron.* **4**, 243.
Solinger, A., Rappaport, S., and Buff, J.: 1975, *Astrophys. J.* **201**, 381.

GAMMA-RAY SOURCES OBSERVED BY COS-B*

W. HERMSEN

Cosmic-Ray Working Group, Huygens Laboratorium, Leiden, The Netherlands

Abstract. The nature of the fine-scale structure in the gamma-ray distribution is not yet disclosed. Considerable debate is going on whether these structures which appear point-like in the data, are mainly diffuse in nature or are genuinely compact objects. Most of the uncertainty is due to the experimental limitations. A status report is presented on the experimental study of the fine-scale structure measured by COS-B in the energy range 50 MeV–5 GeV. All *identified* gamma-ray sources are discussed: (i) the temporal and spectral characteristics of the radio pulsars PSR053 + 21 and PSR0833–45; (ii) COS-B upper limits on the gamma-ray flux from the binary system Cyg X-3; (iii) the ρ Oph molecular cloud now shown to be resolved in gamma rays as is the case for the Orion complex; (iv) the evidence on the detection of 3C273 in three COS-B observations. The 2CG catalogue of high-energy (point-like) gamma-ray sources contains 25 sources, of which 21 are not yet unambiguously identified. Their average properties are discussed. The error region of the unidentified source 2CG195 + 04 (Geminga) is studied at other wavelengths in greatest detail. This search for a counterpart is summarized, showing the possibility that a nearby ($\lesssim 100$ pc) neutron star is the counterpart. In the appendix is presented the cross-correlation method which is applied in the search for gamma-ray sources, as well as the appearances of the sources in the data.

1. Introduction

The European Space Agency's satellite COS-B (Bignami *et al.*, 1975; Scarsi *et al.*, 1977) was launched from NASA's Western Test Range on August 9th, 1975. Being switched off on April 25th, 1982, the COS-B gamma-ray experiment, capable of detecting gamma rays with energies greater than ~ 50 MeV, provided 6.7 years of observation time. The total number of useful gamma-ray events is well over 100 000. Since 1976 a large number of publications appeared in print, based on part of the final data base. After the conclusion of the observations a final, more detailed analysis could be started on the different topics addressed in gamma-ray astronomy, taking advantage of the improved counting statistics, of the knowledge gained in gamma-ray astronomy over the last years and of the progress made meanwhile experimentally and theoretically at other wavelengths.

This paper reports on the status of the study of the fine-scale structure in the large-scale gamma-ray distribution before the final digestion of the data. Given the experimental definition of a gamma-ray source, namely, a gamma-ray source is a statistically significant excess above the surrounding background, the excess exhibiting a spatial distribution consistent with the COS-B pointspread function, a search for these sources resulted in the publication of a catalogue of 25 gamma-ray sources using only about $2\frac{1}{2}$ years of data (Swanenburg *et al.*, 1981; Hermsen, 1980, 1981). At the time of compilation of this catalogue in 1979, 1980 no complete surveys were performed yet at mm wavelengths to reveal the clumpy molecular mass distribution (mainly H_2), which, together with the smoother atomic mass distribution (mainly H I) determines pre-

---

* Proceedings of the XVIII General Assembly of the IAU: *Galactic Astrophysics and Gamma-Ray Astronomy*, held at Patras, Greece, 19 August 1982.

dominantly the structure of the large-scale, diffuse (i.e. originating from the interaction between cosmic rays and interstellar matter) galactic gamma-ray background. This fact made the search for point sources more difficult. To provide an independent estimate of the diffuse galactic background structure, the new results from radio and mm observations will certainly be used in the final search for gamma-ray sources in the COS-B data. However, the sofar reported gamma-ray sources are present in the data as significant excesses above the surrounding background and have to be explained by whatever feature, object or model. Evidence for their presence is given for several sources. Profiles of all sources are shown by Hermsen (1980).

In the following sections mainly experimental results are presented in view of the long list of papers in this volume adressing searches for counterparts of sofar unidentified gamma-ray sources and theoretical models which try to explain the nature of the gamma-ray sources (identified and unidentified ones). An exhaustive review is given by Bignami and Hermsen (1983). For a more complete account of earlier experimental results see also the latter review.

2. Identified Gamma-Ray Sources

Only a few gamma-ray sources detected in the COS-B energy range (50 MeV–5 GeV) are identified with known astronomical objects, which turned out to be of vastly different nature:

– The radio pulsars PSR 0531 + 21 (Crab pulsar) and PSR 0833 – 45 (Vela pulsar) are unambiguously identified because of their characteristic timing signatures. Both radio-pulsars were already seen in the SAS-2 data (Kniffen *et al.*, 1974; Thompson *et al.*, 1975, respectively), while the Crab pulsar was detected even earlier by instruments flown on balloons (see the review of Fazio, 1973).

– The binary system Cyg X-3 was claimed to be detected at high-energy gamma rays by its characteristic period of $\sim$ 4.8 hr ($E > 40$ MeV, Galper *et al.*, 1976; $E > 35$ MeV, Lamb *et al.*, 1977). However, no confirmation is obtained from the COS-B data (Bennett *et al.*, 1977b; Swanenburg *et al.*, 1981).

– The ρ Oph molecular cloud was claimed to be seen in the COS-B gamma-ray data (Mayer-Hasselwander *et al.*, 1980; Bignami and Morfill, 1980) because of a spatial coincidence with one of the catalogued gamma-ray sources.

– The quasar 3C 273 is the first (and sofar only) discovered extragalactic high-energy gamma-ray source (Swanenburg *et al.*, 1978). The proposed identification is again based purely on the spatial coincidence with a gamma-ray source.

Results, derived from the COS-B data, on the above mentioned sources are presented concisely in this section.

2.1. RADIO PULSARS

Vela and Crab. Sofar, gamma-ray emission from radio pulsars has been detected with certainty only from the Vela and Crab pulsars, the only unambiguously identified discrete gamma-ray sources. COS-B had these two sources several times within its field

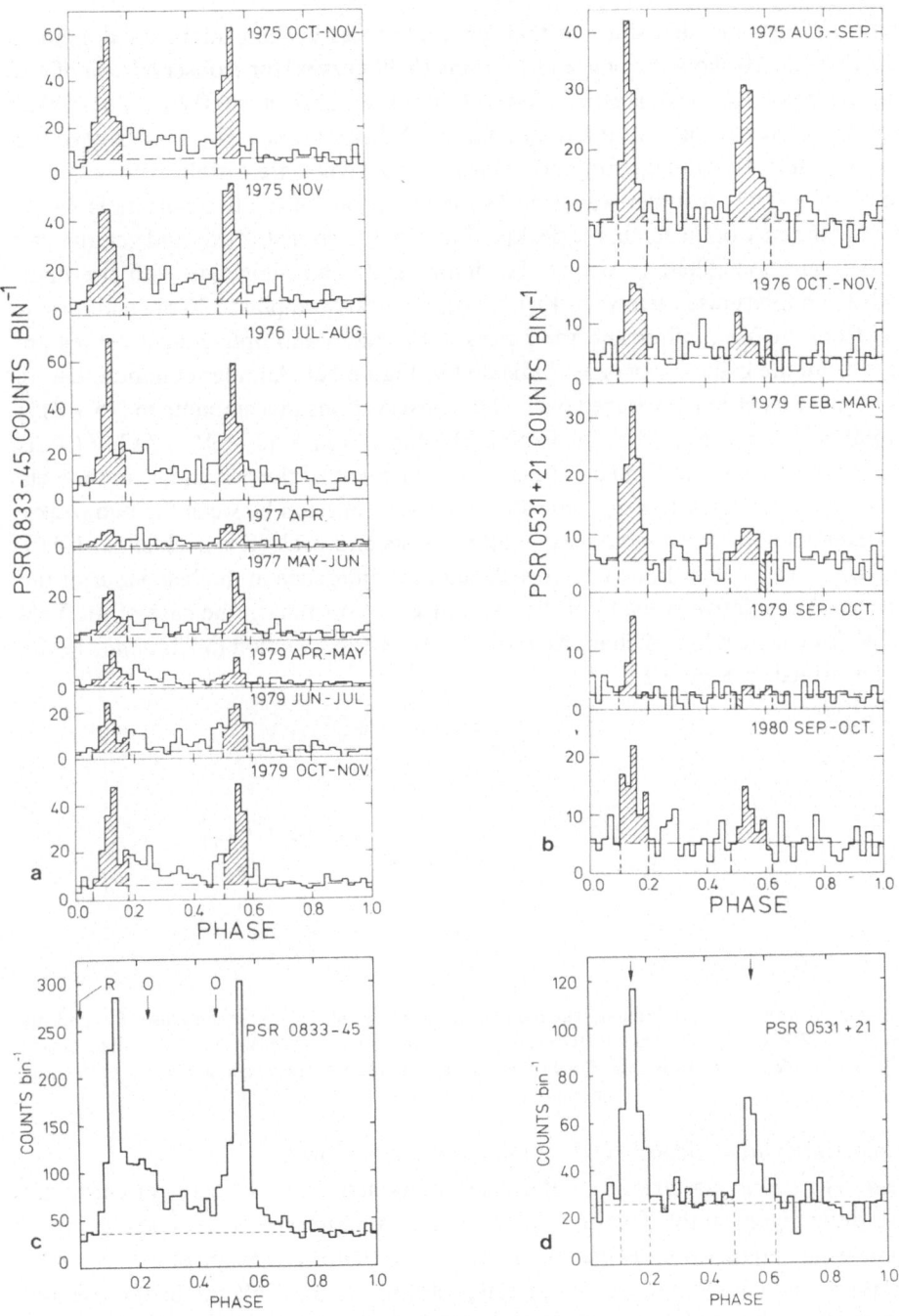

Fig. 1. Gamma-ray light curves of (a) PSR 0833−45 and (b) PSR 0531+21 in the energy interval 50−3000 MeV at the epochs indicated. The shaded areas indicate the phase intervals of the two pulses and the horizontal broken lines show the levels of the background. (c) and (d) present gamma-ray lightcurves obtained by summing the individual curves. PSR 0833−45 (c), the arrows indicate the phases of the radio (R) and optical (O) peaks. PSR 0531+21 (d), the shaded area indicates the interpulse emission and the two arrows indicate the phases of the radio (and optical and X-ray) peaks.

of view, enabling a detailed study of their characteristics in the gamma-ray domain.

Figures 1a and 1b show the phase histograms (light curves) for 8 observations of the Vela pulsar (Wills *et al.*, 1981) resp. 5 observations of the Crab pulsar (Wills *et al.*, 1982) and Figures 1c and 1b present the total (summed) distributions. The similarity of the phase histograms is striking: for both pulsars two peaks are visible with a phase separation of ~ 0.42. At other wavelengths the behaviour of the two pulsars is vastly different: In the case of the Crab, the peakpositions in the optical, X-ray and gamma-ray light curves are coincident to within the timing errors and coincide with the main radiopulse and interpulse (arrows in Figure 1d). Contrary, no pulsed X-ray emission is detected from the Vela pulsar and the pulses in the radio and optical light curves are displaced from the gamma-ray ones (indicated in Figure 1c). Another common feature is the interpulse emission. Averaged over all the observations this amounts to $(15 \pm 4)\%$ of the total pulsed emission from the Crab in the energy range 50 MeV to 3 GeV (Wills *et al.*, 1982; the excess is indicated in Figure 1d), roughly half the percentage in the Vela situation. The Vela light curve not only shows pulsed emission between the two peaks, but also exhibits a tail of pulsed emission after the second peak (Kanbach *et al.*, 1980).

Comparison of Figures 1a and 1b reveals another difference in the behaviour of the two pulsars: The relative strength of the two pulses is constant in the case of the Vela pulsar (Wills *et al.*, 1981) and changed in time in the Crab situation (Wills *et al.*, 1982). This is illustrated in Figure 2.

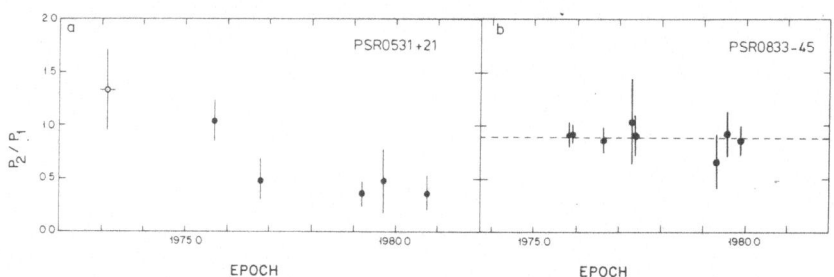

Fig. 2. Variation with time of the ratio of the number of pulsed counts in the second pulse (P2) and the first pulse (P1) of PSR 0531 + 21 (a) and PSR 0833 − 45 (b): open circle, > 35 MeV from SAS-2 data; closed circles, 50–3000 MeV from COS-B data. The broken line in (b) indicates the weighted average of the data points in the Vela case.

The spectral characteristics can be summarized as follows:

– *Vela*. The energy spectra of the different components of the Vela light curve are found to differ significantly. The first pulse exhibits a softer spectrum than the inter-region emission and the second pulse (Kanbach *et al.*, 1980). The shape of the spectrum of the total pulsed emission in the energy range 50 MeV to 3.2 GeV can be represented by a single power law of index 1.89 ± 0.06, with indications of a spectral flattening at low energies and a steepening above a few hundred MeV (Kanbach *et al.*, 1980). The energy spectrum of the total excess at the position of the Vela pulsar is consistent (in flux and shape) with that of the pulsed radiation (Lichti *et al.*, 1980; Hermsen, 1980), indicating that no steady component has been measured.

– *Crab*. No differences in shape are detected sofar between the energy spectra of the emission in the two pulses in the energy range above 50 MeV (Bennett *et al.*, 1977a, using only one observation). The total pulsed spectrum in the energy range 50 MeV to 1 GeV can be represented by a single power law index 2.2 ± 0.2. The total Crab spectrum shows emission in addition to the pulsed component up to energies of about 400 MeV (Lichti *et al.*, 1980; Hermsen 1980).

The total energy spectrum from the radio up to the ultra-high-energy gamma-ray range is presented for the cases of Vela and Crab in Figures 3a and 3b, respectively (from

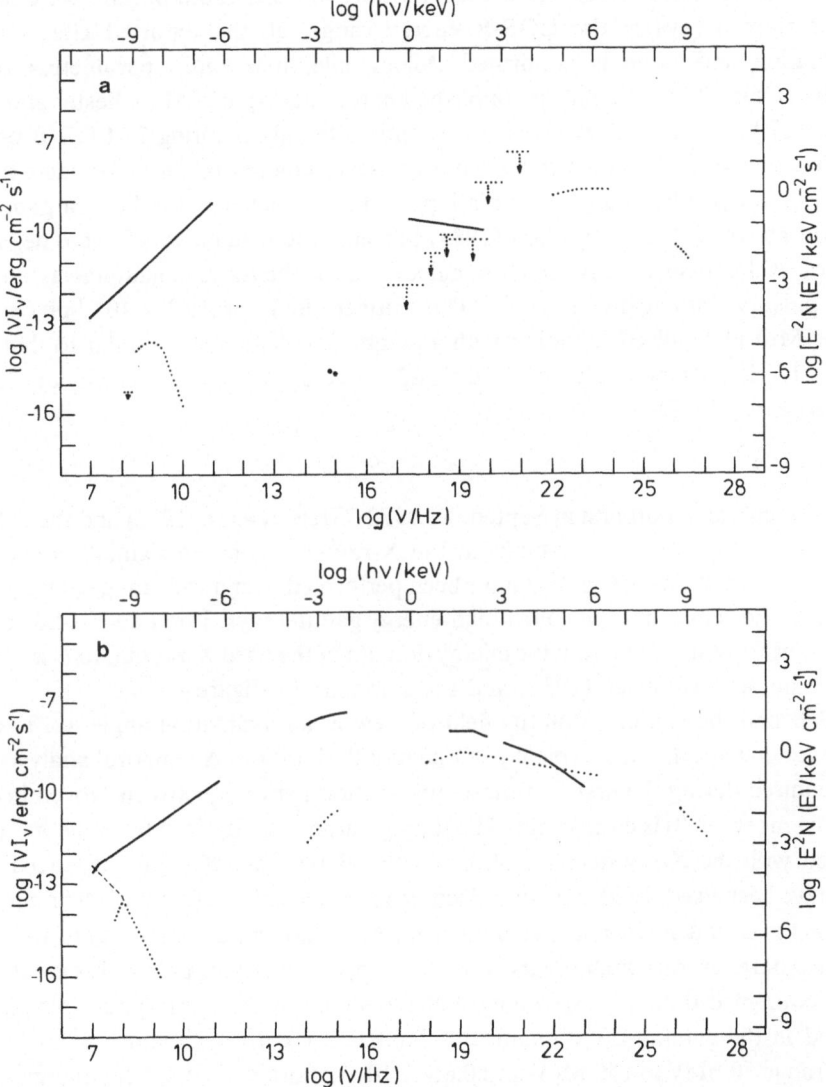

Fig. 3. Distributions of emitted power as a function of frequency for (a) Vela and (b) Crab: Nebula (solid lines), pulsar (dotted lines), compact source (broken line). The data for $E > 500$ GeV are somewhat disputed and show time variability. For data origins see Bignami and Hermsen (1983).

Bignami and Hermsen, 1983). The maximum power of the Vela pulsar is emitted at gamma-ray energies of ~ 1 GeV. The Crab pulsar emits most of its energy in the range of the hard X-rays to the soft gamma rays.

Finally, it is not possible to calculate the intrinsic luminosity of the pulsars exactly, since it depends on the pulsar's beaming geometry; using a conic beam geometry Buccheri (1981) derives for Vela L(50 MeV–10 GeV) $\simeq 4 \times 10^{34}$ erg s^{-1} or a conversion efficiency of $\sim 6 \times 10^{-3}$ and for Crab L(50 MeV–10 GeV) $\simeq 2 \times 10^{35}$ erg s^{-1} or a conversion efficiency of $\sim 4 \times 10^{-4}$.

Other Radio Pulsars. Although a small number of other radio pulsars were reported to emit gamma rays in the COS-B energy range, all the reported effects needed confirmation and none is confirmed. Meanwhile, new radio parameters became available. Of the 330 radio pulsars published in the catalogue of Manchester and Taylor (1981), 117 have parameters measured in time intervals covering 21 COS-B observations. These objects (17 of which have been observed more than once, bringing the total number of independent pulsar observations to 145) have been searched for gamma-ray emission above 50 MeV. Buccheri (1983) presents the results, which were negative in all cases. A 3σ upper limit is therefore calculated for the average gamma-ray flux from these pulsars into gamma rays. This upper limit is 1.2×10^{-7} ph cm^{-2} s^{-1} ($E > 50$ MeV) for pulsed radiation with a gamma-ray light curve similar to that of the Crab and Vela pulsars, implying less than 23% average conversion efficiency (see also Buccheri *et al.*, 1983).

2.2. Cyg X-3

After the great radio outburst in September 1972 (Gregory *et al.*, 1972) and the detection of a periodical variation of ~ 4.8 hr in the X-ray emission (Parsignault *et al.*, 1972), extensive observations of Cyg X-3 have been performed over a wide range of frequencies from the radio range up to the ultra-high-energy gamma rays. For a detailed discussion of the experimental results from the energy domain of the hard X-rays up to $E \geq 10^{16}$ eV, see Bignami and Hermsen (1983) and the summary in Figure 4.

COS-B had the source within its field of view at an inclination angle $< 10°$ during 7 observation periods, each typically of one month duration. A temporal analysis of the data acquired during the first 5 of these observation periods (between November 1975 and November 1980) is carried out. The timing parameters derived from the X-ray data collected with the X-ray detector aboard COS-B (van der Klis and Bonnet-Bidaud, 1981) have been used. Evidence for pulsed gamma-ray emission (at the 4.8 hr periodical variation) is found neither in the data from each individual observation, nor in the combined data set. As a preliminary result a 2σ upper limit to the pulsed flux is calculated for the combined data set. Assuming that the shape of the X-ray phase histogram is expected in the gamma-ray domain, this limit is 2.1×10^{-6} photon cm^{-2} s^{-1} in the energy range 50 MeV to 150 MeV and 0.65×10^{-6} photon cm^{-2} s^{-1} for energies above 150 MeV.

The integral energy spectrum, as summarized by Samorski and Stamm (1983), is given in Figure 4 together with the COS-B 2σ integral upper limits. At present it seems

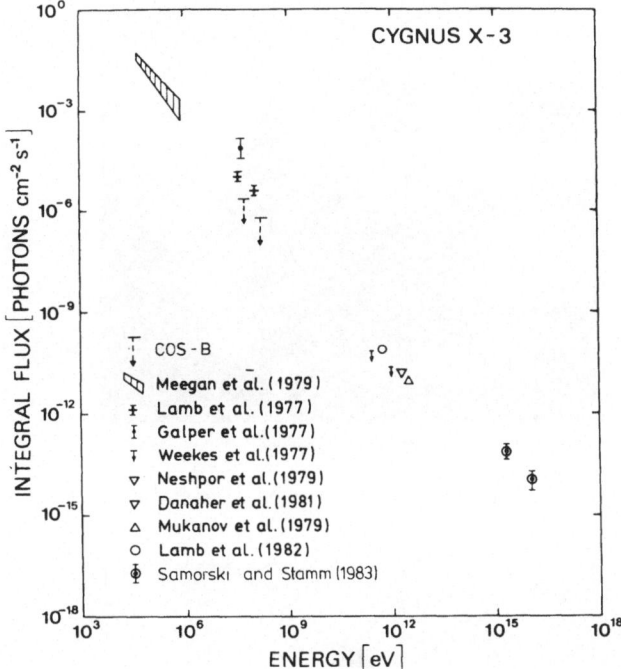

Fig. 4.   Integral gamma-ray spectrum of Cygnus X-3 (after Samorski and Stamm, 1983). The flux values are averaged over the total phase. The COS-B upper limits are included as well to show the experimental uncertainties for energies of ~ 100 MeV.

to be firm that the source is detected at ultra-high gamma-ray energies, while the detection at energies around 100 MeV is uncertain, but, if true, would indicate the presence of an extremely variable strong source, reaching a luminosity of $\gtrsim 10^{37}$ erg s^{-1} for a distance of ~ 11 kpc (Lamb *et al.*, 1977). In this context it is worth noting, that the data point in Figure 4 from Galper *et al.* (1977) is the average value over several years, while the measured flux in 1972 was claimed to be as high as 2×10^{-4} photon cm^{-2} s^{-1} ($E > 40$ MeV), or about an order of magnitude higher than the values claimed by the SAS-2 team.

2.3. MOLECULAR CLOUDS

Interstellar clouds can become visible as gamma-ray sources by interaction of cosmic rays and the matter of the clouds, as was first clearly stated by Black and Fazio (1973). Since the gamma-ray source $2\,CG\,353 + 16$ (Swanenburg *et al.*, 1981) is located in the direction of the ρOph cloud, identification of this source with ρOph was evidently suggested (Mayer-Hasselwander *et al.*, 1980; Bignami and Morfill, 1980). As a result of improved statistics relative to the database of Swanenburg *et al.* (1981), the source is now spatially resolved, making the identification definite. The ρOph region of the sky is given in Figure 5 (from Hermsen and Bloemen, 1983): Figure 5a shows the gamma-ray map, presenting evidence for resolved structure in gamma rays, and Figure 5b

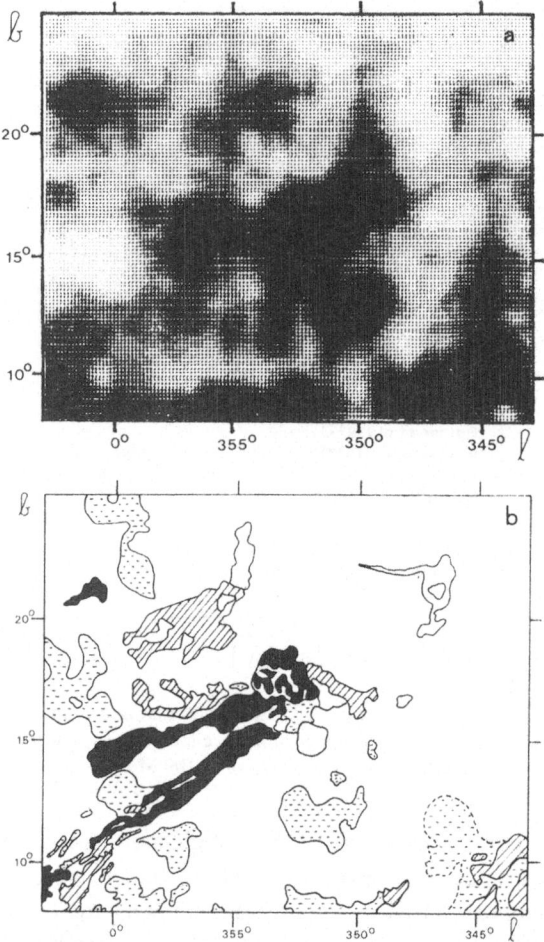

Fig. 5. The ρOph region: (a) half-tone map showing the gamma-ray intensities (100 MeV $< E <$ 5 GeV); (b) sketch from the atlas of dark clouds by Khavtassi (1960). The dotted, shaded and black regions indicate increasing obscuration. The blank features represent bright diffuse nebulae. For a detailed description of the distances of the dark clouds and the molecular gas connected to them see Wouterloot (1981) and references therein. (Figure from Hermsen and Bloemen, 1983.)

presents the distribution of extinction after Khavtassi (1960). The detailed positional correlation is striking, even though the gamma-ray map is measured with appreciably lower resolution. Wouterloot (1981) calculated from his OH measurements a higher mass estimate for the 'upper stream' than for the lower one, both clearly visible in the Khavtassi map (Figure 5b). This is consistent with the measured relative gamma-ray fluxes. A preliminary comparison of the total gamma-ray flux with the rather high mass estimates of Wouterloot (1981), indicates that the gamma-ray flux is within a factor 2 consistent with the gamma-rays being produced by the interaction of cosmic rays with a density equal to that observed in the solar neighbourhood and the estimated total mass (Hermsen and Bloemen, 1983).

The ρOph cloud complex falls now in the same category as the Orion cloud complex, namely a molecular cloud spatially resolved in the COS-B maps. In the case of Orion, the counting statistics obtained during one single observation were sufficient to resolve the structure (Caraveo *et al.*, 1980, 1981). Figure 6 gives a comparison of the gamma-ray map of the Orion region (Figure 6a) and the main part of the complex as seen in CO millimeter emission (Figure 6b). The gamma-ray map includes additional data (compared to Caraveo *et al.*, 1980, 1981) from several observations overlapping partly the Orion complex. The detailed positional correlation is evident, like in the case of ρOph.

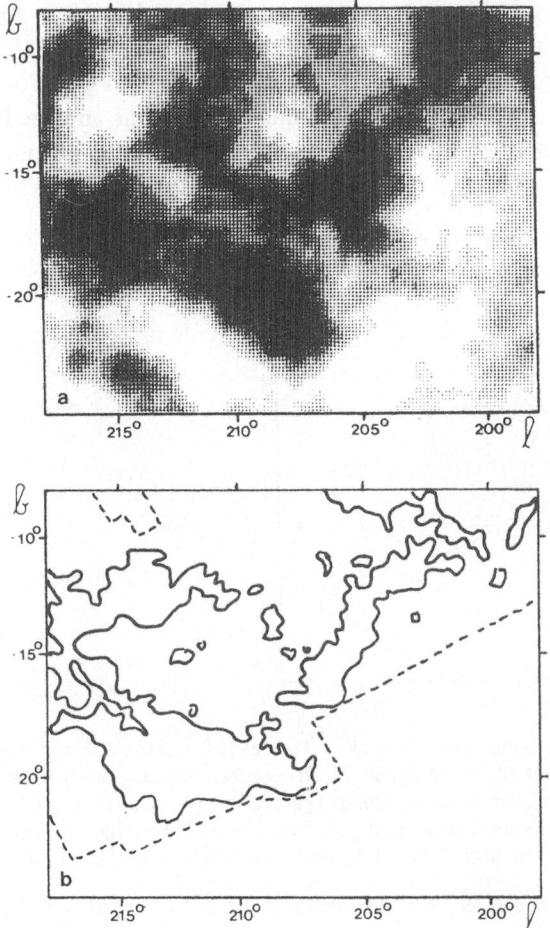

Fig. 6.   The Orion region: (a) half-tone map showing the gamma-ray intensities (100 MeV $< E <$ 5 GeV); (b) CO contour plot of 1 K peak antenna temperature (Morris and Thaddeus). (Figure from Bloemen, 1983.)

The total mass of the Orion complex is determined from the gamma-ray data, assuming that the cosmic-ray density at the complex is equal to the local value, and is found to be well in agreement with radio-astronomical estimates (Caraveo *et al.*, 1980, 1981).

2.4. 3C 273

COS-B observed the Virgo region of the sky three times. An observation in May, June 1976 led to the discovery of the high-energy (> 50 MeV) gamma-ray source CG 291 + 65, which prompted Swanenburg *et al.* (1978) to propose its identification with the quasar 3C 273 (< 1% probability of chance coincidence of the gamma-ray source with the specific object 3C 273). The second measurement (June, July 1978) confirmed the high-energy gamma-ray emission (Bignami *et al.*, 1981) and the position was updated (2CG 289 + 64; Swanenburg *et al.*, 1981). Finally, Hermsen *et al.* (1981) reported on a third detection, namely from data collected in June, July 1980.

Figure 7 presents the combined evidence for the detection of 3C 273 for energies above 100 MeV. The measured arrival directions of the gamma rays of all three observations were sorted into 0.5° × 0.5° bins. The resulting skymap was analyzed using the cross-correlation method described by Hermsen (1980) (see also the appendix). The skymap shown in Figure 7 gives for each bin the number of counts that correlate with the PSF.

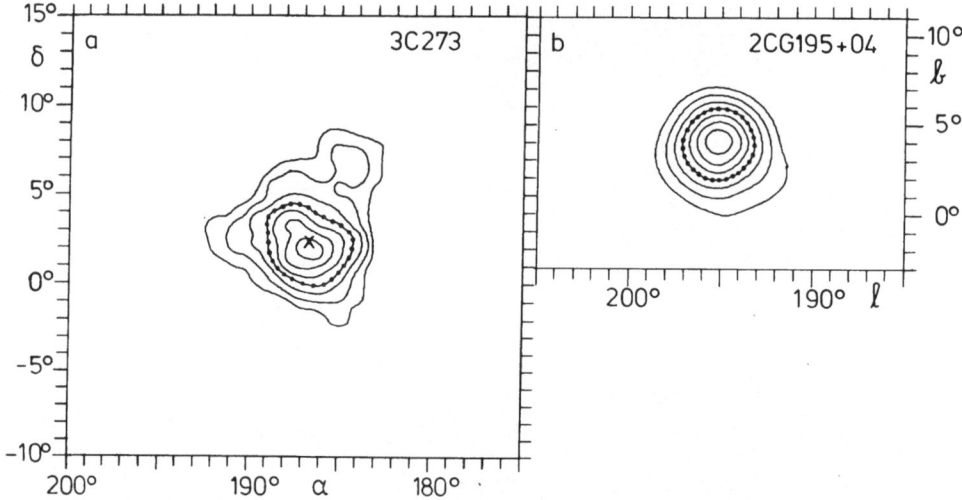

Fig. 7.   (a) Contour plot of the correlation map (*E* > 100 MeV) of an area of the sky containing 3C 273. The contour levels are indicated at 20, 30, 40, 50, 60, 70, 80 correlated counts. The cross indicates the position of 3C 273. (b) For comparison, the strong unidentified gamma-ray source 2CG 195 + 04 is shown in the same representation. Contour levels are indicated at 125, 200, 275, 350, 425, 500, 575 correlated counts. The dotted contours indicate for both figures a level of ~ 56% of the peak values.

In other words, in each bin is given the number of counts that can be attributed to a point source located in that bin. For comparison, the appearance of a point-like gamma-ray source is shown as well, for the same energy range and applying the same cross-correlation analysis. Used is data from the strongest (*E* > 100 MeV), unidentified gamma-ray source, 2CG 195 + 04 (often referred to as Geminga). The gamma-ray source feature at the position of 3C 273 contains ~ 90 counts (a confidence level of $8.2\sigma_0$) in a statistically empty (< 20 correlated counts) field. The Geminga feature contains ~ 620 counts and reaches a significance of $32.5\sigma_0$. For these high numbers of photon counts σ_0 can be considered to follow Gaussian statistics.

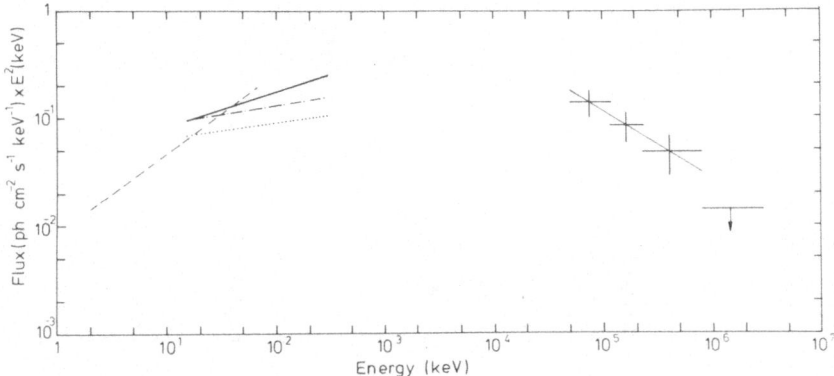

Fig. 8. High-energy spectrum of 3C273 including data from HEAO-1, experiment A2 (broken line), experiment A4 (solid line), the MIT/Leiden balloon experiment, 1979 (broken, dotted line) and 1980 (dotted line) and the COS-B data points with the best fit between 50 and 800 MeV. For references, see Hermsen *et al.* (1981).

The total energy spectrum from the soft X-rays up to the high-energy gamma rays is given in Figure 8 (from Hermsen *et al.*, 1981). Between 50 MeV and 800 MeV the spectrum can be well fitted by the power law:

$$dN/dE = (3.9 \pm 1.2) \times 10^{-6} \, [E\,(\mathrm{GeV})/0.15]^{-2.6 \pm 0.4} \, \text{photon}$$
$$\text{cm}^{-2} \, \text{s}^{-1} \, \text{GeV}^{-1} \, .$$

The data suggest that the peak luminosity of 3C273 could be expected in the 1 MeV to 10 MeV range. Neither the problem to decide on an appropriate gamma-ray production model, nor that to decide on the location of production in the QSO are sofar solved (see for example the discussion in Bignami *et al.*, 1981).

3. Unidentified Gamma-Ray Sources

3.1. THE 2CG CATALOGUE OF GAMMA-RAY SOURCES

The Caravane Collaboration published in 1977 the first CG Catalogue of high-energy gamma-ray sources (Hermsen *et al.*, 1977a), which contained 13 sources for a ~ 40% coverage of the galactic disc. The list of sources was the result from a search following a relatively simple procedure (Hermsen *et al.*, 1977b). This first success stimulated the development of a more sensitive and firm analysis procedure, based on a cross-correlation method, which was presented in full detail by Hermsen (1980). In this method the measured distribution of the photon arrival directions is correlated with the distribution expected for a point source. This latter distribution, the intrinsic point-spread function of the instrument, was determined by pre-launch calibration and confirmed by the actual flight data for the strong Vela gamma-ray source. A gamma-ray source is thus defined as a significant excess which has a spatial distribution consistent with the point-spread function. In an appendix, added to this paper, an extended abstract is presented, which

TABLE I

The 2CG catalogue of gamma-ray sources

Source name	No. of observations	Position l	Position b	Error radius	Flux[b] E > 100 MeV (10^−6 photons cm^−2 s^−1)	Spectral[a] parameter	Comments	CG source (Hermsen et al., 1977)	Identification
2CG006−00	3	6°.7	−0°.5	1°.0	2°.4	0.39±0.08		–	–
2CG010−31	1	10°.5	−31°.5	1°.5	1°.2	–		–	–
2CG013+00	4	13°.7	+0°.6	1°.0	1°.0	0.68±0.14		–	–
2CG036+01	3	36°.5	+1°.5	1°.0	1°.9	0.27±0.07		–	–
2CG054+01	3	54°.2	+1°.7	1°.0	1°.3	0.20±0.09		–	–
2CG065+00	4	65°.7	0°.0	0°.8	1°.2	0.24±0.09	could be an extended feature	CG64+0	–
2CG075+00	5	75°.0	0°.0	1°.0	1°.3	–		CG75−0	–
2CG078+01	5	78°.0	+1°.5	1°.0	2°.5	–		CG78+1	–
2CG095+04	3	95°.5	+4°.2	1°.5	1°.1	–		–	–
2CG121+04	3	121°.0	+4°.0	1°.0	1°.0	0.43±0.12		CG121+3	–
2CG135+01	3	135°.0	+1°.5	1°.0	1°.0	0.31±0.10		CG135+1	–
2CG184−05	4	184°.5	−5°.8	0°.4	3°.7	0.18±0.04		CG185−5	–
2CG195+04	3	195°.1	+4°.5*	0°.4	4°.8	0.33±0.04	γ195+5	CG195−4	–
2CG218−00	3	218°.5	−0°.5	1°.3	1°.0	0.30±0.08		–	–
2CG235−01	2	235°.5	−1°.0	1°.5	1°.0	–		–	–
2CG263−02	4	263°.6	−2°.5	0°.3	13°.2	0.36±0.02	could be an extended feature	CG263−2	PSR 0833−45
2CG284−00	1	284°.3	−0°.5	1°.0	2°.7	–		–	–
2CG288−00	1	288°.3	−0°.7	1°.3	1°.6	–		–	–
2CG289+64	2	289°.3	+64°.6	0°.8	0°.6	0.15±0.07		CG291+65	3C273
2CG311−01	2	311°.5	−1°.3	1°.0	2°.1	–		CG312−1	–
2CG333+01	3	333°.5	+1°.0	1°.0	3°.8	–		CG33+0	–
2CG342−02	5	342°.9	−2°.5	1°.0	2°.0	0.36±0.09		–	–
2CG353+16	4	353°.3	+16°.0	1°.5	1°.1	0.24±0.09		–	ρOph
2CG356+00	1	356°.5	+0°.3	1°.0	2°.6	0.46±0.12	prob. variable	–	–
2CG359−00	3	359°.5	−0°.7	1°.0	1°.8	–		–	–

[a] Assuming E^{-2} spectra.

[b] Intensity ($E > 300$ MeV)/Intensity ($E > 100$ MeV), assuming E^{-2} spectra calculating both intensities.

* See text.

describes the method, the performed Monte-Carlo simulations and experimental results (including data on the average extent of the published sources).

Based on the data acquired during the first $2\frac{1}{2}$ years of observation time the second COS-B catalogue of gamma-ray sources has been compiled (Swanenburg et al., 1981; Hermsen, 1980) and is shown in Table I. It is noted that, although the sources in the Carina region (2CG 284 – 00 and 2CG 288 – 00) and in the Cygnus region (2CG 075 + 00 and 2CG 078 + 01) are quoted in the catalogue as individual entries, the corresponding structures could also be interpreted as extended features. Also, it has been recognized that the flux from the Cygnus region shows differences between several observations of this region, suggesting a contribution from variable sources, not related to Cyg X-3 (Bloemen et al., 1981). One probably variable source (2CG 356 + 00) is contained in the catalogue. This source has been clearly seen only in one of four observations and presents a hint for time variability at the 99% confidence level.

After publication of the 2CG catalogue the position of one source has been updated, namely that of 2CG 195 + 04. New data became available on this source and the updated position is $l = 195.1°$, $b = 4.2°$ with an error radius of $0.4°$ (Masnou et al., 1981; see also Figure 9).

The four identified sources in the catalogue have been discussed already in Section 2. Strictly taken, the source 2CG 353 + 16, identified with ρOph, should be deleted from the catalogue. Using more data than was available at the time of compilation of the catalogue, the source is now resolved (see Section 2.3). All remaining sources have been the subject of searches for counterparts measured at other wavelengths. Some have been studied in great depths. (e.g. 2CG 135 + 01 and 2CG 195 + 04; see the review by Bignami and Hermsen, 1983). Still, 21 sources lack firm identification.

3.2. AVERAGE PROPERTIES OF THE UNIDENTIFIED GAMMA-RAY SOURCES

The vast majority of gamma-ray sources is galactic, as is evident from their distribution over the sky. In fact, 20 unidentified sources are closely aligned with the galactic-disc line emission. Most discussions on the nature of the unidentified sources are concentrated on these low-latitude sources. We do the same here.

Prime characteristics of a source are total luminosity and size. The lack of identification, and hence the lack of individual distances, prohibits the direct derivation of these parameters from the observations. In order to enable us to estimate the order of magnitude of these parameters we assume that all unidentified sources, or at least the majority, belong to one class of objects or population. With this assumption the main characteristics of this population follow from the observed longitude and latitude distribution (Swanenburg et al., 1981). The average properties are summarized in Table II. Many other authors have discussed this topic (see e.g. Bignami and Hermsen, 1983, and reference therein). Among them Buccheri et al. (1981), who propose that the 2 kpc lower limit to the average distance is probably a 'typical' distance. This would imply that the typical luminosity of the 2CG sources is about 4×10^{35} erg s^{-1}, while they consider 2×10^{36} erg s^{-1} to be an upperlimit. These values would support their

TABLE II

Average properties of the unidentified galactic gamma-ray sources

Average $\|b\|$	$\sim 1.5°$
Angular size	$0° - 2°$
Photon-flux range (100 MeV $< E <$ 1 GeV)	$\sim (1-5) \times 10^{-6}$ ph cm^{-2} s^{-1}
Distance range	2–7 kpc
Scale height $\langle \|z\| \rangle$ ($300° < l < 60°$)	$\lesssim 130$ pc
Spectral shape	different for individual objects, average dN/d$E \sim E^{-2}$
Average photon energy	~ 250 MeV in the 100 MeV–1 GeV range
Energy-flux range (100 MeV $< E <$ 1 GeV)	$\sim (4-20) \times 10^{-10}$ erg cm^{-2} s^{-1}
Luminosity range (100 MeV $< E <$ 1 GeV)	$\sim (0.4-5) \times 10^{36}$ erg s^{-1}

suggestion that young (yet undetected) radio pulsars are the counterparts for most of the unidentified gamma-ray sources.

The proposed identification of the unidentified gamma-ray sources with radio pulsars is one of the many suggestions made to solve the problem of the puzzling nature of the sources. This is an example of a comparison between different populations of sources, using the average characteristics of the populations. Other attempts are made for individual sources by making detailed experimental investigations of the region of the sky covered by the COS-B error box. This way we derive constraints on source models which try to explain the emission from an individual source. An example is given below for 2CG 195 + 04 (Geminga). Also, many authors propose first theoretical scenario's for gamma-ray production which can then be tested against the experimental constraints or can stimulate the search for a particular object or astrophysical setting in the general direction of a gamma-ray source. In the introduction it was already noted, that a general discussion of these topics is beyond the scope of this paper. The reader is referred to the many papers on this topic in this volume and e.g. the review by Bignami and Hermsen (1983).

3.3. 2CG 195 + 04 (Geminga)

The source 2CG 195 + 04 was discovered by the SAS-2 experiment (Kniffen *et al.*, 1975) and called γ 195 + 5. In Figure 9 the SAS-2 error box (Thompson *et al.*, 1977) is indicated in a contour map of the gamma-ray intensities measured by COS-B for energies in the range 500 MeV to 5 GeV. The data are from four COS-B observations of typically 5 weeks duration each, between August 1975 and October 1980. The high-energy range is selected in order to exploit the best possible angular resolution obtainable with COS-B. A sharp source profile, representing ~ 190 counts, is nicely located in the SAS-2 error box (actually, the intrinsic source profiel has been smoothed to produce the contour map). The COS-B position is indicated in the figure, together with its uncertainty (Masnou *et al.*, 1981). The feature at $l \simeq 185°$, $b \simeq 6°$ is the Crab gamma-ray source; the position of PSR 0531 + 21 is indicated.

This unidentified gamma-ray source Geminga is a mystery, still 8 years after its discovery! Its gamma-ray spectrum is vastly different from that of Crab or Vela. In

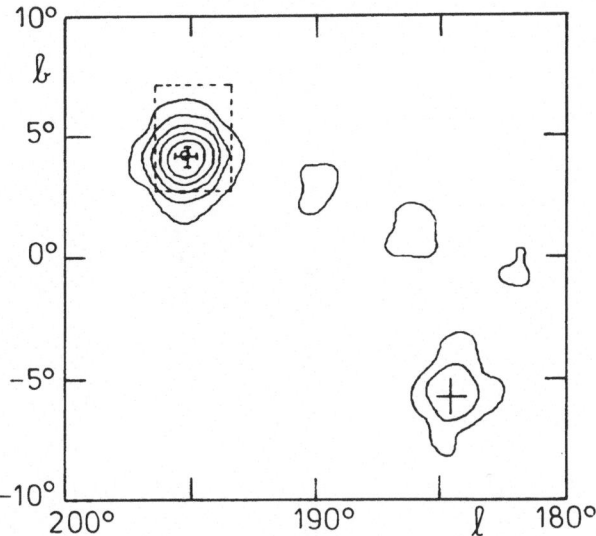

Fig. 9. Contour map of the gamma-ray intensities measured by COS-B for energies between 500 MeV and 5 GeV of the sky region containing Geminga and Crab. The COS-B exposure used to produce this map is for both sources $\sim 1.8 \times 10^8$ cm^2 s. The dotted lines indicate the SAS-2 error box for $\gamma 195 + 5$. Also is indicated the COS-B position with its uncertainty and the position of PSR 531 + 21.

Figure 9 it is evident that the strength of the source for energies between 500 MeV and 5 GeV is twice the strength of the total Crab, while at energies between 50 MeV and 100 MeV Geminga reaches only half the strength of the Crab pulsar! Sofar no unambiguous counterpart has been proposed for this enigmatic source; Bignami and Hermsen (1983) present a résumé of the experimental status in the search for counterparts, summarized in Figure 10. However, a recent and very tempting identification has been proposed by Bignami *et al.* (1983) after a deep survey of the Geminga error box in several wavelengths bands. Caraveo (1981) reported already on the discovery of an Einstein soft-X-ray source in the Geminga error box. This source (1 E0630 + 178) turned out to be a unique object. The detailed work by Bignami *et al.* (1983) resulted in the following expose: An identification was obtained with a faint blue object exhibiting a unique luminosity ratio $L_X/L_V \simeq 1300$, while no source has been found at 6 cm down to ~ 1 mJy using the VLA. The X-ray source (flux $\sim 2 \times 10^{12}$ erg cm^{-2} s^{-1}) has a very soft spectrum and little evidence has been found for interstellar absorption, leading to a distance estimate of 100 pc or less. Although no temporal evidence for the presence of a pulsar has been found, Bignami *et al.* (1983) argue that the extreme parameters characterizing this source, point at the presence of a neutron star.

Figure 10 shows the total energy spectrum from the radio range up to the ultra-high-energy gamma rays. The data point indicated at X-ray energies and the upper limit in the radio range, are drawn with the assumption that 1 E0630 + 178 is to be associated with the gamma-ray one. However, if the radio flat-spectrum source of which the

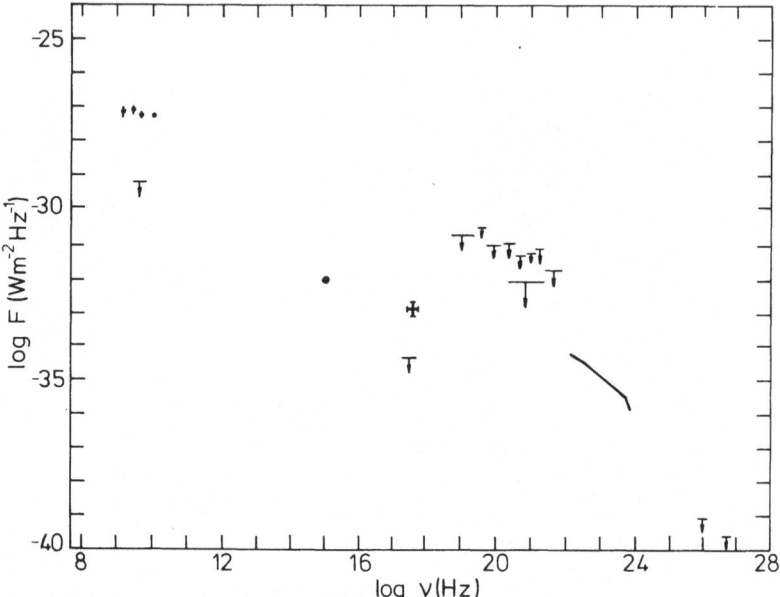

Fig. 10.   Composition of spectral data for 2CG 195 + 04 (Geminga), drawn in the assumptions: (1) that the newly discovered X-ray source is to be associated with the gamma-ray one, (2) that the radio flat-spectrum source is the counterpart (see text). Data origins: see Bignami and Hermsen (1983).

spectrum is given in the figure, is the counterpart (as was suggested by Sieber and Schlickeiser, 1982), the upper limit at soft X-ray energies should be regarded. This latter suggestion is now less likely to be the reality, given the very interesting identification of 1 E0630 + 178. The COS-B gamma-ray spectrum is taken from Masnou *et al.* (1981).

Geminga is certainly the prototype of the unidentified gamma-ray sources. The energy flux (F) emitted by this source in the gamma-ray domain dominates the energy output in all other frequency bands by several orders of magnitude, e.g.: (1) if the radio flat-spectrum source is the counterpart, $\nu F_\gamma / \nu F_{radio} \approx 2 \times 10^5$ and $\nu F_\gamma / \nu F_x > 10^4$; and (2) if the Einstein X-ray source is the counterpart, $\nu F_\gamma / \nu F_{radio} > 10^7$ and $\nu F_\gamma / \nu F_x \approx 3 \times 10^3$. Most likely the mystery around the Geminga gamma-ray source has been solved now with the identification of the unique X-ray source 1 E0630 + 178.

Acknowledgements

The results presented in this paper are obtained, thanks to the continuing efforts of many colleagues in the Caravane Collaboration for the COS-B satellite. I thank Hans Bloemen for reading the manuscript and for some valuable contributions to this paper.

Appendix: Search for Point Sources

In the search for point sources in the COS-B data we used a cross-correlation method. The main advantage of such a method is that the presence in the data of the signature of a point source, i.e. the point-spread function (PSF), is exploited, yielding a favourable combination of sensitivity, positional accuracy, and low susceptibility to spurious effects. On the other hand, the method imposes certain limitations. For instance, the effective angular resolution is slightly degraded, which impacts on the ability to distinguish between a real point source and an extended region of emission and on the sensitivity in confused regions. Also, if a candidate source is superimposed on a structured background, the excess should be fitted. This fitting procedure itself is not unique, but introduces a certain amount of interpretation. Therefore it is usefull to get insight in the applied method and to see examples from the data. What follows is an extended abstract from the presentation by Hermsen (1980) of the method and the data, with some additional figures and remarks.

A. CROSS-CORRELATION FUNCTION

In looking for point sources we are examining the data for statistically significant excesses above the surrounding background, the excess having a spatial distribution consistent with the PSF. Therefore we search for a signal of predictable shape super-imposed on the background. The general method available for separating signal and background in such a case is the cross-correlation method. With this method the distribution of the photon arrival directions is correlated with the PSF. A gamma-ray source is thus defined as an excess for which the cross-correlation analysis results in a significant positive correlation. From the amount of correlation also the strength of the source can be derived, or in case of lack of correlation an upper limit to the source strength.

We explain now the basis and the technical execution of this method. The data in the skymaps are grouped in skybins (D_{ij}) and may be presented in counts (N_{ij}) or in flux values (F_{ij}). The indexes i and j represent the bin coordinates in the skymap. When position $\mathbf{r}$ is taken as a trial position, the PSF, $f(\mathbf{r}, i, j)$, predicts the contribution of a source at $\mathbf{r}$ to bin (i, j). The correlation function $C(\mathbf{r})$ is then defined as

$$C(\mathbf{r}) = \frac{A}{m} \sum_{ij} (D_{ij} - \overline{D})(f(\mathbf{r}, i, j) - \overline{f}), \qquad (1)$$

where A is a normalization constant, yet to be chosen, m is the number of bins over which Σ_{ij} is accumulated, $\overline{D}$ is the average of D_{ij} and $\overline{f}$ is the average of $f(\mathbf{r}, i, j)$ over the area of the m bins considered. In the computations square arrays containing m bins are used. The distribution of $C(\mathbf{r})$ over the skymap is called the correlation map. If for a point $\mathbf{r}$ the deviations of D_{ij} with respect to $\overline{D}$ have predominantly the same sign as the deviation of $f(\mathbf{r}, i, j)$ with respect to $\overline{f}$, $C(\mathbf{r})$ is positive. We would like to underline the importance of the inclusion in this correlation function of the average values $\overline{D}$ and $\overline{f}$. The result is, that structures with angular sizes lager than the PSF are partly suppressed, while a uniform background, or a background linear with r, gives even zero correlation. A source superimposed on such a background causes non zero correlations as a function of $\mathbf{r}$.

Equation (1) may be rewritten in the form

$$C(\mathbf{r}) = A \left[\frac{\Sigma f D}{m} - \frac{(\Sigma f)(\Sigma D)}{m^2} \right],$$ (2)

where all subscripts are omitted for clarity. The normalization constant A can be chosen in such a way that, if we take for D_{ij} the number of counts

$$N_{ij} = \frac{N_s}{\Sigma' f} \; f(\mathbf{r}, i, j) + B,$$ (3)

where B is a uniform background level, N_s is the total number of counts (= strength) of the source, and $\Sigma' f$ is the sum over the total PSF, the maximum value of the correlation function equals the source strength. This gives

$$C_n(\mathbf{r}) = [m \, \Sigma \, fN - (\Sigma f)(\Sigma \, N)] \; \frac{\Sigma' f}{m \, \Sigma f^2 - (\Sigma f)^2} \; .$$ (4)

Similarly, if we take the flux values F_{ij} instead of the number of counts:

$$C_f(\mathbf{r}) = [m \, \Sigma \, fF - (\Sigma f)(\Sigma \, F)] \; \frac{\Sigma' f}{m \, \Sigma f^2 - (\Sigma f)^2} \; .$$ (5)

In principle the correlation functions may be determined for *any* point in a skymap by accumulating the sums over the *entire* map.

For practical reasons three restrictions are made:

(i) Sky bins of size $0.5 \times 0.5°$ are used.

(ii) The correlation function is determined only at the centre of bins. This is justified because a step size of $0.5°$ is fine enough in view of the available statistics.

(iii) The sums are accumulated over a smaller array, usually a matrix of about $10° \times 10°$ or $15° \times 15°$. This is done to restrict the range over which significant structure (e.g. strong source, or galactic plane) affects the value of the correlation.

Examples of the response of the cross-correlation analysis to a strong point source (PSR 0833 – 45, Vela) are given in Figure 11. The derived correlation maps peak at the position of the pulsar and it is verified that the correct number of source counts is retrieved by using the PSF as determined by the pre-launch calibration. The negative 'ring' around the maximum reflects the expected anti-correlation.

B. PROBABILITY DENSITY FUNCTION

A good insight into the statistical behaviour of the cross-correlation function is essential, both for estimating the accuracy of intensities and positions, and even more so for deriving firm conclusions about the statistical significance of the results. Therefore, Hermsen (1980) determined the probability density function (PDF) of the cross-correlation function. The shape of the PDF depends on the used PSF, the number of counts in the matrix and the matrix size. For the COS-B profile and counting statistics the

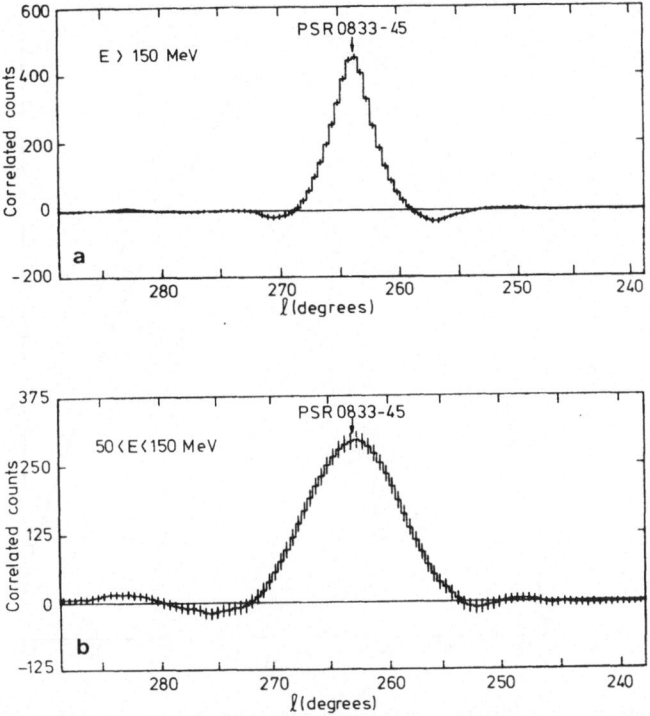

Fig. 11.   Longitude profiles across PSR 0833−45 in the correlation map for (a) energies above 150 MeV and (b) energies between 50 MeV and 150 MeV. In both figures the diffuse galactic gamma-ray background is suppressed because the events are selected in the peaks of the light curve of the pulsar.

cross-correlation function does not follow Gaussian statistics. Two examples of the PDF are shown in Figure 12 for parameters typical for COS-B high-latitude observations and the energy range $E > 150$ MeV. The peak at negative values of C (or C_n) and the tail towards positive values are apparent. For a field without sources (flat background, high-latitude observations) the parent mean is 0 and the parent standard deviation of the distribution of values for C found, is given by

$$\sigma_0(C) = \frac{1}{m^2} \sqrt{(m \Sigma f^2 - (\Sigma f)^2) \Sigma N} , \tag{6}$$

where again the subscripts are deleted.

The ratio

$$C/\sigma_0 = \frac{m(\Sigma fN) - (\Sigma f)(\Sigma N)}{\sqrt{(m \Sigma f^2 - (\Sigma f)^2) \Sigma N}} \tag{7}$$

gives the correlation value in number of parent standard deviations for a random distribution of the counts. A sufficiently large value of C/σ_0 indicates that the data set probably does *not* represent a random distribution of counts, possibly indicating the

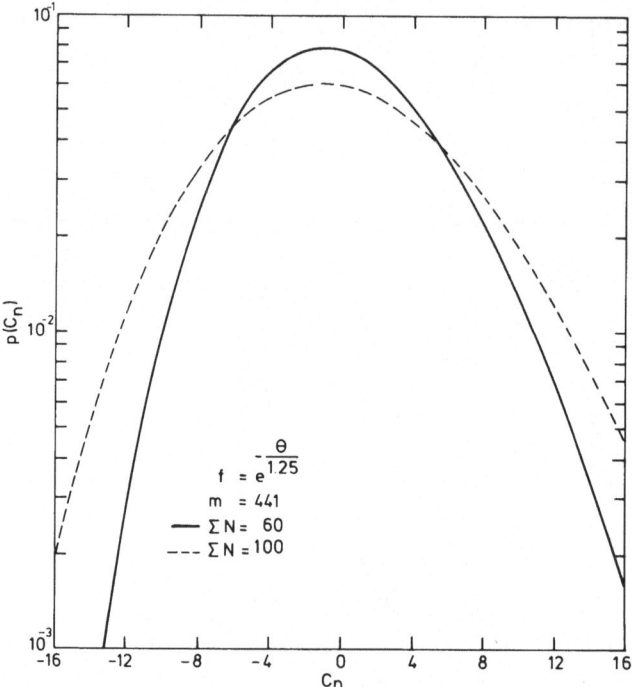

Fig. 12.   Two probability density functions of the cross-correlation function for different ΣN as a function of C_n (Equation (4)) for the energy range $E > 150$ MeV. The point-spread function, f, applicable in this energy range is given.

presence of a source. The probability can be worked out numerically, as was done to produce Figure 12. For large ΣN the PDF may be approximated by the normal distribution.

If the values of C/σ_0 indicate that we do have a significant correlation, we claim the detection of a source with strength given by (4) (correlated counts) or (5) (correlated flux). Then we wish to attribute a variance to the correlation (or source strength) actually measured. For this we have to derive the sample variance of the measured C or C_n. These are respectively (Hermsen, 1980):

$$\sigma(C) = \frac{1}{m^2} \sqrt{[m^2 \, \Sigma \, (f^2 N) + (\Sigma f)^2 \, \Sigma \, N - 2m(\Sigma f)(\Sigma fN)]}, \tag{8}$$

$$\sigma(C_n) = \sigma(C) \, \frac{m^2 \, \Sigma f}{m \, \Sigma f^2 - (\Sigma f)^2} \,. \tag{9}$$

C. Monte Carlo simulations

Although the PDF of the cross-correlation function can be determined numerically and the response of the correlation analysis to fine structure in the data can be calculated, representative simulations of complicated situations encountered in the flight data, including the effect of a non-uniform background, are needed. Expecially near the Galactic Centre the background is in the form of a strong line source and systematically yields a positive correlation. Another aspect which can only be assessed by simulations is the evaluation of the probability to detect spurious sources as a result of a *search* for sources by analysing points of maximum correlation. The significance of each maximum is reduced because in the search a large number of trials is made, namely one for each bin in the correlation map. Therefore, the probability to find a maximum of given significance anywhere in the map is proportional to the number of *independent* trials. By the nature of the correlation function a certain degree of correlation exists between individual trials. The derivation of the degree of correlation of individual bins in the correlation map cannot be done analytically. Therefore simulations are needed.

In the case of a uniform background region (applicable in the COS-B data for high-latitude observations) the simulation is unambiguous and a sufficiently accurate approximation for the probability calculation is determined (Hermsen, 1980). A probability threshold, of finding a value C_n or larger, of $< 10^{-5}$ is adopted in the flight data analysis. With this threshold the probability of accepting a spurious source in a single high latitude observation is $\sim 3\%$.

The situation is more complex if the candidate source is superimposed on a structured background, because the expectation value of C is not zero in this case. This situation occurs in the flight data along the galactic plane for $|b| < 5°$, where the line source introduces a positive excess in the correlation map. Figure 13 presents a one-dimensional

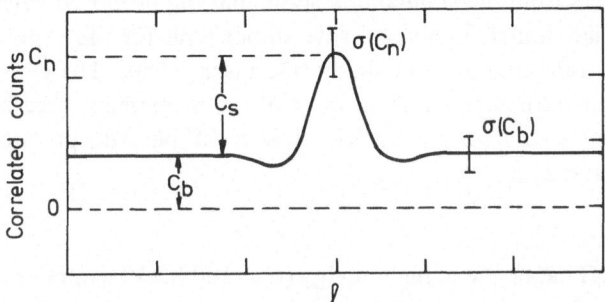

Fig. 13.   One-dimensional sketch of a correlation map for a source superimposed on a line source. Explanation is given in the text.

sketch of such a situation, i.e. a point source superimposed on a line source which is parallel to the longitude direction. The values of C_n and $\sigma(C_n)$ at the source position are not a measure of the correlation and significance of the source, but of the combination of source *and* background (i.e. the line source). In Figure 13 the number of counts attributed to the source is C_s and to the background C_b. The values of C_s and C_b are derived using a fitting procedure. Since no independent background estimate is available (still at least for the major fraction of the galactic plane region), each candidate source is fitted and subtracted, in such a manner that the remaining background connects smoothly with the surroundings. The fitted source strength (C_s) together with the parameters $\sigma(C_n)$ and ΣN at the fitted position are then used to express the significance of the excess in number of standard deviations (σ'_0) of the underlying background (Hermsen, 1980). Since it cannot be assumed that the statistical behaviour of the values obtained by the fit is identical to that of the correlation values in the single bins, and because the fitting procedure itself is not unique but introduces a certain amount of interpretation, the entire method is calibrated by analysing representative simulated skymaps. This calibration also removes the ambiguity resulting from the unknown number of independent trials made in a search.

Because the shape and strength of the galactic gamma-ray distribution varies with longitude, nine longitude intervals, each $10°$ wide, are selected to give a representative sample of plane shapes (in latitude direction), plane to background ratios and counting statistics. In total 99 skymaps, $40°$ wide in longitude each, are simulated, namely 11 for each latitude distribution. No structure as a function of longitude is introduced in the simulation. The correlation maps have been analysed in the latitude range $|b| < 5°$. The entire search thus covers 1.584×10^5 bins of size $0.5° \times 0.5°$, equivalent to a surface 11 times the measured galactic plane distribution in the same latitude range. Candidate (spurious) sources are fitted and judged by eye using an interactive computer display terminal, identical to the procedure followed for the flight-data analysis. The significance in C_s/σ'_0 is calculated for each analysed excess. The number of counts in the region along the galactic plane is high. Therefore, Gaussian statistics are initially assumed to apply and the corresponding probability for random occurence for the analysed excess is calculated. The results from the simulations show that the proposed calculation of the significance (Gaussian statistics) gives a save upper limit for the expected numer of random excesses as a function of probability (Hermsen, 1980). Therefore, in the flight data analysis of the structured region $|b| < 5°$ a probability threshold of 10^{-6} $(\sim 4.75\sigma'_0)$ is adopted, so that for $|b| < 5°$ (i.e. 1.44×10^4 bins) the probability to accept a spurious source is $< 2\%$.

D. RESULTS

In Section 3, the 2CG catalogue of high-energy $(E > 100 \text{ MeV})$ gamma-ray sources has been introduced and the average properties of the sources have been discussed. Below, the appearances in the data of several sources contained in the catalogue are shown in order to give a good impression of the actual measurements.

Most results obtained for high-latitude regions are shown already: 2CG 353 + 16, now being resolved and unambiguously identified with ρ Oph (Section 2.3) and 2CG 289 + 64 or 3C 273 (Section 2.4). One more source is detected away from the galactic plane: 2CG 010 − 31. It is measured at about the significance level obtained for 3C 273 in a single observation: $5.7\sigma_0$ for 33 counts, giving $P(C_n) = 4.6 \times 10^{-8}$. Confirmation is needed in this case.

The search for sources along the galactic plane encounters two main problems: (i) sources are superimposed on a structured background, (ii) source confusion occurs due to a high surface density of possible sources.

In the region of the plane $50° \lesssim 1 \lesssim 310°$ source confusion is minimal. Here, this problem mainly plays a role in the Cygnus and the Carina region. For both cases two sources are contained in the catalogue: the measured distributions are best explained with the presence of these sources, but it cannot be excluded that they are part of more extended features (as is stated in the catalogue). For other regions in this longitude interval confusion doesnot play an important role. Therefore, selection of data with the best possible angular resolution, namely for energies above 300 MeV or 500 MeV, is not required. All reported sources are detected for energies above 100 MeV or 150 MeV. Figure 14a, 14b, and 14c show examples of the profile of a gamma-ray source in the

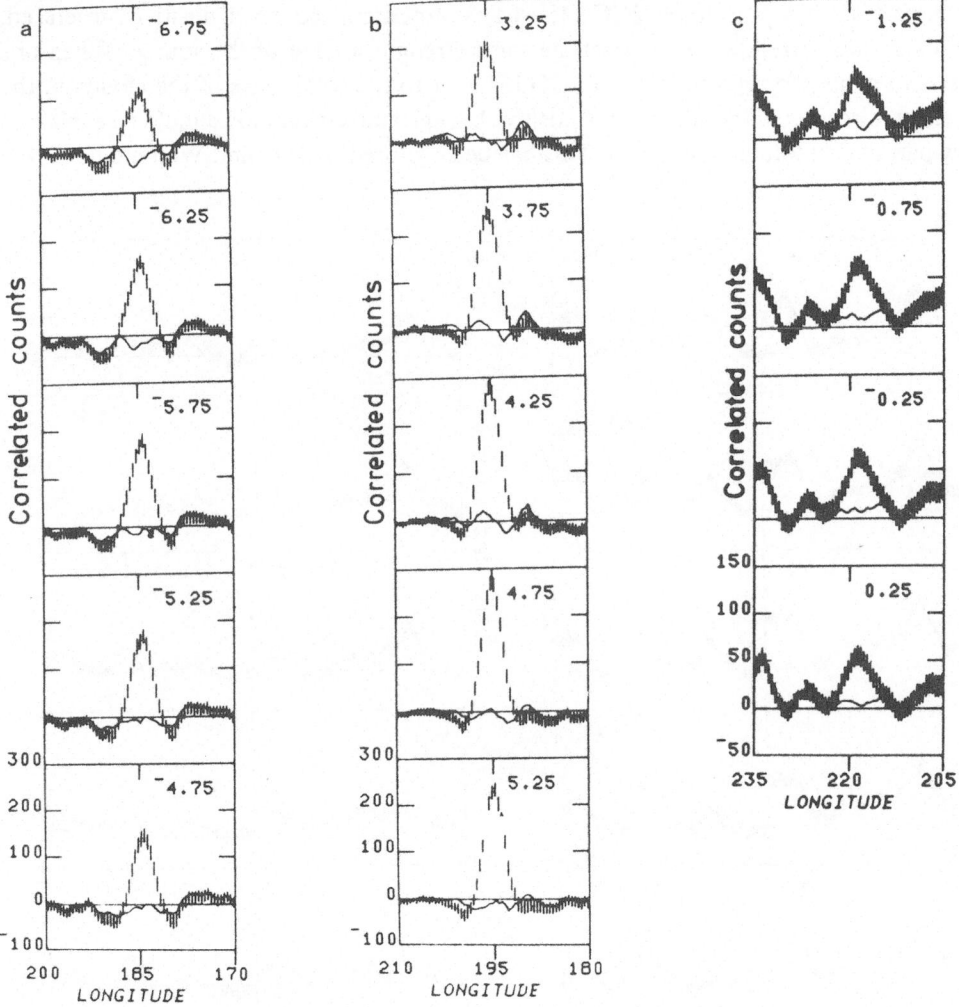

Fig. 14. Longitude profiles of the correlation maps showing for energies above 150 MeV (a) 2CG 184−05 (Crab), (b) 2CG 195 + 04 (Geminga), and for energies above 100 MeV (c) 2CG 218−00. The latitude values are given for each profile. The solid lines show the distributions which remain after subtraction of the cross-correlated point-spread functions at the positions of the sources.

correlation map for: (a) a genuine point source (PSR 0531 + 21, Crab, 2CG 184 − 05; $E > 150$ MeV), (b) a strong unidentified source (Geminga, 2CG 195 + 04; $E > 150$ MeV), and (c) a weak unidentified source (2CG 218 − 00; $E > 100$ MeV). In each figure the solid line shows the distribution which remains after subtraction of the cross-correlated PSF at the position of the source. Evidently, all three excesses have a shape compatible with the PSF. Hardly any plane component is evident in the correlation maps, except in the case of 2CG 218 − 00, for which a relatively small positive correlation remains after subtraction of the source profile. This can be quite different in other regions, such as that around 2CG 135 + 01. Figure 15a presents, for energies above 100 MeV, the correlation map containing this source and 2CG 121 + 04. The source feature representing 2CG 135 + 01 is superimposed on a positive structured, background correlation of at least the same strength as that of the source. When one tries to relate the reported flux of 2CG 135 + 01 to the total mass of the clouds in the general direction of the source, this positive background correlation should be explained by part of that total mass as well. This has been ignored by Issa and Wolfendale (1981)

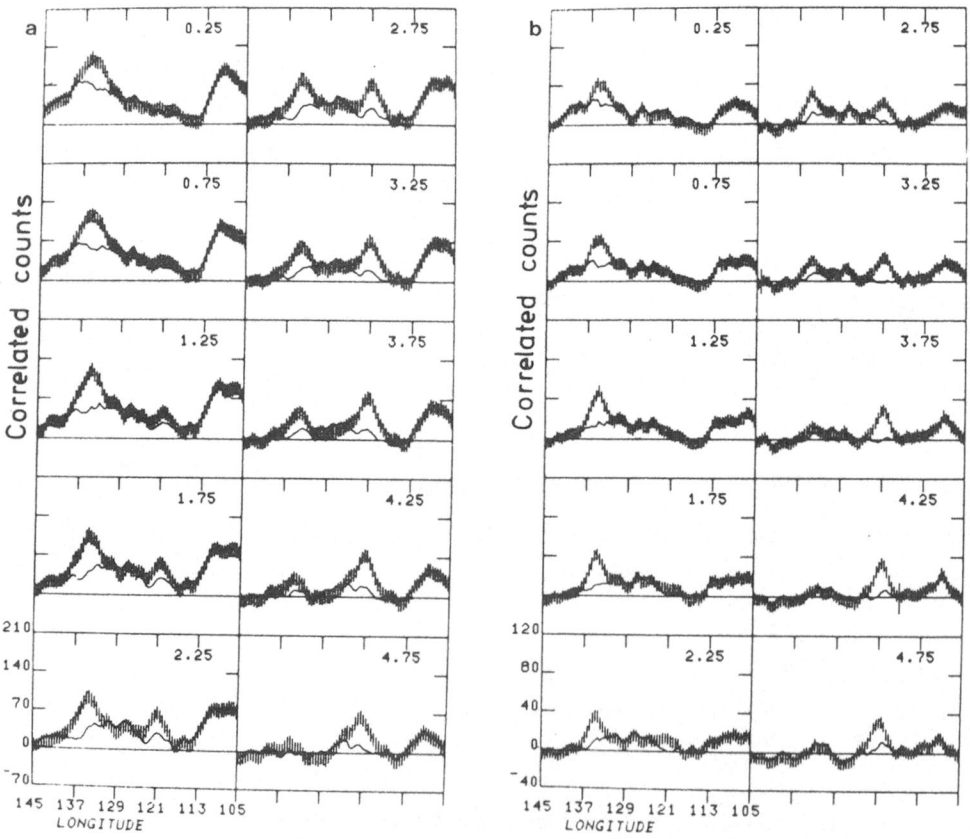

Fig. 15.   Longitude profiles of the correlation maps showing 2CG 135 + 01 and 2CG 121 + 04 for energies above (a) 100 MeV and (b) 300 MeV. See legend of Figure 14.

in the case of this source and several others. The source 2CG 121 + 04 is well away from the plane, therefore, the background level is lower. Both sources show up nicer in Figure 15b for energies above 300 MeV. The background level in the correlation map is practically reduced to zero at the position of 2CG 121 + 04 and is also reduced relative to the source strength for 2CG 135 + 01. In addition the profiles are sharper.

In the longitude interval encompassing the direction of the Galactic Centre, $310° \lesssim l \lesssim 50°$, source confusion dominates in the integral energy range above 100 MeV. Given the COS-B PSF, counting statistics and applied selection criteria it is practically impossible to resolve source features from the general galactic structure at a sufficiently high significance level. In fact, all sources reported in the 2CG catalogue

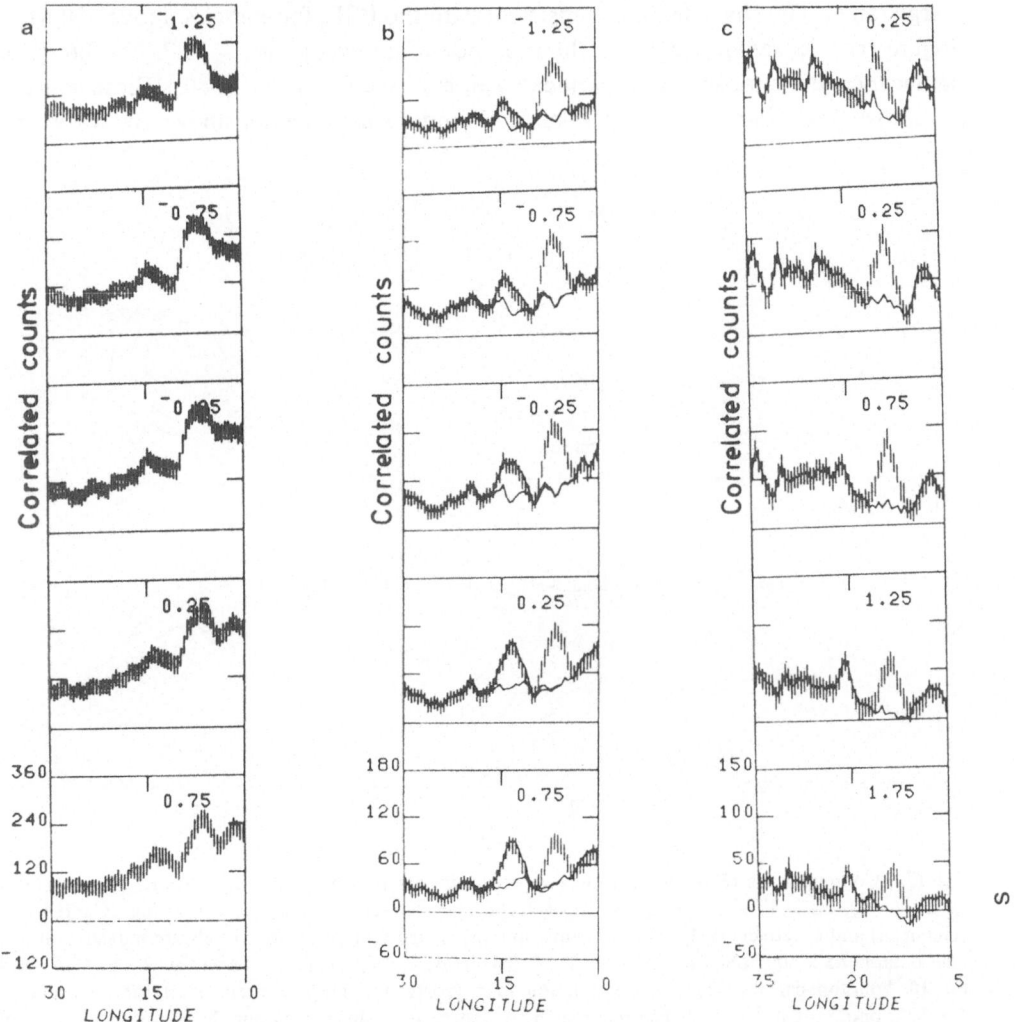

Fig. 16. Longitude profiles of the correlation maps showing the region containing 2CG 006−00 and 2CG 013 + 00 for energies above (a) 100 MeV, (b) 300 MeV, and (c) 500 MeV. See legend of Figure 14.

which are located within this longitude interval are detected at energies above 300 MeV (seven sources), with the single exception of 2CG 311 + 01 (detected for $E > 150$ MeV), which is located just at the boundary of this interval and of the intense ridge of gamma radiation from the inner Galaxy. The following examples are representative for the confusion problem in this longitude range. First two examples from the data base used to compile the 2CG catalogue. Figures 16a, 16b, and 16c present the gamma-ray sources 2CG 006 – 00 and 2CG 013 + 00 in the correlation map. For all energies above 100 MeV (Figure 16a) practically all information is lost on the fine structure, in particular no evidence at all can be claimed for a source at $l = 13.7$, $b = 0.6$ (2CG 013 + 00). However, the source feature of 2CG 006 – 00 can easily be resolved at energies above 300 MeV (Figure 16b). In fact, also a significant correlation is found for 2CG 013 + 00 in this energy range. The sharp feature, consistent with the PSF, for energies above 500 MeV (Figure 16c), leaves no doubt on the presence of an excess at $l = 13.7$, $b = 0.6$. (For significance levels, observation periods used, etc., see Hermsen, 1980.) These sources are shown two dimensionally in Figures 17 and 18 for energies above 100 MeV and

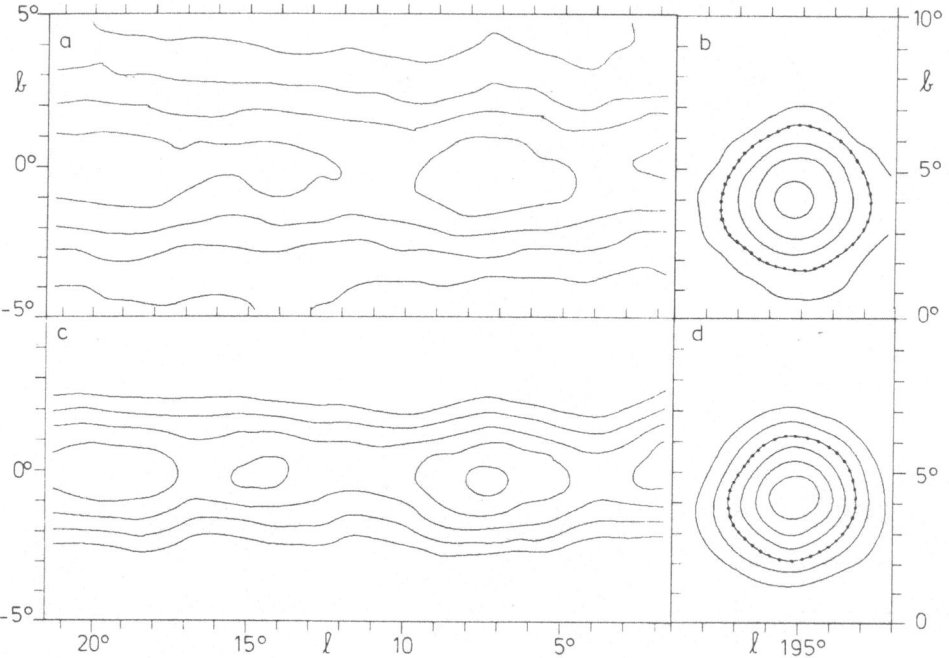

Fig. 17. Contour maps ($E > 100$ MeV) of two very different regions of the sky: (a) and (c) containing 2CG 006 – 00, 2CG 013 + 00 and the strong galactic 'line' source; (b) and (d) containing 2CG 195 + 04 (Geminga) and effectively no background emission. (a) intensity map; contour levels are indicated at 3.0, 4.5, 6 and 7.5×10^{-4} photon $cm^{-2} s^{-1} sr^{-1}$. (b) intensity map; contour levels at 2, 3, 4, 5, and 6×10^{-4} photon $cm^{-2} s^{-1} sr^{-1}$. (c) correlation map (correlated flux); contour levels at multiples of 6×10^{-7} photon $cm^{-2} s^{-1}$. (d) correlation map (correlated flux); contour levels at multiples of 5×10^{-7} photon $cm^{-2} s^{-1}$. The dotted contour levels in (b) and (d) indicate the level of $\sim 45\%$ of the peak value of the source profiles. Evidently, no feature of the same angular extent in latitude and exhibiting closed contour lines, is present in (a) or (c).

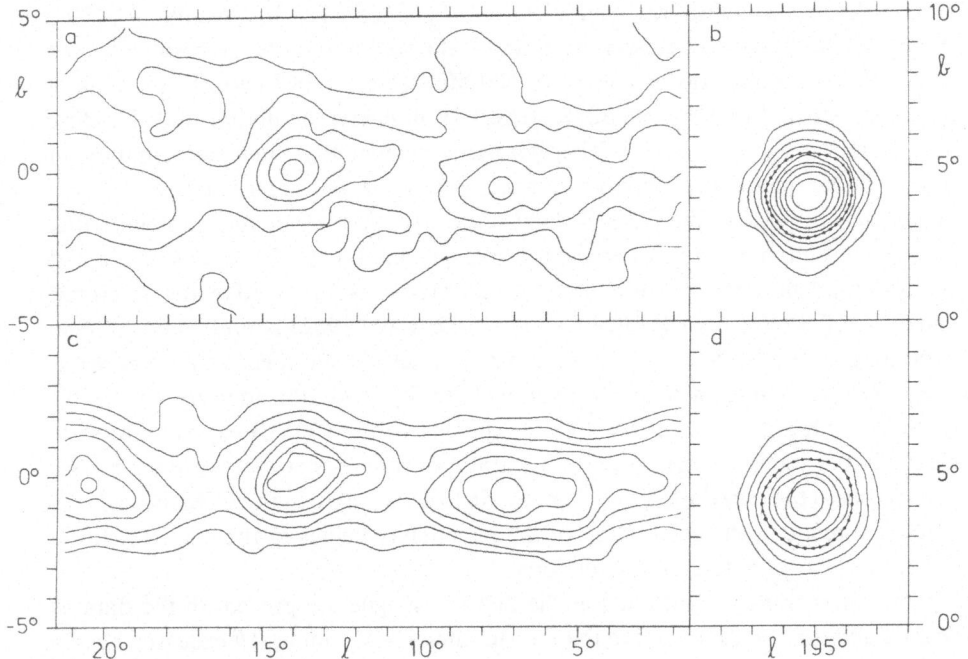

Fig. 18.   Contour maps like the ones shown in Figure 17 but for $E > 500$ MeV. Contour levels are indicated at multiples of: (a) 5×10^{-5} photon $\mathrm{cm}^{-2}\,\mathrm{s}^{-1}\,\mathrm{sr}^{-1}$, (b) 2.5×10^{-5} photon $\mathrm{cm}^{-2}\,\mathrm{s}^{-1}\,\mathrm{sr}^{-1}$ with the first contour at 5×10^{-5} photon $\mathrm{cm}^{-2}\,\mathrm{s}^{-1}\,\mathrm{sr}^{-1}$, (c) 1×10^{-7} photon $\mathrm{cm}^{-2}\,\mathrm{s}^{-1}$, and (d) 1×10^{-7} photon $\mathrm{cm}^{-2}\,\mathrm{s}^{-1}\,\mathrm{sr}^{-1}$ with the first contour at 2×10^{-7} photon $\mathrm{cm}^{-2}\,\mathrm{s}^{-1}$.

500 MeV, respectively (some additional data are used to produce these figures). For both energy ranges contour maps are given for intensity maps (a), correlated-flux maps (c), and the corresponding maps showing the Geminga source feature for comparison (b) and (d). Again it is evident that in the maps for energies above 100 MeV no source feature can be resolved. At energies above 500 MeV the fine structure is apparent with source features near to $b = 0°$ (the shapes of the sources are somewhat elongated due to the underlying line source).

Li and Wolfendale (1982) discuss the nature of the gamma-ray sources and claim that the majority of the catalogued sources are 'apparent sources' due to the irradiation of clumpy interstellar medium by cosmic ray electrons and nuclei. This conclusion is reached from a Monte Carlo analysis for the longitude interval $10° < l < 100°$. For that interval they predict the gamma-ray intensity distribution for energies above 100 MeV starting from the available information on the distribution of gas (CO and H I data) and using the assumption of a uniform cosmic-ray distribution. Furthermore, they claim that the characteristics of the COS-B detector are imposed in the Monte Carlo analysis as well as the finite number of gamma-ray counts detected experimentally, and, finally, that the COS-B technique of source identification is adopted. However, in their five simulations of the gamma-ray distribution for energies above 100 MeV, they 'detect' at least 4 out of 5 times a gamma-ray source at $l \simeq 12°$, $l \simeq 25°$, $l \simeq 38°$, $l \simeq 49°$, $l \simeq 79°$, and $l \simeq 93°$. In all these cases it is in the actual flight data impossible to resolve source features for the integral energy range above 100 MeV (see e.g. in Figure 17 the case of 2CG 013 + 00, which Li and Wolfendale 'detected' in every simulation and therefore classified as an 'apparent source'). Therefore it is evident that their combination of (subjective) fitting procedure and acceptance criteria is *not* representative for the analysis actually applied to the COS-B flight data. As a result, one may conclude that there is no sound basis to apply their significance calculations to the 2CG sources and no conclusions are allowed to be drawn on the nature of the sources. Also their reference to an analysis of SAS-2 data (Houston and Wolfendale, 1982) is ambiguous, since: (1) the binning of the SAS-2 data ($2.5° \times 0.8°$) results in an effectively even worse angular resolution, (2) the low SAS-2 exposure of the sky makes the analyses impossible (taking into account the ratio of the SAS-2 exposure to the COS-B exposure used to compile the 2CG catalogue, on average ~ 8 counts per source are expected in the SAS-2 data), (3) no fits to the correlation maps are made like was done in the COS-B analysis, (4) errors were made in the calculation of the probability values of the claimed effects (A. W. Wolfendale, private communication).

The gamma-ray sources contained in the 2CG catalogue are present in the data as *statistically significant* excesses above the surrounding background. *All* excesses have a spatial distribution consistent with the PSF. This means that the lower limit which can be set on the angular extent for *each* source is $0°$. The high lower limits claimed for some of the sources by Li and Wolfendale (1981) are not in agreement with the experimental data (all reported sources *do* fit the PSF) nor is the claimed high accuracy on the derived extent for some sources (e.g. $6.3° \pm 1.6°$ for 2CG 218 − 00, which is shown in Figure 14c to fit nicely the PSF). The average extent of the sources has been claimed by Hermsen (1980) to be less than $1°$ to $2°$. Li and Wolfendale (1981) tried to fit published COS-B data (by Hermsen, 1980) and arrived at values of $\sim 4°$ (the diameter of an equivalent sphere of uniform volume emissivity). In order to give an impression of the most likely average extent the following procedure has been worked out: For the region along the Galactic plane (containing 22 sources) skymaps in the energy range 300 MeV to 5 GeV are produced, such that in each skymap a source is placed in the central bin. High-energy data is selected since the better angular resolution enables a more accurate determination of the angular extent. Then, a point-summation technique is applied, namely, all skymaps are summed to obtain one total map with the total source signal again in the central bin. Finally the total correlation map is produced. Excluded from the analysis are the 'pair sources' in the Cygnus and Carina regions, to avoid confusion. In addition the three strongest sources, the genuine point sources PSR 0833 − 45 and PSR 0531 + 21 and Geminga, are also excluded. The first two, because their angular extent is known, the latter, because its large number of counts would dominate the result. Of the 15 sources which remain, no source contributes more than 10% to the total number of counts. Like in the analysis of Li and Wolfendale, a geometrically simple source model has been assumed: spheres of uniform volume emissivity. The total source feature, shown in Figure 19a, is fitted with the correlation profiles for source diameters (D) of

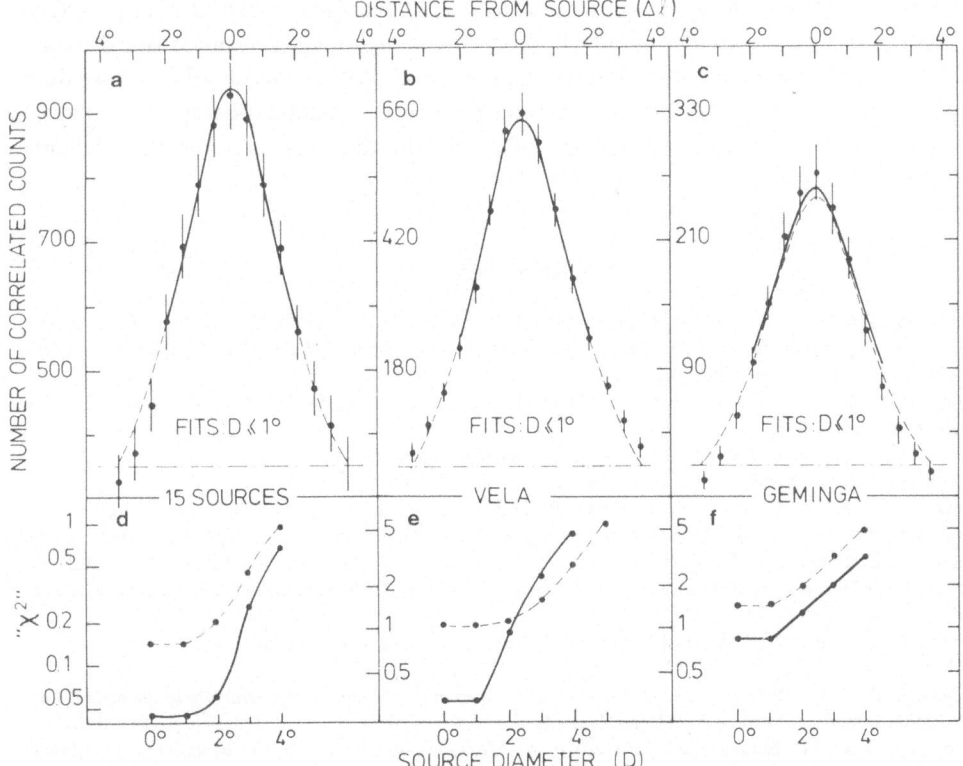

Fig. 19. The upper figures show longitude profiles of the correlation maps across the position of (a) the maximum of the summed profiles of 15 2CG sources (see text), (b) Vela, and (c) Geminga (2CG 195 + 04). The lower figures present the dependence of the minimum, reduced χ^2 values on source diameter (see text). The solid lines give the results obtained for fits to 9 points of the correlation profile, the broken lines for fits to 15 data points.

$0°$ up to $4°$ and 'χ^2' values are calculated to judge the fits. Since the values in Figure 19a are not independent for each bin, it is not possible to calculate directly a probability for the goodness of fit. However, one obtains a valid indication for the most-probable average angular extent. The best fit to the data points is indicated in Figure 19a and the obtained χ^2 values are given as a function of D in Figure 19d. The whole excercise has been performed also for PSR 0833 – 45, to show the appearance of a genuine point source, and for Geminga being a strong unidentified source (Figure 19b, c, e, f). It is evident that the three cases shown, are quite similar. One should note that the uncertainties in the source positions will broaden the distribution of the total summed signal somewhat. In the case of Geminga the number of counts is significantly lower than for PSR 0833 – 45. The conclusion from these figures is that the average extent of the sources is indeed between $0°$ and $2°$. As far as the individual sources are concerned: some of the weak sources and sources detected only in the integral energy range above 100 MeV certainly have a higher *upper limit* to the angular extent. However, as remarked upon before, for all sources the lower limit to D remains $0°$. Still, in the search for counterparts of the gamma-ray sources, the indicated average extent of the sources does not a priori rule out molecular clouds, bombarded by high-energy cosmic rays, to be the counterparts of some individual sources. Therefore, if a dense gas cloud (or more than one) is detected in the direction of a gamma-ray source a detailed comparison should be made, between the total gamma-ray emission from that direction and the available astronomical data.

References

Bennett, K., Bignami, G. F., Boella, G., Buccheri, R., Hermsen, W., Kanbach, G., Lichti, G. G., Masnou, J. L., Mayer-Hasselwander, H. A., Paul, J. A., Scarsi, L., Swanenburg, B. N., Taylor, B. G., and Wills, R. D.: 1977a, *Astron. Astrophys.* **61**, 279.

Bennett, K., Bignami, G. F., Hermsen, W., Mayer-Hasselwander, H. A., Paul, J. A., and Scarsi, L.: 1977b, *Astron. Astrophys.* **59**, 273.

Bignami, G. F. and Hermsen, W.: 1983, *Ann. Rev. Astron. Astrophys.* **21**, 67.

Bignami, G. F. and Morfill, G. E.: 1980, *Astron. Astrophys.* **87**, 85.

Bignami, G. F., Boella, G., Burger, J. J., Keirle, P., Mayer-Hasselwander, H. A., Paul, J. A., Pfefferman, E., Scarsi, L., Swanenburg, B. N., Taylor, B. G., Voges, W., and Wills, R. D.: 1975, *Space Sci. Instr.* **1**, 245.

Bignami, G. F., Bennett, K., Buccheri, R., Caraveo, R., Hermsen, W., Kanbach, G., Lichti, G. G., Masnou, J. L., Mayer-Hasselwander, H. A., Paul, J. A., Sacco, B., Scarsi, L., Swanenburg, B. N., and Wills, R. D.: 1981, *Astron. Astrophys.* **93**, 71.

Bignami, G. F., Caraveo, P. A., and Lamb, R. C.: 1983, *Astrophys. J. Letters*, in press.

Black, J. H. and Fazio, G. G.: 1973, *Astrophys. J. Letters* **185**, L7.

Bloemen, J. B. G. M.: 1983, in W. L. H. Shuter (ed.), *Kinematics, Dynamics and Structure of the Milky Way*, D. Reidel Publ. Co., Dordrecht, Holland, p. 31.

Bloemen, J. B. G. M., Bennett, K., Caraveo, P. A., Hermsen, W., Kanbach, G., Masnou, J. L., Mayer-Hasselwander, H. A., Paul, J. A., Sacco, B., Strong, A. W., and Wills, R. D.: 1981, *Proc. 17th Int. Cosmic Ray Conf. Paris* **1**, 181.

Buccheri, R.: 1981, in W. Sieber and R. Wielebinski (eds.) 'Pulsars', *IAU Symp.* **95**, 241.

Buccheri, R.: 1983, *Bull. Am. Astron. Soc.* **14** (4), 966.

Buccheri, R., Morini, M., and Sacco, B.: 1981, *Phil. Trans. Roy. Soc. London* **A301**, 495.

Buccheri, R., Bennett, K., Bignami, G. F., Bloemen, J. B. G. M., Boriakoff, V., Caraveo, P. A., Hermsen, W., Kanbach, G., Manchester, R. N., Masnou, J. L., Mayer-Hasselwander, H. A., Özel, M. E., Paul, J. A., Sacco, B., Scarsi, L., and Strong, A. W.: 1983, *Astron. Astrophys.*, in press.

Caraveo, P. A.: 1981, *Phil. Trans. Roy. Soc. London* **A301**, 523.

Caraveo, P. A., Bennett, K., Bignami, G. F., Hermsen, W., Kanbach, G., Lebrun, F., Masnou, J. L., Mayer-Hasselwander, H. A., Paul, J. A., Sacco, B., Scarsi, L., Strong, A. W., Swanenburg, B. N., and Wills, R. D.: 1980, *Astron. Astrophys.* **91**, L3.

Caraveo, P. A., Barbareschi, L., Bennett, K., Bignami, G. F., Hermsen, W., Kanbach, G., Lebrun, F., Masnou, J. L., Mayer-Hasselwander, H. A., Sacco, B., Strong, A. W., and Wills, R. D.: 1981, *Proc. 17th Int. Cosmic Ray Conf., Paris* **1**, 139.

Danaher, S., Fegan, D. J., Porter, N. A., and Weekes, T. C.: 1981, *Nature* **289**, 568.

Fazio, G. G.: 1973, in H. Bradt and R. Giacconi (eds.), *X- and Gamma-Ray Astronomy*, D. Reidel Publ. Co., Dordrecht, Holland, p. 303.

Galper, A. M., Kirillov-Ugryumov, V. G., Kurochkin, A. V., Leikov, P. G., Luchkov, B. I., and Yurkin, Yu. T.: 1976, *Pisma v. Astron. Zh.* **2**, 254.

Galper, A. M., Kirillov-Ugryumov, V. G., Kurochkin, A. V., Leikov, P. G., Luchkov, B. I., Yurkin, Yu. T., Fomin, V. P., Neshpor, Yu. I., Stepanian, A. A., and Vladimirsky, B. M.: 1977, *Proc. 15th Int. Cosmic Ray Conf., Plovdiv* **1**, 131.

Gregory, P. C., Kronberg, P. P., Seaquist, E. R., Hughes, V. A., Woadsworth, A., Viner, M. R., and Retallack, D.: 1972, *Nature* **239**, 440.

Hermsen, W.: 1980, 'Gamma-Ray Sources', Ph.D. Thesis Univ. of Leiden, Holland.

Hermsen, W.: 1981, *Phil. Trans. Roy Soc. London* **A301**, 519.

Hermsen, W. and Bloemen, J. B. G. M.: 1983, in W. B. Burton and F. P. Israel (eds.), *Proc. Leiden Workshop on Southern Galactic Surveys*, D. Reidel Publ. Co., Dordrecht, Holland, in press.
Hermsen, W., Swanenburg, B. N., Bignami, G. F., Boella, G., Buccheri, R., Scarsi, L., Kanbach, G., Mayer-Hasselwander, H. A., Masnou, J. L., Paul, J. A., Bennett, K., Higdon, J. C., Lichti, G. G., Taylor, B. G., and Wills, R. D.: 1977a, *Nature* **269**, 494.
Hermsen, W., Bennett, K., Bignami, G. F., Boella, G., Buccheri, R., Higdon, J. C., Kanbach, G., Lichti, G. G., Masnou, J. L., Mayer-Hasselwander, H. A., Paul, J. A., Scarsi, L., Swanenburg, B. N., Taylor, B. G., and Wills, R. D.: 1977b, in R. D. Wills and B. Battrick (eds.), *Recent Advances in Gamma-Ray Astronomy*, ESA SP-124, p. 13.
Hermsen, W., Bennett, K., Bignami, G. F., Bloemen, J. B. G. M., Buccheri, R., Caraveo, P. A., Kanbach, G., Masnou, J. L., and Wills, R. D.: 1981, *Proc. 17th Int. Cosmic Ray Conf., Paris* **1**, 230.
Houston, B. and Wolfendale, A. W.: 1982, *Proc. Roy Irish Acad.*, in press.
Issa, M. R. and Wolfendale, A. W.: *Nature* **292**, 430.
Kanbach, G., Bennett, K., Bignami, G. F., Buccheri, R., Caraveo, P., D'Amico, N., Hermsen, W., Lichti, G. G., Masnou, J. L., Mayer-Hasselwander, H. A., Paul, J. A., Sacco, B., Swanenburg, B. N., and Wills, R. D.: 1980, *Astron. Astrophys.* **90**, 163.
Khavtassi, J. Sh.: 1960, *Atlas of Galactic Dark Nebulae*, Tiblis, Abastumani Astrophys. Obs.
Kniffen, D. A., Hartman, R. C., Thompson, D. J., Bignami, G. F., Fichtel, C. E., Tümer, T., and Ögelman, H.: 1974, *Nature* **25**, 397.
Kniffen, D. A., Bignami, G. F., Fichtel, C. E., Hartman, R. C., Ögelman, H., Thompson, D. J., Özel, M. E., and Tümer, T.: 1975, *Proc. 14th Int. Cosmic Ray Conf. Munchen* **1**, 100.
Lamb, R. C., Fichtel, C. E., Hartman, R. C., Kniffen, D. A., and Thompson, D. J.: 1977, *Astrophys. J. Letters* **212**, L63.
Lamb, R. C., Godfrey, C. P., Wheaton, W. A., and Tümer, T.: 1982, *Nature* **296**, 543.
Li, T. P. and Wolfendale, A. W.: 1981, *Astron. Astrophys.* **100**, L26.
Li, T. P. and Wolfendale, A. W.: 1982, *Astron. Astrophys.* **116**, 95.
Lichti, G. G., Buccheri, R., Caraveo, P., Gerardi, G., Hermsen, W., Kanbach, G., Masnou, J. L., Mayer-Hasselwander, H. A., Paul, J. A., Swanenburg, B. N., and Wills, R. D.: 1980, in R. Cowsik and R. D. Wills (eds.), *Non-Solar Gamma Rays (COSPAR)*, Advances in Space Exploration, Vol. 7, p. 49, Pergamon Press, Oxford and New York.
Manchester, R. N. and Taylor, J. H.:1977, *Pulsars*, Freeman Publ. Co., San Francisco, U.S.A.
Masnou, J. L., Bennett, K., Bignami, G. F., Bloemen, J. B. G. M., Buccheri, R., Caraveo, P. A., Hermsen, W., Kanbach, G., Mayer-Hasselwander, H., Paul, J. A., and Wills, R. D.: 1981, *Proc. 17th Int. Cosmic Ray Conf. Paris* **1**, 177.
Mayer-Hasselwander, H. A., Bennett, K., Bignami, G. F., Buccheri, R., D'Amico, N., Hermsen, W., Kanbach, G., Lebrun, F., Lichti, G. G., Masnou, J. L., Paul, J. A., Pinkau, K., Scarsi, L., Swanenburg, B. N., and Wills, R. D.: 1980, *Ann. NY Acad. Sci.* **336**, 211.
Meegan, C. A. and Fishman, G. J.: 1979, *Astrophys. J. Letters* **234**, L123.
Mukanov, J. B., Nesterova, N. M., Stepanian, A. A., and Fomin, V. P.: 1979, *Proc. 16th Int. Cosmic Ray Conf., Kyoto* **1**, 143.
Neshpor, Yu. I., Stepanian, A. A., Fomin, V. P., Gerasimov, S. A., Vladimirsky, B. M., and Ziskin, Yu. L.: 1979, *Astrophys. Space Sci.* **61**, 349.
Parsignault, D., Gursky, H., Kellog, E., Matilsky, T., Murray, S., Schreier, E., Tananbaum, H., Giacconi, R., and Brinkman, A.: 1972, *Nat. Phys. Sci.* **239**, 129.
Samorksi, M. and Stamm, W.: 1983, *Astroph. J. Letters*, in press.
Scarsi, L., Bennett, K., Bignami, G. F., Boella, G., Buccheri, R., Hermsen, W., Koch, L., Mayer-Hasselwander, H. A., Paul, J. A., Pfeffermann, E., Stiglitz, R., Swanenburg, B. N., Taylor, B. G., and Wills, R. D.: 1977, in R. D. Wills and B. Battrick (eds.), *Recent Advances in Gamma-Ray Astronomy*, ESA SP-124, p. 3.
Sieber, W. and Schlickeiser, R.: 1982, *Astron. Astrophys.* **113**, 314.
Swanenburg, B. N., Bennett, K., Bignami, G. F., Caraveo, P., Hermsen, W., Kanbach, G., Masnou, J. L., Mayer-Hasselwander, H. A., Paul, J. A., Sacco, B., Scarsi, L., and Wills, R. D.: 1978, *Nature* **275**, 298.
Swanenburg, B. N., Bennett, K., Bignami, G. F., Buccheri, R., Caraveo, P., Hermsen, W., Kanbach, G., Lichti, G. G., Masnou, J. L., Mayer-Hasselwander, H. A., Paul, J. A., Sacco, B., Scarsi, L., and Wills, R. D.: 1981, *Astrophys. J. Letters* **243**, L69.
Thompson, D. J., Fichtel, C. E., Kniffen, D. A., and Ögelman, H. B.: 1975, *Astrophys. J. Letters* **200**, L79.

Thompson, D. J., Fichtel, C. E., Hartman, R. C., Kniffen, D. A., and Lamb, R. C.: 1977, *Astrophys. J.* **213**, 252.

Van der Klis, M. and Bonnet-Bidaud, J. M.: 1981, *Astron. Astrophys.* **95**, L5.

Weekes, T. C. and Helmken, H. F.: 1977, in R. D. Wills and B. Battrick (eds.), *Recent Advances in Gamma-Ray Astronomy*, ESA SP-124, p. 39.

Wills, R. D., Bennett, K., Bignami, G. F., Buccheri, R., Caraveo, P. A., Hermsen, W., Kanbach, G., Masnou, J. L., Mayer-Hasselwander, H. A., Paul, J. A., and Sacco, B.: 1981, *Proc. 17th Int. Cosmic Ray Conf. Paris* **1**, 22.

Wills, R. D., Bennett, K., Bignami, G. F., Buccheri, R., Caraveo, P. A., Hermsen, W., Kanbach, G., Masnou, J. L., Mayer-Hasselwander, H. A., Paul, J. A., and Sacco, B.: 1982, *Nature* **296**, 723.

Wouterloot, J. G. A.: 1981, 'The Large-Scale Structure of Molecular Clouds', Ph.D. Thesis Univ. of Leiden, Holland.

Electromagnetic Induction in the Earth and Moon

Invited Review Papers presented at the 5th Workshop, Istanbul, August, 1980

A. M. ISIKARA and S. R. MALIN

1982, 184 pp.
Paper Dfl. 42,50 / US $ 22.50   Order ref. GEOP 00404
A Special Issue of the Journal, *Geophysical Surveys*, Vol. 4, No. 4

D. Reidel Publishing Company

P.O. Box 17, 3300 AA Dordrecht, the Netherlands
190 Old Derby St., Hingham, MA 02043, U.S.A.

The Source Region of the Solar Wind

Invited Review Papers, IX Lindau Workshop, November 1981

Edited by
WOLFGANG SCHMIDT and HEINER GRÜNDWALDT
Max-Planck Institut für Aeronomie, Lindau

1982, 275 pp.
Cloth Dfl. 115,– / US $ 49.50   ISBN 90-277-1537-8
Reprinted from the journal, *Space Science Reviews*, Vol. 33, Nos. 1 and 2

In the sense that the study of solar wind and that of coronal effects have developed into different branches of research in the past because of their differing experimental techniques, the meeting reported here could be seen as an interdisciplinary one. Emphasis is placed on research results rather than instrumentation although a series of observational methods are presented in order to give an overview of the large variety of methods that can contribute data to coronal and solar wind research.

Contents

D. Reidel Publishing Company

P.O. Box 17, 3300 AA Dordrecht, the Netherlands
190 Old Derby St., Hingham, MA 02043, U.S.A.

SPACE SCIENCE REVIEWS

Volume 36 No. 2 1983

Published monthly.
Subscription prices, per volume: Institutions $92.00, Individuals $30.00.
Second-class postage paid at New York, N.Y. USPS No. 509–100.
U.S. Mailing Agent: Expediters of the Printed Word Ltd., 527 Madison Avenue (Suite 1217), New York, NY 10022.
Space Science Reviews is published by D. Reidel Publishing Company, Voorstraat 479–483, P.O. Box 17, 3300 AA Dordrecht, Holland, and 190 Old Derby Street, Hingham, MA 02043, U.S.A.
Postmaster: please send all address corrections to: c/o Expediters of the Printed Word Ltd., 527 Madison Avenue (Suite 1217), New York, NY 10022, U.S.A.

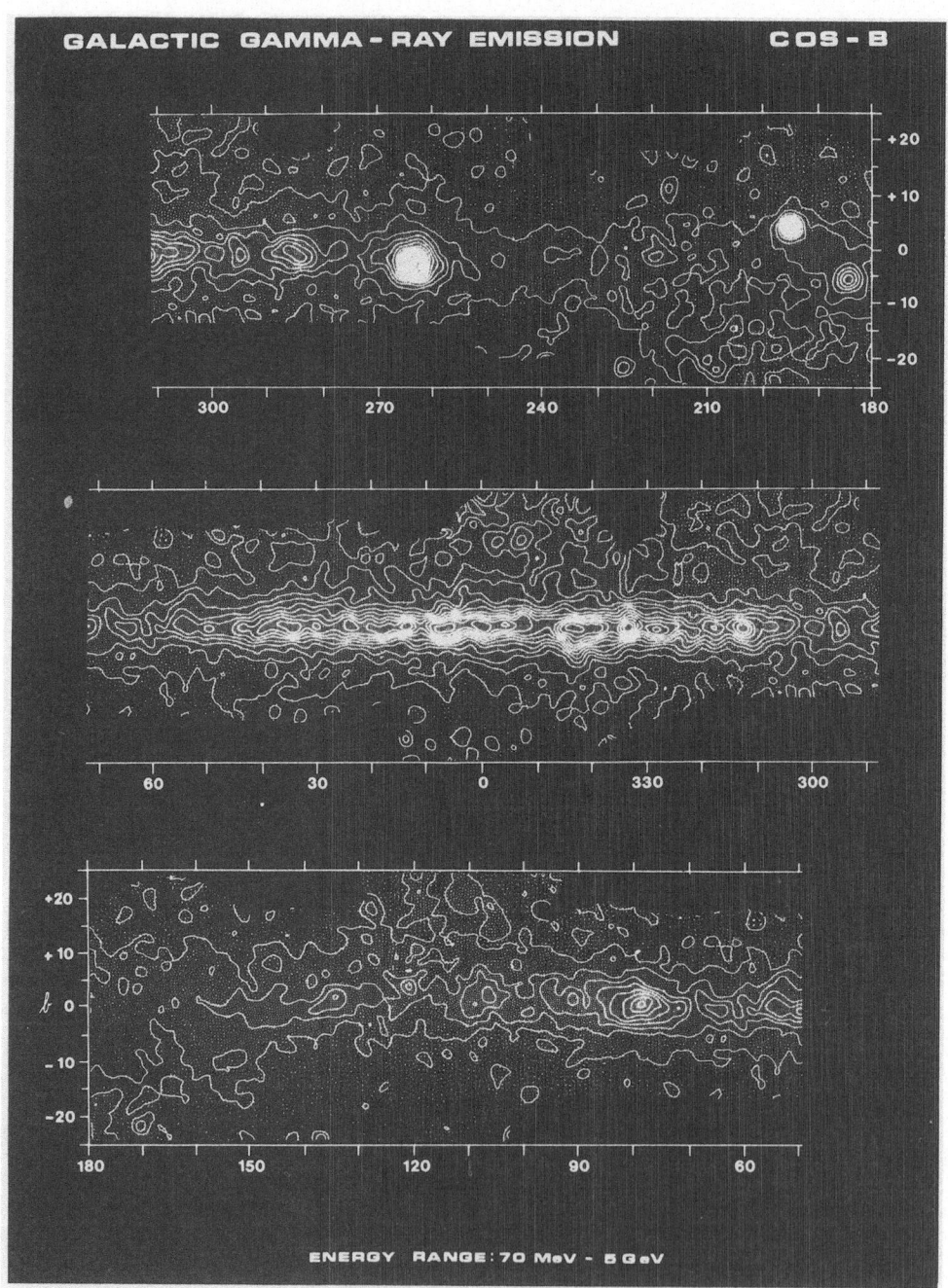

Presentation in galactic coordinates of the structure of the galactic gamma-ray emission as measured by the gamma-ray experiment of the 'Caravane' collaboration aboard the ESA satellite COS-B. In the map the surface fitted to the data is indicated by contour lines and a grey scale. Regions outside the accepted field of view are left blank. The contour levels are indicated at multiples of 3×10^{-3} *on-axis* counts $s^{-1} sr^{-1}$. A detailed discussion can be found in H. A. Mayer-Hasselwander *et al.*: 1982, *Astron. Astrophys.* **105**, 164–175.

POINT-LIKE GAMMA-RAY SOURCES*

G. E. MORFILL

*Max-Planck-Institut für Physik und Astrophysik, Institut für extraterrestrische Physik,
8046 Garching, F.R.G.*

and

G. TENORIO-TAGLE

Max-Planck-Institut für Physik und Astrophysik, Institut für Astrophysik, 8046 Garching, F.R.G.

Abstract. Various models are examined, which could give rise to point-like gamma-ray sources, at the present time indistinguishable, experimentally, from true point sources. These models involve energetic processes associated with interstellar clouds, e.g. supernova-cloud interactions, neutron star accretion inside interstellar clouds, cloud collisions, etc. The dynamical evolution of such systems is discussed and physical processes are described in a mathematical framework which can be solved. Statistical arguments are presented, where possible, on the likelihood that the scenarios may actually occur in our Galaxy. The visibility of the systems at other wavelengths, e.g. infrared, X-ray, radio etc., and further consequences, e.g. gamma-ray line emission, special radio emission line features, absorption features etc. are also discussed. Finally, a limited attempt at identification of some 'gamma-ray objects' is made based on the theoretical predictions.

1. Introduction

From an analysis of several years of COS-B gamma-ray data, it was concluded that there exist ∼ 25 point or point-like sources of sufficient strength to satisfy some stringent statistical constraints (for references, see e.g., Hermsen, 1980; Swanenburg *et al.*, 1981; Hermsen, 1983). Most of these sources have not yet been identified, the Crab and Vela pulsars being notable exceptions as well as the quasar 3C273 and possibly the ρ-Oph dark cloud. The bulk of the sources may constitute a new class of objects (Gamma-Ray Objects – GRO's) which may be genetically similar – or we may be dealing with a number of phenomena that happen to have exceptionally strong gamma-ray emission.

Clearly the identification of GRO's is very important for galactic astrophysics as a whole, since they are individually very energetic and contribute a sizeable fraction of the total galactic gamma-ray luminosity (see e.g. Bignami and Hermsen, 1982). The role of theoretical astrophysics in this context is to aid the search for counterparts at other wavelengths by calculating the consequences of certain models.

In this review we concentrate on three aspects: the energy source, the physical processes which convert available energy into gamma rays and the consequences for other branches of astronomy.

To start with, we make a few simple estimates based on the observations of GRO's, and the energetics of some galactic objects.

* Proceedings of the XVIII General Assembly of the IAU: *Galactic Astrophysics and Gamma-Ray Astronomy*, held at Patras, 19 August 1982.

Space Science Reviews **36** (1983) 93–143.   0038–6308/83/0362–0093$07.65.

GRO's have a typical gamma-ray energy output of $\lesssim 10^{36}$ ergs s^{-1} and it was concluded that most of these sources lie within ~ 5 kpc of the Sun (e.g. Bignami *et al.*, 1978). Extrapolation to the Galaxy as a whole implies then some 150 GRO's.

Energetic events in the Galaxy are e.g. supernovae, which release $\sim 10^{50}$ to 10^{51} ergs instantaneously. Most of this energy ends up in the form of kinetic bulk motion of the ejected matter. A slower release of energy, but of much longer duration, is found in e.g. mass losses from stars. Of these, the Wolf–Rayet stars are the strongest (see e.g., Maeder, 1981, and references therein) with typical mass loss rates of 1.5–$6 \times 10^{-5}\ M_\odot$ yr^{-1}, terminal wind speeds of 1–4×10^8 cm s^{-1} and life time (of this mass loss stage) of $\sim 2.5 \times 10^5$ yr. The total mechanical energy of these objects is then $E \approx 3.7 \times 10^{49}$ – 2.4×10^{51} ergs, and the mechanical energy output rate is 5×10^{36} ergs s^{-1} – 3.2×10^{38} ergs s^{-1}. Total mass loss during this stage is between 3.7–$15\ M_\odot$. (The upper limits chosen here may perhaps be somewhat extreme.)

The determination of the efficiency of conversion of this energy into gamma rays requires detailed examination of processes – here we will just assume that it is of the order 1% or less.

We then have the simple relationship

$$\dot{N}/\dot{E} = N/E, \tag{1}$$

where $\dot{N}$ is the production rate of GRO's, N the present number in our Galaxy, $\dot{E}$ the typical gamma-ray luminosity of GRO's, and E the total gamma-ray energy available. Setting $E = 10^{48}$ ergs as a reasonable upper limit for $\geqslant 100$ MeV gamma rays, gives

$$\dot{N} = 4.5 \times 10^{-3} \text{ GRO's yr}^{-1} \tag{2}$$

in order to account for the observations. We will not delve into the uncertainties involved in deriving (2), and simply take it as a 'baseline' value.

In the subsequent review, we shall estimate the GRO production rate, for each scenario considered, as a rough feasibility test.

2. Models for GRO's

Supernovae (SN) were already mentioned in the introduction as being interesting candidates for an association with GRO's. This association suggests itself for energy reasons, and it remains to be tested whether physical processes, geometry etc. allow the formation of GRO's. In addition to an energy supply, we also require in general an energy absorber (e.g. a large amount of matter in the vicinity) so that some of the energy may be converted into gamma rays. The principal processes are: collisional excitation (leading to gamma-ray line production in the region 0.5 to 10 MeV), acceleration of nucleons to > 1 GeV energies and subsequent π° production by inelastic collisions (leading to gamma-ray production in the region $\gtrsim 70$ MeV) and finally acceleration of electrons to high energies and subsequent Bremsstrahlung losses (leading to a gamma-ray spectrum roughly identical with that of the accelerated electrons), or inverse Compton scattering.

Hence a common feature of non-compact GRO models is the existence of an energy source, a process for accelerating particles with a reasonable conversion efficiency, and a geometry which allows gamma-ray production by one of the above processes. (Collisional excitation is irrelevant for the COS-B GRO's.)

3. Supernovae in Interstellar Clouds

The possibility that many supernovae may explode deep inside molecular clouds has been discussed recently in several papers (Wheeler *et al.*, 1980; Shull, 1980a, b; Morfill and Drury, 1981).

The number of such objects may well be comparable with the number of 'free' supernovae.

3.1. DYNAMICAL EVOLUTION OF THE SUPERNOVA REMNANT

In this section we consider supernova remnants (SNRs) caused by explosions occuring deep inside clouds, in such a way that the remnant does not break through the boundary of the cloud, and thus it diffuses while acquiring an inner pressure comparable to that of the cloud. A case of especial interest is that of a supernova explosion with an energy comparable to the binding energy of the recipient cloud. If allowances for cooling are

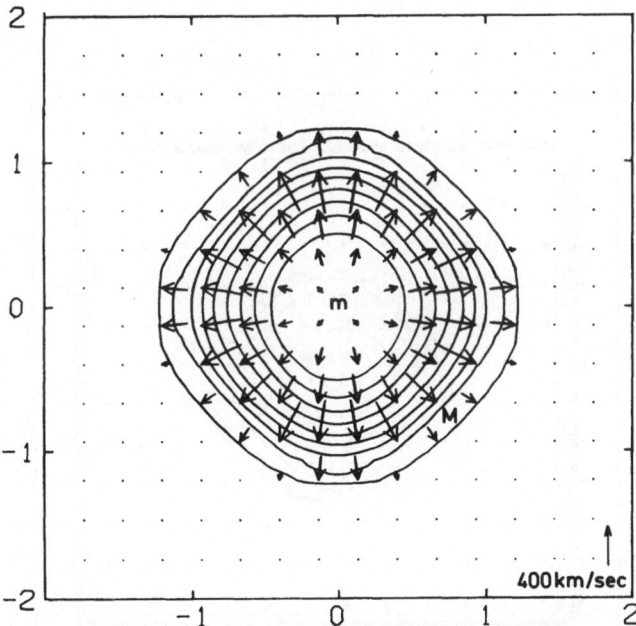

Fig. 1.   Supernova explosion inside an interstellar cloud at the cloud centre. The SN energy equals the cloud gravitational binding energy = 10^{51} ergs. Initial cloud parameters were: density 10^3 cm^{-3}, radius 15 pc, height 30 pc, temperature 10 K. The figure corresponds to a time 710 yr after the SN explosion. Density contours (solid lines) spaced logarithmically are shown between the minimum ($m = 10^{-22}$ g cm^{-3}) and the maximum ($M = 10^{-20.6}$ g cm^{-3}) with log $\Delta\rho = 0.2$. The velocity field is denoted by the arrows.

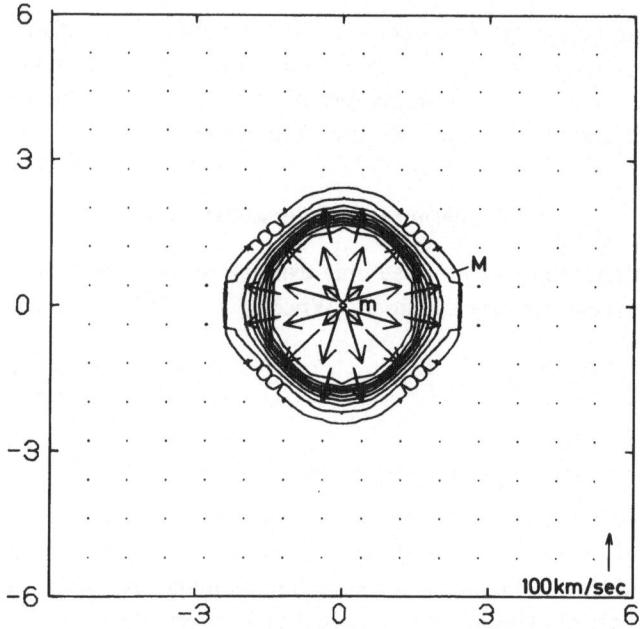

Fig. 2.   As in Figure 1, except for a time 8.28×10^3 yr after the SN explosion ($m = 10^{-24}$ g cm^{-3}, $M = 10^{-20.6}$ g cm^{-3}, log $\Delta\rho = 0.5$.).

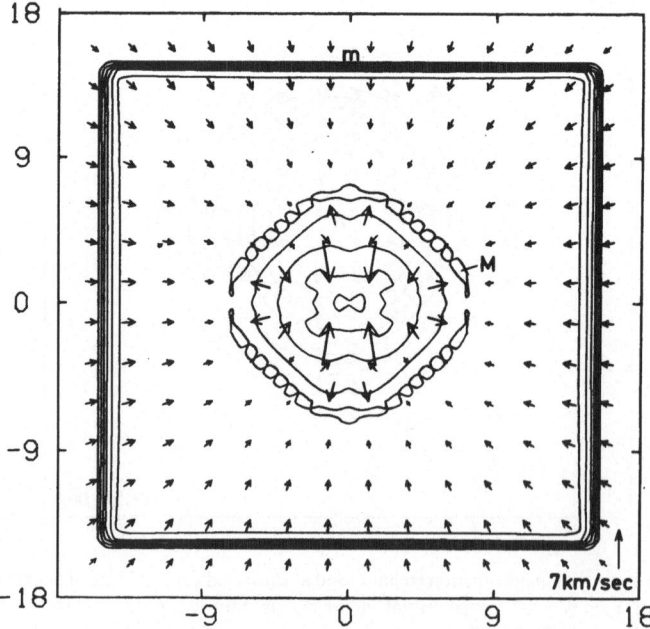

Fig. 3.   As in Figure 1, except for a time 3.37×10^5 yr after the SN explosion ($m = 10^{-23.5}$ g cm^{-3}, $M = 10^{-20.5}$ g cm^{-3}, log $\Delta\rho = 0.5$.).

made a large fraction of the supernova energy would be radiated away mainly in the infra-red, and the cloud is then well able to contain the remnant. Figures 1–4 show the gas dynamical evolution of such a case (see also Tenorio–Tagle *et al.*, 1983). The energy of the explosion, 10^{51} ergs, was deposited in a small (< 1 pc in radius) box along the axis of symmetry of a cylindrical cloud of radius $R = 15$ pc and height = $2R$. A cross section of this can be seen in Figure 3. The density contrast between cloud and intercloud medium $N_{\text{cloud}}/N_{\text{ic}}$ was set equal to 10^3 and an initial pressure balance was established between the cool cloud and a warm intercloud gas. The calculations are based on the two-dimensional hydrodynamical code described by Black and Boden-heimer (1975). The equations are solved explicitly on an Eulerian grid in a cylindrical (R, Z, ϕ) coordinate system, with symmetry axis Z and with derivatives in ϕ suppressed.

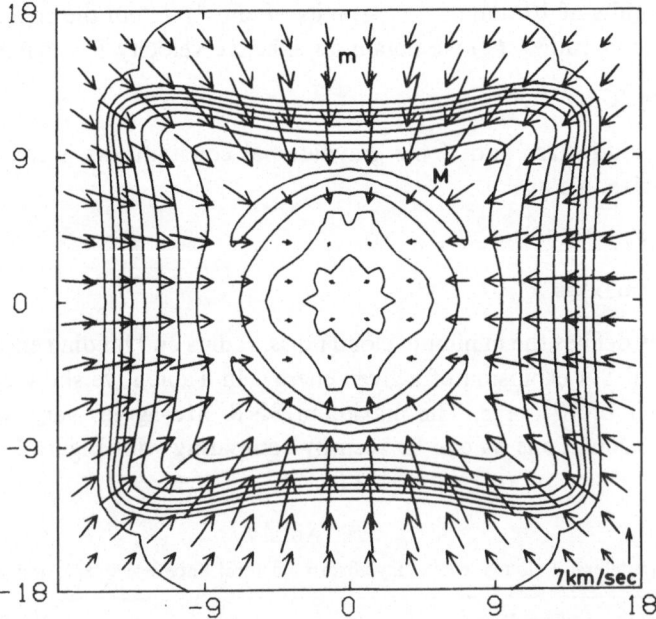

Fig. 4.   As in Figure 1, except for a time 1.15×10^6 yr after the SN explosion. The SNR begins to collapse again, as does the cloud ($m = 10^{-24}$ g cm^{-3}, $M = 10^{-20}$ g cm^{-3}, log $\Delta\rho = 0.5$).

The calculations shown here have a fixed grid in space and 60×120 or 150×500 grid points in the R and Z directions, respectively. The energy equation was rewritten as

$$\frac{\partial}{\partial t}(\rho E) + \nabla \cdot (\rho E \mathbf{v}) = -P\nabla \cdot \mathbf{v} + L,$$

where L is the energy loss rate per unit volume due to cooling. L was taken from Cox and Daltabuit's (1974) law.

Figures 1–4, where the scale lengths are measured in pc, show the remnant at various stages of the calculation. Here one can clearly appreciate the maximum dimension reached by the SNR (~ 9 pc) before the collapse of the cloud reverses the flow to refill the cavity generated during the early expansion. The SNR is slowed down quite rapidly (within a few 10^5 yr) and the velocity field becomes dominated by the free collapse of the cloud. The final frame shows the inevitable collapse of the cloud, despite the initial explosion of 10^{51} ergs.

We can estimate the required minimum binding energy of the cloud in the following way:

From McKee and Hollenbach (1980) we use the expression for the SNR size in the radiative phase

$$R_{SNR}(PC) = 27E_{51}^{0.32}/V_{S7}^{0.45}n_0^{0.36}, \tag{3a}$$

where the subscripts denote the size of the quantities in cgs units (e.g. $E_{51} \equiv$ units of 10^{51} ergs, $V_{S7} \equiv$ units of 10^7 cm s^{-1}, $n_0 \equiv$ units of cm^{-3}). From the selfgravity of the cloud (homogeneous, radius R_c) we obtain an effective velocity V_*, defined by

$$\tfrac{1}{2}V_*^2 \equiv \tfrac{3}{5}GM/R_c, \tag{3b}$$

where M is the cloud mass and G the gravitational constant. As a criterion for the maximum SNR size, we use

$$V_s \equiv V_*, \tag{4a}$$

$$R_{SNR}(\text{max}) \equiv R_c. \tag{4b}$$

This then uniquely defines the minimum cloud mass, radius and binding energy required to absorb a central SN explosion of a given energy. In Table I we show actual values obtained for a SN of 10^{51} ergs. The maximum SNR size agrees very well with the hydrodynamical calculations as can be seen by comparing with Figure 3.

TABLE I

Parameters of clouds which may absorb a central Supernova explosion of 10^{51} ergs

Density (cm^{-3})	Cloud mass ($M_\odot$)	Cloud radius ($R_{SNR}(\text{max})$) (pc)	Binding energy (ergs)
100	8×10^4	20.7	1.65×10^{49}
300	6.4×10^4	13.3	1.62×10^{49}
1000	4.9×10^4	8.2	1.58×10^{49}
3000	3.9×10^4	5.3	1.56×10^{49}
10000	3×10^4	3.2	1.53×10^{49}

For our purposes of investigating particle acceleration and gamma-ray production, we have to examine the initial situation more carefully:

The supernova is thought to explode in the pre-supernova stellar wind cavity. This cavity contains very little mass, and the supernova ejecta expand freely until they encounter the edge of the stellar wind cavity. The subsequent evolution of the supernova remnant (SNR) can be described by the snowplough solution (Oort, 1951). We have:

(1) *energy equation*

$$\frac{dE}{dt} = -2V\frac{E}{r} ; \tag{5a}$$

(2) *momentum equation*

$$\frac{d}{dt}(MV) = \frac{2E}{r} , \tag{5b}$$

where $M(t) = \frac{4}{3}\pi\rho_0(r(t)^3 - R_w^3)$ is the instantaneous displaced mass. ρ_0 is the ambient cloud density, R_w the original stellar wind cavity radius, $r(t)$ is the size of the SNR, and $V(t)$ the expansion speed. From these equations it is easy to show that the expansion time scale of the SNR is given by

$$\tau_{ex} \approx R_W(3M_D/2E_{SN})^{0.5}. \tag{6}$$

Here, $M_D = \frac{4}{3}\pi\rho_0 R_w^3$ is the originally displaced mass. For a SN energy $E_{SN} \equiv 10^{51}$ ergs, a cloud density $n_0 = 10^4$ cm^{-3}, and R_{SW} measured in pc we obtain

$$\tau_{ex} = 6.4 \times 10^3 \, (E_{51}^{-1/2}n_4^{1/2}R_{pc}^{5/2}) \text{ yr}. \tag{7}$$

This time scale agrees very well with the hydrodynamical calculations, which did not have a stellar wind cavity, if we substitute for R_{pc} the instantaneous SNR radius. τ_{ex} is also the adiabatic energy loss time scale of the plasma and any trapped accelerated particles inside the SNR. This result implies that the initial stages (stellar wind cavity, possible Sedov phase, etc.) are unimportant, quantitatively, for the SNR evolution once the snowplough phase has been reached.

3.2. COSMIC RAY PRODUCTION

During the initial free propagation of the SN shock through the stellar wind cavity, energetic particles can be produced by repeated diffusive shock acceleration (see e.g. Scholer and Morfill, 1975; Axford *et al.*, 1977; Blandford and Ostriker, 1978; Bell, 1978a, b). This process has been observed in the near-earth environment, e.g. the Earth's bow shock, and in the interplanetary medium (see e.g., Jokipii, 1983) and the theory has been tested. According to Drury (1983, and references therein), the process may be very efficient (conversion of the order $\sim\frac{1}{3}$ of the available mechanical energy into energetic particles). The SN shock runs through the stellar wind cavity with a constant velocity $V_S = 1.19 \times V_E$ (Parker, 1963) where $\frac{1}{2}M_E V_E^2 = E_{SN}$ and M_E is the ejecta mass from the SN. In general $M_E \gg M_W$, the total mass of stellar wind in the cavity. M_W is given by

$$M_W \approx R_W \dot{M}/V_W, \tag{8}$$

where $\dot{M}$ is the mass loss rate in the pre-SN phase and V_W is the stellar wind speed. The flux of mechanical energy into the shock is

$$2\pi r^2 \rho_W(V_S - V_W)^3 = \frac{1}{2}\frac{\dot{M}}{V_W}(V_S - V_W)^3. \tag{9}$$

If the shock compression ratio is 4, the mechanical energy flux out of the shock is 1/16th of the inflow. The power available for heating the gas, exciting MHD waves and accelerating energetic particles is

$$\frac{15}{32} \frac{\dot{M}}{V_W} (V_S - V_W)^3. \tag{10}$$

We shall take the fraction of the energy flux, which goes into energetic particles, to be $\frac{1}{3}$. Integrating over the time $t = R_W/V_S$ the shock takes to cross the cavity, and allowing for the adiabatic deceleration of particles in the SNR, gives

$$E_{CR} \approx \int_0^{R_W/V_S} \frac{5}{32} \frac{\dot{M}}{V_W} (V_S - V_W)^3 \frac{tV_S}{R_W} \, dt. \tag{11}$$

This is an estimate of the energy which is put into energetic particles (cosmic rays). Evaluating the integral yields

$$E_{CR} \approx 0.5 \left(1 - \frac{V_W}{V_S}\right)^3 \frac{M_W}{M_E} E_{SN}. \tag{12}$$

As an example, for an ejecta mass $M_E = 10 \, M_\odot$, a stellar wind mass loss $\dot{M} = 10^{-5} \, M_\odot \, \text{yr}^{-1}$; a cavity radius $R_W = 1$ pc, a stellar wind velocity $V_W = 10^8$ cm s^{-1} and a SN energy $E_{SN} = 10^{51}$ ergs, we obtain for (12)

$$E_{CR} \approx 8 \times 10^{-5} E_{SN} \approx 10^{47} \text{ ergs}. \tag{13}$$

A necessary requirement for the shock acceleration process to work is that the diffusion length scale K/V_S should be smaller than the shock radius r_S. Here K is the diffusion coefficient which depends in a complicated way on the MHD wave spectrum, which in turn may even be generated by the accelerated particles themselves. In the extreme limit of strong diffusion, which may be reached in the situation we are dealing with, we have a diffusion mean free path $\lambda \equiv 3 \, \text{Km} \, p^{-1} \approx$ particle gyroradius $\equiv pc/ZeB$, where p, Z, and m are particle momentum, charge number, and mass respectively, and B is the ambient magnetic field strength, i.e. the field frozen into the stellar wind. Hence the requirement for shock acceleration becomes

$$p^2 c/3ZemBV_S < r_S. \tag{14}$$

The magnetic field in the stellar wind cavity at a distance $r = r_S$ from the star (stellar radius R_*, angular velocity Ω_*) is roughly

$$B \approx B_* R_*^2 \Omega_*/r_S V_W, \tag{15}$$

where B_* is the field at the stellar surface. Substituting this in Equation (14) yields, using rigidity $R = pc/Ze$

$$R < 3\sqrt{3} \, R_* \left(\frac{A}{Z} B_* \Omega_* \frac{V_S}{V_W}\right)^{1/2}. \tag{16}$$

Using a stellar radius $R_* = 10^{11}$ cm, a stellar rotation $\Omega_* = 10^{-5}$, a surface magnetic field strength $B_* = 10$ G, a stellar wind velocity $V_W = 10^8$ cm and a shock speed (as before) corresponding to a 10^{51} erg SN, we obtain

$$R < 10 \sqrt{\frac{A}{Z}} \text{ GV} . \tag{17}$$

Thus we expect acceleration of particles up to mildly relativistic energies via this process.

When the SN shock reaches the edge of the stellar wind cavity, the expansion is stopped abruptly. A forward shock propagates into the cloud, the snowplough solution applies almost immediately (as was mentioned earlier) and a reverse shock propagates back into the cavity. This reverse shock thermalises the SN ejecta, converting most of the 10^{51} ergs of kinetic energy into thermal energy. The role of the reverse shock in accelerating particles is probably small, since the SN ejecta has swept out the stellar magnetic field during its initial expansion, leaving a practically 'field-free cavity'. From Equation (16), we see that acceleration to relativistic energies is then impossible. (A stellar surface field B_* frozen into the SN ejecta will be reduced by a factor $\sim (R_*/R_W)^2$.)

Thus a total energy in the form of mildly relativistic particles of $\sim 10^{47}$ ergs is produced. This number is in good agreement with the upper limit derived by Morfill and Drury (1981) from the galactic gamma-ray luminosity.

3.3 TRANSPORT AND LOSSES OF COSMIC RAYS, GAMMA-RAY PRODUCTION

The accelerated particles may now interact with the surrounding cloud. Confinement in the cavity seems unlikely, rather we expect the cosmic rays to leak into the cloud immediately. Inside the cloud the particles propagate freely, without scattering, since all waves of relevant frequencies are damped by ion-neutral friction (Kulsrud and Pearce, 1969). The typical cloud traversal time is ~ 100 yr for a cloud of radius $R_c \approx 10$ pc. When the cosmic rays leave the cloud, their streaming velocity is probably limited to the Alfvén speed, $V_A \approx 100$ km s^{-1}, in the hot interstellar medium by selfexcited waves (e.g. Lerche, 1967; Wentzel, 1974). This implies an adiabatic loss time of

$$\tau_{ad} \approx R_c/V_A \approx 10^5 \text{ yr}, \tag{18}$$

i.e. much larger than the cloud filling time of $\sim 10^2$ yr.

An interstellar cloud of mass $10^5 \, M_\odot$ and radius 10 pc has a mean density $n_c \approx 10^3$ cm^{-3}. As discussed by Morfill and Drury (1981) this implies a total inelastic nuclear loss rate of $\tau_N^{-1} \approx 4 \times 10^{-16} \, n_c = 4 \times 10^{-13}$ s^{-1}, of the same order as $1/\tau_{ad}$. A fraction of the order 0.3 of the energy lost this way is converted into gamma rays. The gamma luminosity of the cloud is then

$$\dot{E}_\gamma \approx 10^{34} \text{ ergs s}^{-1}. \tag{19}$$

This estimated gamma luminosity is on the low side when compared with typical values for GRO's. Improvements may be achieved both in the amount of energy which goes into cosmic rays, R_{CR}, and the upper limit of individual particle energies (or rigidities) by reducing the stellar wind speed. For instance, putting $V_W = 4 \times 10^7$ cm s^{-1} (the solar value) increases E_{CR} by a factor ~ 4.5 and the rigidity, R, by a factor 1.6.

At this stage we should discuss the uncertainties of the estimate. E_{CR} is probably quite well determined within a factor 10 of the quoted result of 10^{47} ergs. R, the upper limit for the particle rigidity is very uncertain, since in particular B_* is a poorly known quantity. However, we believe that our estimate is conservative rather than optimistic. The gamma-ray luminosity, finally, depends on the gas density in the cloud. For a more compact cloud, e.g. $R_c = 5$ pc, we obtain $\tau_{ad} \approx 5 \times 10^4$ yr, $n_c = 8 \times 10^3$ cm^{-3}, $\tau_N \approx 10^4$ yr, and $\dot{E}_\gamma \approx 8 \times 10^{34}$ ergs s^{-1}. Hence there is some reason to believe that perhaps a few buried SN in interstellar clouds may be detectable as point-like GRO's.

Very little can be said about the rate of occurrence of buried supernovae. For a discussion see e.g. Shull (1980a), Morfill and Drury (1981).

3.4. CORRELATIONS WITH OTHER ASTRONOMICAL OBSERVATIONS

Since all SN-cloud interactions are expected to have a common range of other signatures, here is a brief summary:

First, one would expect the radiation from the hot SNR to escape in some form. The confined remnant is very hot, and emits large amounts of X-rays and UV radiation. This penetrates into the cloud, is absorbed (at least below a few keV) and subsequently re-emitted by the dust in the infra-red. Luminosities as a function of time were calculated by Wheeler *et al.* (1980) and Shull (1980a). These authors concluded that apart from an initial 'flash phase', a buried SN was essentially indistinguishable from a buried O, B star. Hard X-rays may escape through the cloud, but this gets progressively more difficult to observe as the SNR expands and the emission softens.

Second, it was suggested by Morfill and Meyer (1981) that such objects might be gamma-ray line emitters. The lines are produced by collisional excitation of the confined hot SN ejecta material in the SNR. Detection of such lines would be particularly valuable, as it would provide us with an opportunity to sample the elemental abundances of SN ejecta, and thus learn something at first hand about explosive nucleosynthesis.

Third, the signature of bulk gas motion should be observable in the radio line emissions if the cloud can be resolved (see Figures 1 to 4), in particular ~ 100 km s^{-1} 'shocked' CO lines should be searched for.

Fourth, enhanced ionisation should be indicated for the whole cloud (or large portions of the complex) due to the increased cosmic-ray intensity. In particular, the ionising particle flux should have been substantially enhanced by the shock acceleration process, a fact that should become visible in, for instance, the HCO$^+$ radio line data.

Fifth, the hot SNR is surrounded by a thin H II shell at $\sim 10^4$ K, which may appear as a compact H II region at radio wavelengths. Many such objects are observed (Mezger, 1978), most of which undoubtedly are associated with O, B stars, as can be verified by correlations with infrared spectra. For the identification of a buried SNR, the *absence* of an O, B star must be searched for – a problem that is much more difficult to solve, convincingly.

Sixth, the SNR should be surrounded by a C II, S II, etc. region. The existence of recombination lines is another feature which, in principle, is also measurable.

4. Supernovae Partially Embedded in Interstellar Clouds

When a SN explodes too near the edge of an interstellar cloud (but initially still embedded) the situation is qualitatively different only after the SN has 'broken out'. The initial evolution proceeds as described in the last chapter, including the initial burst of energetic particle production. The SNR expansion then slows down (after reaching R_W); over a time period of a few 100 yr we may expect the cosmic rays to leak into the cloud, etc.

4.1. DYNAMICAL EVOLUTION OF THE SUPERNOVA REMNANT

The SN shock reaches the edge of the cloud at some point, and accelerates into the more tenuous intercloud medium. The shock is not driven by the kinetic energy of the SN ejecta at this stage. The SNR is either in the Sedov or in the snowplough phase.

In working its way to the surface the SNR has lost a large amount of energy by radiation (mainly infrared) and the appearance is 'older'.

The available left over energy in a SNR which has just 'burst' from an interstellar cloud depends on its size R_{SNR} = distance from SN origin to cloud surface. It equals approximately the kinetic energy, E_K, of the displaced cloud mass, M_D. This expands with a velocity $V_D = V_S/1.19$ (Parker, 1963). Using again relationship (3a) of a SNR in the radiative phase, we obtain

$$E_K = \text{Min} \begin{Bmatrix} E_{SN} \\ 1.47 \times 10^{52} \, E_{51}^{1.42}/n_0^{0.6} R_{SNR}^{1.44} \end{Bmatrix} \text{ergs,} \tag{20}$$

where n_0, the cloud density, is in cm^{-3} and R_{SNR} is in pc.

In Table II we have computed E_K for the case of a SN exploding 2 pc below the cloud surface. As can be seen, the effective energy of the SN after it breaks out of the cloud ($= E_K$) is significantly reduced, leading to a falsified picture when using standard surface brightness-distance relationships in order to find the approximate location of the SNR.

We now give some examples of hydrodynamical calculations of such a situation:

The first calculations of this kind are due to Falle and Garlick (1982) who attempted to match the appearance of the Cygnus Loop with such a model. They were able to show the development of a shell moving with a large velocity (~ 300 km s^{-1}) into the low density medium, while a slower ($v \sim 100$ km s^{-1}) one concurrently expanded into the cloud. Their model, although restricted by an adiabatic assumption, seems to fit the optical observations of Kirschner and Taylor (1976) which showed an Hα emitting component with velocities up to 300 km s^{-1} as well as the low velocity filaments. The fact that the X-ray temperature is consistent with a shock velocity of about 400 km s^{-1} (Rappaport et al., 1974) can be explained with a radiative SNR where the X-ray temperature soon becomes unrelated to the outer shock velocity.

TABLE II

Available energy of a SNR after it breaks out of
a cloud. Explosion took place 2 pc from cloud
edge, initial SN energy was 10^{51} ergs

Density (cm^{-3})	Supernova energy (ergs)
100	3.4×10^{50}
300	1.8×10^{50}
1 000	8.6×10^{49}
3 000	4.4×10^{49}
10 000	2.2×10^{49}

New further calculations (see Yorke *et al.*, 1982; Tenorio–Tagle *et al.*, 1983) have confirmed and given additional information related to the hydrodynamics of breakout. One of the main features is the fact that the part of the shock moving into the cloud is unaffected by the other part breaking through the cloud's edge, i.e. it continues to move according to the well known Sedov and/or snowplough relations, therefore it is easy to trace. On the other hand, the part of the shell that breaks into the intercloud gas, although it accelerates at first due to the density drop, soon fragments and is overtaken by the hotter gas generated during the early evolution. This is true at least in the case of a radiative flow. Estimates of the amount of cloud material dispersed by the explosion show rather low values. These are of the order of the mass overtaken by the remnant shock during the Sedov phase. The dependence of this mass on the ambient density is such, that it lowers the ejected mass for denser clouds. The 'Sedov gas', i.e. the gas overtaken by the shock during the Sedov phase, having the largest sound speed in the system, is the first to react to the pressure gradient produced by breaking out of the cloud. It streams away following the path of least resistance, while modifying the radio, X-ray and optical appearance of the remnant (see Yorke *et al.*, 1982). Figures 5 and 6 give two examples of hydrodynamical calculations of such a situation. For both calculations the SN energy (= 10^{51} ergs) was placed in a small box, along the axis of symmetry of a cylindrical cloud, at a distance of two pc from the top. The density gradient N_{cloud}/N_{ic} was set equal to 10^3. Figure 5 shows an adiabatic calculation while for the case in Figure 6 cooling was taken into account. The reader may note that the velocity at breakout is about four times smaller in the radiative case (Figure 6) and almost a factor of 10 difference is shown by the time the computational grid was filled (see Figures 6, 5g, and 5h). Therefore, the evolutionary times are also rather different. In their morphology the major discrepancy between the two cases occurs in the part of the shell that breaks into the intercloud medium. Note how in the adiabatic calculation the hottest gas continuously pushes away the shell while in the radiative case it is able to overtake the dense matter until it becomes an embedded condensation within the remnant. The original shape of the cloud was drastically modified in both calculations,

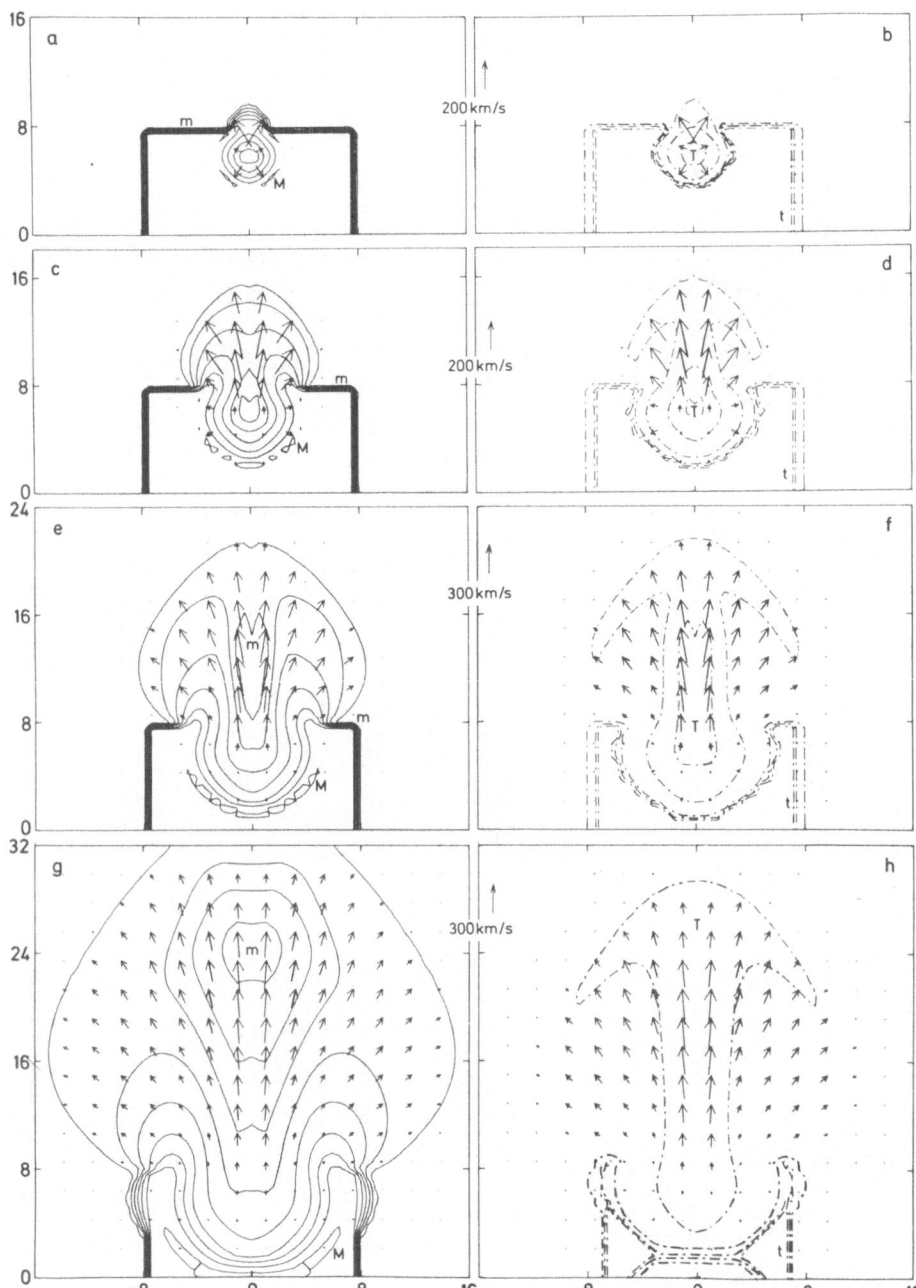

Fig. 5.   Supernova explosion inside an interstellar cloud, 2 pc from the edge. The SN energy is 10^{51} ergs. Initial cloud parameters were: density 10^3 cm^{-3}, radius 8 pc, height 16 pc, temperature 100 K. The intercloud medium is assumed to have a density 1 cm^{-3}, temperature 10^4 K. The evolution of the SNR is shown at various times. The expansion was assumed to be adiabatic. Solid contours (left panel) refer to density (in logarithmic decrements of 0.5) dashed contours (right panel) refer to temperature (log $\Delta T = 0.5$). We have a minimum temperature of $t = 10^2$ K for all sequences. (a) Time 6.64 × 10^2 yr, $m = 10^{-23.5}$ g cm^{-3}, $M = 10^{-20.5}$ g cm^{-3}, $T = 10^7$ K. (b) Time 1.96 × 10^4 yr, m, M, T as in (a). (c) Time 4.09 × 10^4 yr, m, M as in (a), $T = 10^6$ K. (d) Time 8.66 × 10^4 yr, M, T as in (c), $m = 10^{-24}$ g cm^{-3}.

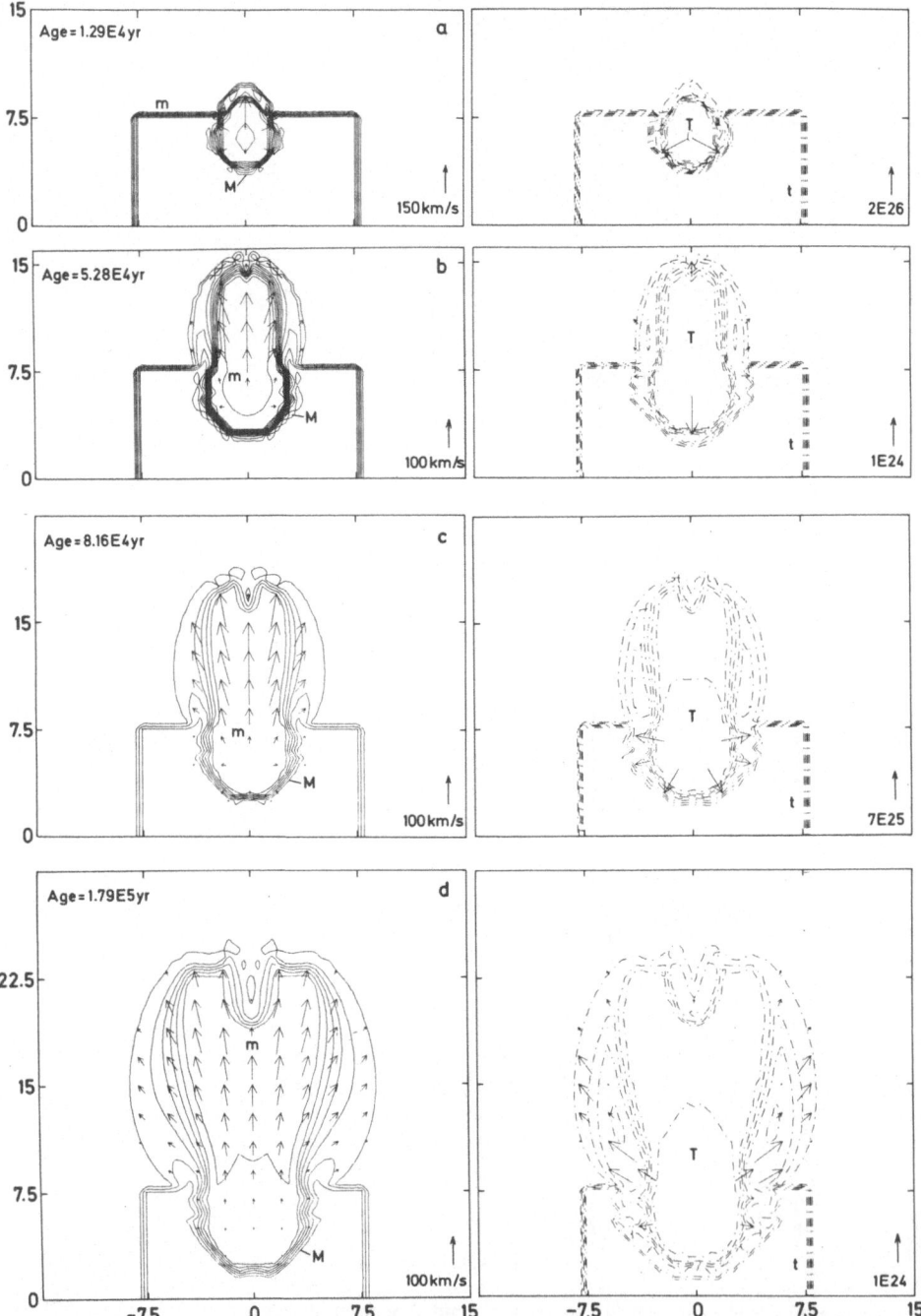

Fig. 6.   As in Figure 5, except that cooling was considered. For a description see text. We have a minimum temperature of $t = 10^2$ K for all sequences. (a) Time 8.77×10^3 yr, $m = 10^{-24.5}$ g cm^{-3}, $M = 10^{-20.5}$ g cm^{-3}, $T = 10^6$ K. (b) Time 4.8×10^4 yr, $m = 10^{-26}$ g cm^{-3}, $M = 10^{-21}$ g cm^{-3}, $T = 10^5$ K. (c) Time 1.65×10^5 yr, m, M as in (b), $T = 10^{4.5}$ K. (d) Time 5.86×10^5 yr, m, M as in (b), $T = 10^4$ K.

but only a small fraction of it was dispersed by the SN explosions. One should conclude that SN explosions tend to condense matter rather than disperse it.

4.2. COSMIC RAY PRODUCTION AND TRANSPORT

The SN, when it bursts from the cloud, reaches shock speeds of ~ 300 km s^{-1} in the adiabatic case. In the radiative case the shock speed is even less, of the order ~ 70 km s^{-1}. The Alfvén speed, V_A, in the 'warm' intercloud medium, into which the SNR expands, is of the order 20 to 30 km s^{-1}. (This assumes a magnetic field ~ 3 μG, an ion density ~ 0.1 cm^{-3} and H$^+$ as the dominant ion.) At the same time, the medium outside the shock is not fully ionised ($n_i/n \approx 0.1$), which implies significant wave damping by ion-neutral friction (Kulsrud and Pearce, 1969).

Shock acceleration in such a situation is only possible through selfexcited waves, provided the residual wave damping is not too large. One necessary requirement for shock acceleration to be effective is that the acceleration time scale τ_{acc} should be smaller than the wave damping time scale τ_D.

The acceleration time is given by (Forman and Morfill, 1979)

$$\tau_{\text{acc}} = 4K/V_S^2, \tag{21a}$$

where K is the diffusion coefficient. The wave damping time is given by (Kulsrud and Pearce, 1969)

$$\tau_D = 2/\nu_{\text{in}}, \tag{21b}$$

where $\nu_{\text{in}} \approx 1.68 \times 10^{-8} n$ is the ion-neutral collision frequency at a gas temperature of 10^4 K, with n being the neutral gas density (see Kulsrud and Cesarsky, 1971). In terms of the diffusion mean free path $\lambda \equiv 3$ Km p^{-1} (see earlier) we then obtain from the condition $\tau_{\text{acc}} < \tau_D$, for relativistic particles

$$\lambda < V_S^2/170n \tag{22}$$

(all quantities in cgs. units). For $V_S = 7 \times 10^6$ cm s^{-1} and $n = 1$ cm^{-3}, this yields $\lambda < 2.9 \times 10^{11}$ cm. By comparison, the gyroradius in a 3 μG field is

$$r_g = 10^7 R(\text{GV})/3B \approx 10^{12} R(\text{GV}) \text{ cm}, \tag{23}$$

where $R(\text{GV})$ is the cosmic-ray rigidity in GV. If the magnetic field is completely turbulent, we expect λ to approach its minimum value $= r_g$, but in no circumstance may we expect $\lambda < r_g$. Comparison of (22) and (23) shows that the mean-free-path constraint demands an unphysically small λ, at least for the typical parameters of the background medium which we have chosen for our numerical example. Thus it is difficult to see how shock acceleration may take place during the expansion of an initially buried SNR into the 'warm' intercloud medium (or cloud 'halo'). The situation may be somewhat better when the SNR reaches the 'hot' intercloud medium (McKee and Ostriker, 1977), provided the SN energy has not been spent already. This has been discussed by Völk (1981, 1983). It implies that the cloud 'halo' must not be too large, or in a more quantitative sense $\ll 10$ pc. We restrict ourselves in this analysis to massive clouds with correspondingly large 'haloes'.

Whilst the mean-free-path constraints (Equations (22) and (23)) depend on local values (V_S, n_i, n, B), which may vary from situation to situation and we thus cannot rule out relativistic cosmic-ray production in general, there exist two major problems with respect to 'late' cosmic-ray and gamma-ray production.

The first problem is the following: we noted earlier that the shock accelerated particles must produce their own waves by streaming in the 'warm' intercloud medium. The wave growth rate is given by

$$\frac{1}{\tau_+} \approx \frac{\pi}{8} \Omega \frac{n}{n_i} \eta \frac{\langle E_G \rangle}{\langle E_{CR} \rangle} \left(\frac{|V_R|}{V_A} - 1 \right). \tag{24}$$

This equation has been derived from Kulsrud and Cesarsky (1971). Ω is the gyro-frequency of the cosmic rays, $\eta \geqslant 1$ is the cosmic-ray enhancement factor, $\langle E_G \rangle = 1$ eV is the mean-gas energy (10^4 K) in the 'warm' intercloud medium, $\langle E_{CR} \rangle$ is the mean cosmic-ray energy above the π° production threshold of ~ 0.5 GeV (we use $\langle E_{CR} \rangle = 1$ GeV), and V_R is the cosmic-ray streaming velocity = shock velocity, V_S. A cosmic-ray spectrum $\propto p^{-2}$ has been assumed, in keeping with the shock acceleration theory (Axford et al., 1977). In order that waves may grow (and thus make the acceleration process feasible), we need

$$\frac{1}{\tau_+} > v_{in}/2. \tag{25}$$

From Equation (25) we can estimate how large the cosmic-ray enhancement factor η (above the π° production threshold) must be (since all other quantities are determined). We obtain, for $V_R = V_S = 70$ km s^{-1}, $V_A = 25$ km s^{-1}, $n_i = 0.1$ cm^{-3}, $B = 3$ μG:

$$\eta \geqslant 40. \tag{26}$$

This result implies that for an initially lower cosmic-ray enhancement, waves cannot grow (damping dominates) and the shock acceleration process cannot get started. Since, on the other hand, the usual assumption relies on the existence of background test particles (which are then re-accelerated), or the elevation of such particles from the thermal plasma (to ever increasing energies), requirement (26) seems to be extremely difficult to fulfill.

The second problem is the following: the shock accelerated particles have to diffuse from the shock through the SNR back into the cloud against the convection, which tries to keep them out (see Figures 5 and 6). In a single one dimensional diffusion-convection picture, we have for the cosmic-ray density n_c in the SNR:

$$\frac{\partial n_c}{\partial t} + V \frac{\partial n_c}{\partial X} - K \frac{\partial^2 n_c}{\partial X^2} = Q\delta(X - X_0), \tag{27}$$

where the convection speed $V \approx V_S$, and Q is the particle source at the shock (distance

X_0 from the cloud). Taking the time independent result as an upper limit for cloud penetration, gives the density (at the cloud) as

$$n_c (X = 0) \propto \exp (- VX_0/K). \tag{28}$$

Inside the SNR we may assume a fully ionised medium (at least in the initial 10^4 yr), in which case waves produced by the accelerated particles streaming ahead of the shock are *not* effectively damped. (On the contrary, they are even enhanced at the shock (Morfill and Scholer, 1978; McKenzie and Westphal, 1968)). This means that K may be taken as constant everywhere. From condition (22) we have $X_0 V_S/K > 5.1 \times 10^{10} nR_{pc}/V_S$, which, using $V_S = 70$ km s^{-1} becomes $0.7 \times 10^4 nP_{pc} \gg 1$.

Hence we see from (28) that cosmic rays, even if they can be formed by the SN shock, cannot propagate effectively back through the expanding SNR into the cloud and supply energy for a possible GRO. The gamma-ray intensity is then dominated by the initial cosmic-ray burst, the life time should be similar, possibly even a little less than the buried SN case.

4.3. CORRELATIONS WITH OTHER ASTRONOMICAL OBSERVATIONS

The obvious feature, different from and additional to the buried SN situation is the existence of a 'visible' SNR (in X-rays and radio). We have already discussed that this SNR may appear 'older' and would be placed at too large a distance away from us on the basis of the normal surface brightness-distance ($\Sigma - D$) relationship. This fact must be borne in mind.

A possible candidate for such a situation is the Carina system. A nonthermal radio source (G 286.8–0.5) has been identified by Elliott (1979) as a SNR. From the $\Sigma - D$ relationship, a distance of ~ 15 kpc was inferred. On the other hand, the SNR is in front of the Carina Nebula (NGC 3372), which is estimated to be only ~ 2 kpc away (Thé and Vleming, 1971). This difference can be resolved if the SNR in question is partially buried in the interstellar cloud, and has a diameter, at the present time, of only ~ 2 pc. A possible identification with the GRO 2CG 288–00 cannot be ruled out on the basis of the COS-B error analysis.

However, there is another suggestion regarding η Car, which we will mention later.

5. Supernovae Outside Interstellar Clouds

The obvious, and in some sense the easiest interaction to verify observationally, is the interaction between a visible SNR and an interstellar cloud. A relatively 'free' SNR, which may expand at large magnetosonic Mach numbers for a long time, can convert a much larger fraction of its energy into cosmic rays than a 'confined' SNR. The process is again diffusive shock acceleration, the most advantageous regions for this process to operate is the 'hot' interstellar medium (see e.g., Blandford, 1979; Axford, 1980, 1981a, b; Völk et al., 1981) and the energy conversion efficiency is high (Drury et al., 1982; Eichler, 1979). In the 'warm' medium, there is almost total pre-ionisation, ahead of the shock, provided the shock speed exceeds ~ 100 km s^{-1} (Shull and McKee, 1979; Raymond, 1979). Thus for all practical purposes we can ignore ion-neutral friction for the strong shocks which we discuss here.

5.1. DYNAMICAL EVOLUTION OF THE SUPERNOVA REMNANT

In Figures 7–9 we show the interaction of an external SN with a nearby cloud. The distance to the explosion from the cloud was chosen to be 2 pc in order to compare it with the earlier results shown in Figures 5 and 6. When the SNR interacts strongly with the cloud the swept up matter is of the order of 90 $M_\odot$ (for a warm intercloud medium with a gas density of 1 cm^{-3}), and the remnant is well into the Sedov phase. The SN energy was chosen again to be 10^{51} ergs and cooling was taken into account. Note, however, when comparing with Figure 6 how different the resultant flows are. The hotter gas looks always for the path of least resistance and is able to engulf the cloud. In such a case no cloud disruption takes place. We argued previously that external SN can produce cosmic rays by diffusive shock acceleration relatively easily, and transport this energetic particle population into the cloud by convection with the expanding SNR. The hydrodynamical calculations show that this process works quite well for that portion of the shock, which runs into the cloud. Subsequent cosmic-ray production by shock portions which run away from the cloud appears to be unimportant again, for the reason mentioned previously, i.e. the severe convection losses inside the SNR.

5.2. TIME DEPENDENT COSMIC-RAY ACCELERATION

In order to estimate how long the cosmic-ray convection into the cloud may last, we have to look at time dependences. Such calculations have been made by Krymsky *et al.* (1979), Forman and Morfill (1979), Axford (1981a), and will not be repeated here. Figures 10 and 11 show calculations made by Forman and Morfill (1979). Figure 10 shows the evolution of the cosmic-ray distribution function $f(x = 0, p, t)$, where the shock is kept at position $x = 0$. The plasma streams into the shock (from the right) with velocity V_S, and leaves the shock with velocity V in this frame. The initial distribution function was a 'square box' of the type

$$f(x = 0, p < p_0, t = 0) \equiv 1, \qquad f(x = 0, p \geqslant p_0, t = 0) \equiv 0, \tag{29}$$

where p_0 is some threshold momentum. (This may be regarded as a crude approximation to a thermal spectrum with $kT = p_0^2/2m$). As can be seen, the accelerated particles obtain a spectrum $\propto p^{-4}$ with a time constant $\tau_{\text{acc}} \approx 4K_1/V_S^2$, where K_1 is the diffusion coefficient upstream of the shock. (Note that the shock accelerated cosmic-ray *intensity* spectrum, $N(E)$, is related to the *distribution function, f,* in phase space, by $I(E) = fp^2$). The factor 4 comes from $3/(V_S - V)$, where for strong shocks $V_S = 4$ V (compression ratio 4).

Figure 11 shows the spatial evolution of the accelerated particles in the frame of the shock. We have plotted the distribution function, $f(p, x, t)$, using the initial distribution (29), at a value $p = 3p_0$ both in x and t. Again we see the approach to steady state, with the time constant $\tau_{\text{acc}} = 4K_1/V^{2s}$. The length scale of the accelerated particles ahead

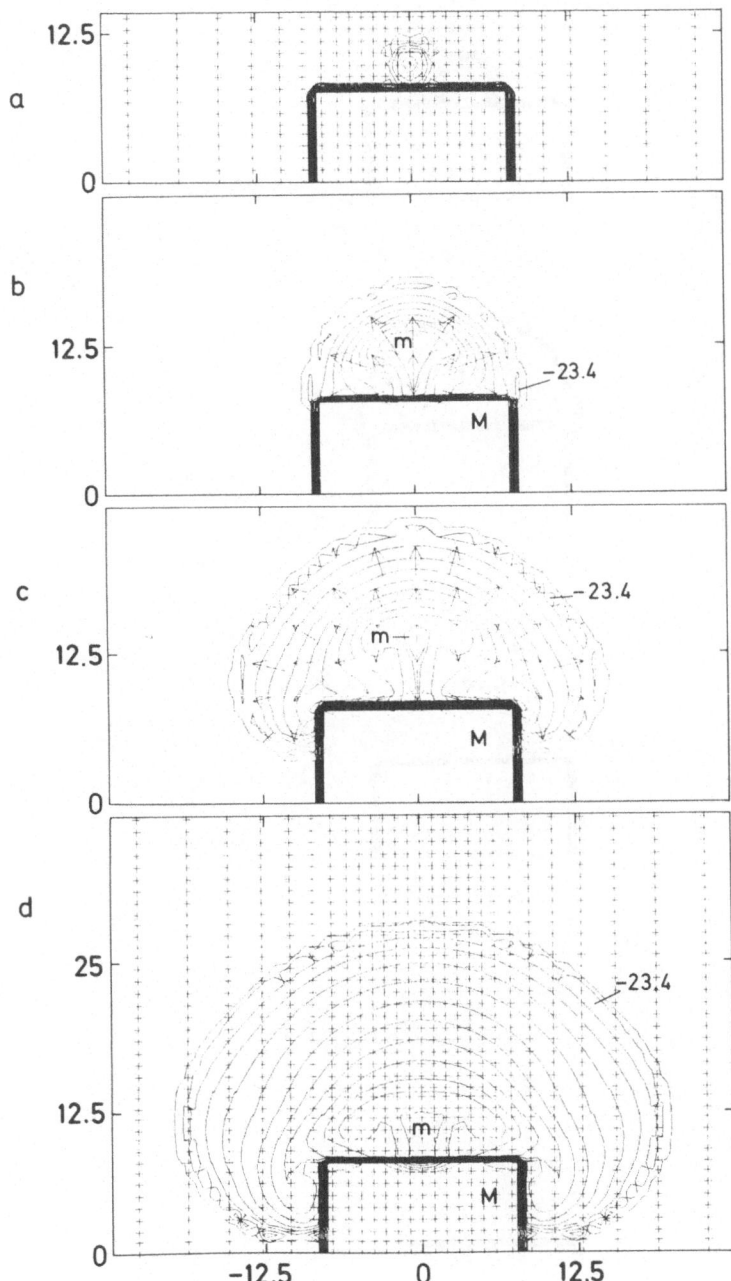

Fig. 7. Supernova explosion in the 'warm' medium (density (cm^{-3}, temperature 10^4 K) surrounding an interstellar cloud. SN energy is 10^{51} ergs, distance from the cloud surface is 2 pc. Cooling was considered. Initial cloud parameters are those of Figures 5 and 6. The evolution of the SNR is shown at different times. (100 yr, 3.06×10^3 yr, 1.19×10^4 yr, and 2.65×10^4 yr). Only density contours are shown between the limits. (a) $m = -$g cm^{-3}, $M = 10^{-21}$ g cm^{-3}. (b) $m = 10^{-25.5}$ g cm^{-3}, $M = 10^{-21}$ g cm^{-3}. (c) $m = 10^{-26.1}$ g cm^{-3}, $M = 10^{-21}$ g cm^{-3}. (d) $m = 10^{-26.4}$ g cm^{-3}, $M = 10^{-21}$ g cm^{-3}.

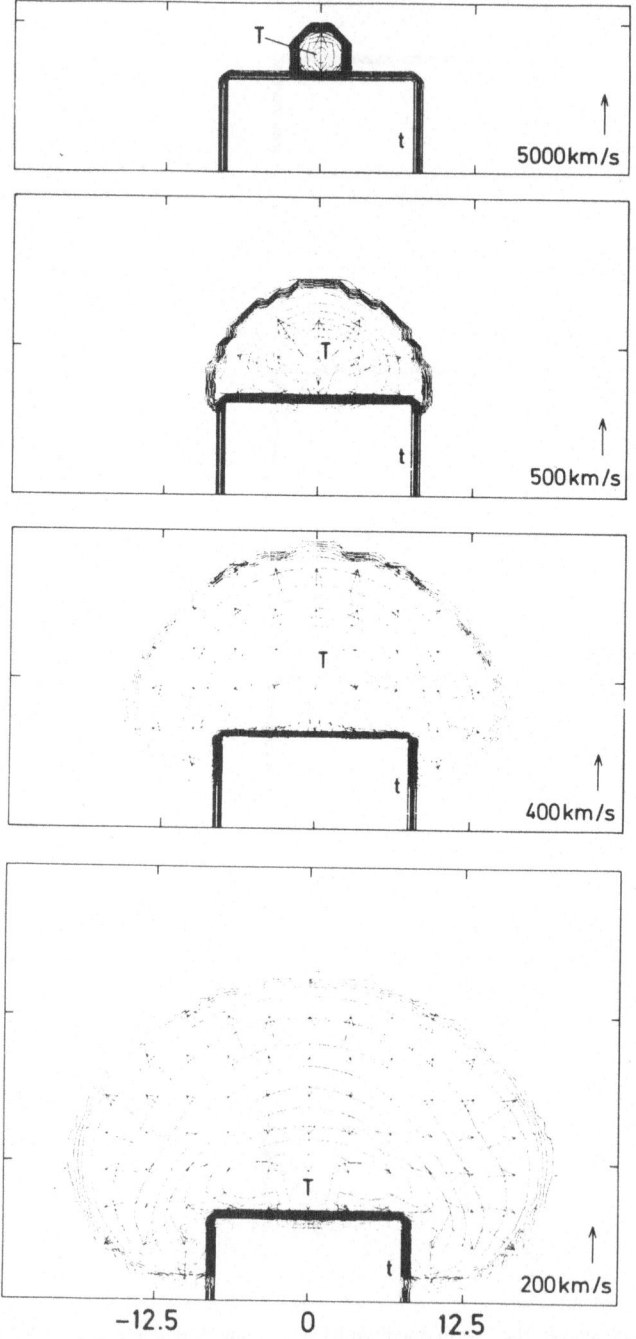

Fig. 8.   As in Figure 7, except that the temperature contours are shown between the limits. (a) $t = 10^{1.8}$ K, $T = 10^{9.9}$ K. (b) $t = 10^{1.8}$ K, $T = 10^9$ K. (c) $t = 10^{1.8}$ K, $T = 10^{8.7}$ K. (d) $t = 10^{1.8}$ K, $T = 10^{8.4}$ K.

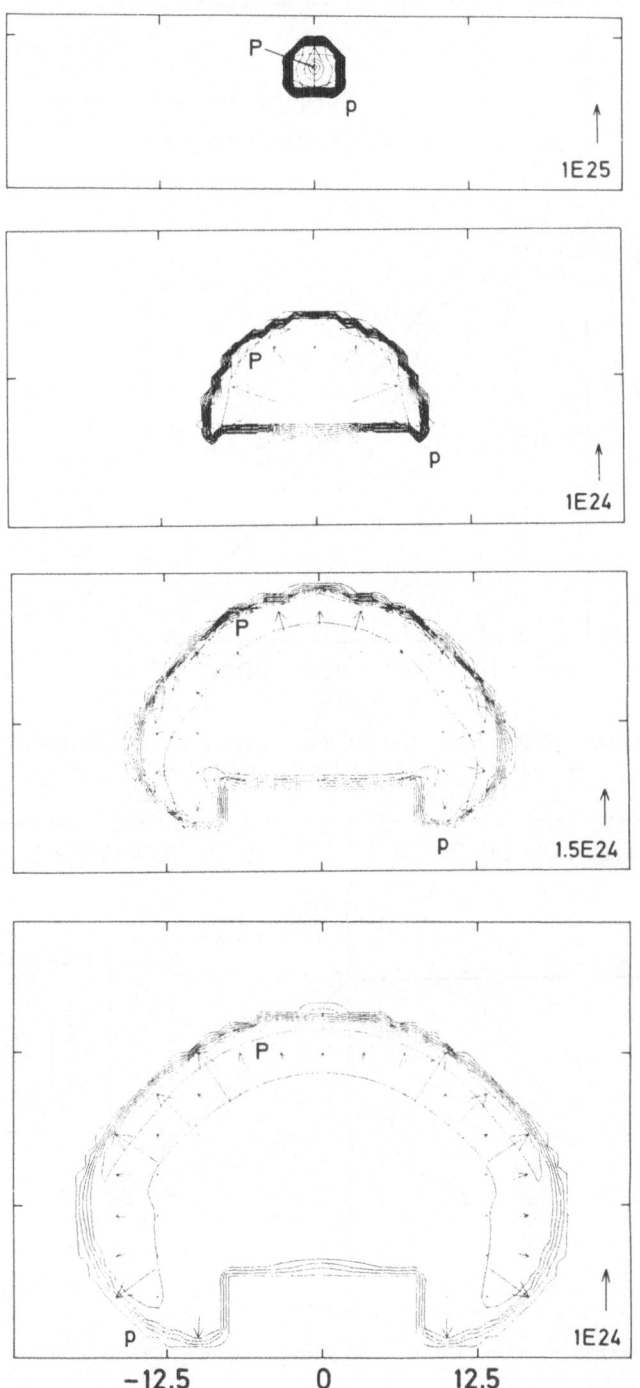

Fig. 9. As in Figure 7, pressure contours are shown between the limits. (a) $p = 10^{-10.8}$ ergs cm^{-3}, Pp = $10^{-6.9}$ ergs cm^{-3}. (b) $p = 10^{-10.8}$ ergs cm^{-3}, $P = 10^{-8.1}$ ergs cm^{-3}. (c) $p = 10^{-10.8}$ ergs cm^{-3}, $P = 10^{-8.7}$ ergs cm^{-3}. (d) $p = 10^{-10.8}$ ergs cm^{-3}, $P = 10^{-9.3}$ ergs cm^{-3}.

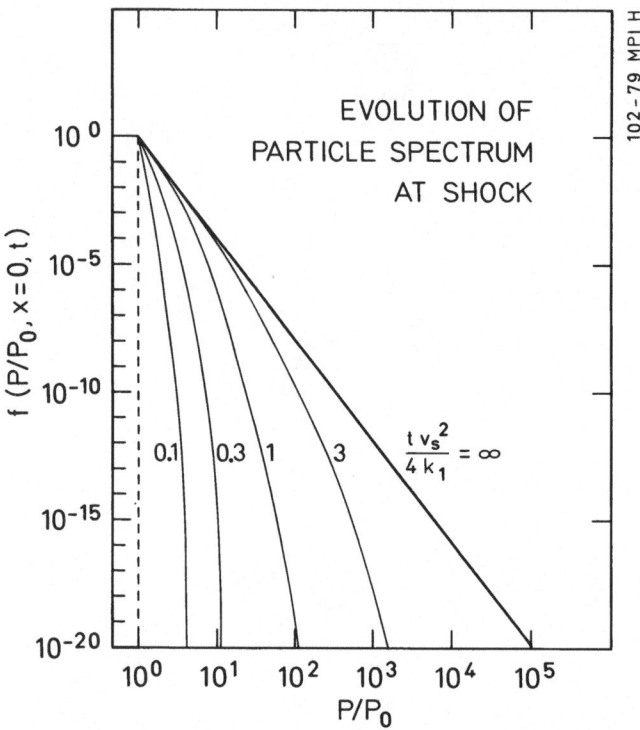

Fig. 10.   Temporal evolution of the cosmic-ray distribution function at the shock. The initial distribution was $f(x = 0, p, t = 0) = 1$ if $p < p_0$ and 0 if $p > p_0$.

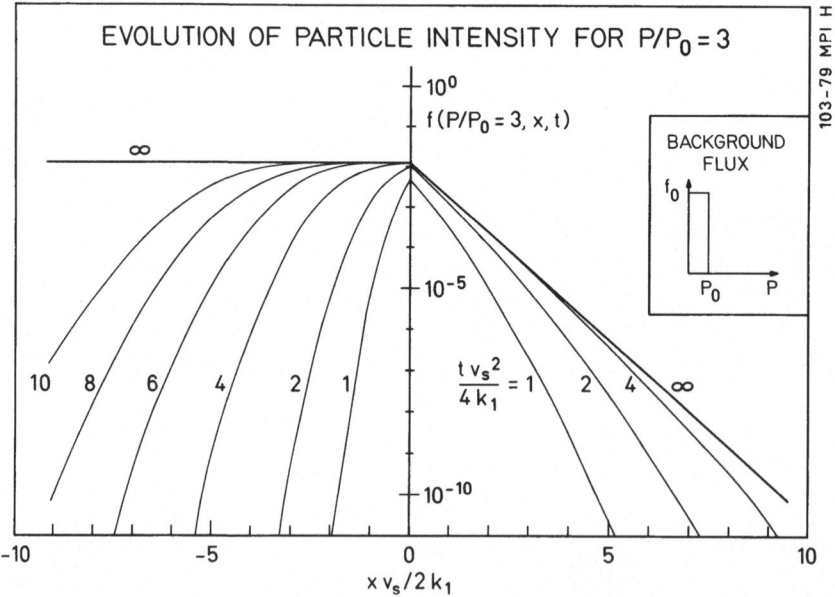

Fig. 11.   Temporal and spatial evolution of the cosmic-ray distribution function around the shock. The shock is located at $x = 0$, the initial distribution function is shown in the insert (it is the same as that used in Figure 10) and the contours of Figure 11 are shown at a momentum $p/p_0 = 3$.

of the shock is $\sim 2K_1/V_S$. In addition, it is seen that the particles propagate downstream, away from the shock, at about the same speed as the plasma.

Special assumptions which went into the calculations for Figures 10 and 11 were: (1) the problem is one dimensional; (2) the diffusion coefficient K_1 was taken to be constant, independent of x, p and t; (3) no loss or production processes were considered, e.g. by nuclear or ionising interactions; (4) the calculations assumed a strong shock with a compression ratio 4, unmodified by the accelerated particles; (5) $K_1/V_S^2 \equiv K_2/V^2$ where K_2 is the diffusion coefficient downstream and V the plasma velocity downstream (this was done in order to use tabulated Laplace transforms, however, it is also not unphysical to propose that scattering in the post-shock region is enhanced, because waves are shock amplified – see Morfill and Scholer, 1978). More general calculations (see e.g. Drury, 1983, and references therein) give

$$\tau_{\text{acc}} \equiv p \left/ \frac{dp}{dt} \right. = \frac{3}{V_S - V} \left(\frac{K_1}{V_S} + \frac{K_2}{V} \right). \tag{30}$$

5.3. ENERGY CONSIDERATIONS

In order to obtain a net particle acceleration at the SNR shock, the acceleration time must be shorter than the age of the SNR, t_{SN}. As a criterion we shall use

$$3\tau_{\text{acc}} \leqslant t_{\text{SN}} \tag{31}$$

(see Morfill et al., 1983). Substituting t_{SN}, as obtained from the Sedov (1959) similarity solution for SNR expansion into a homogeneous medium, we obtain after some algebra

$$K < 4.6 \times 10^{26} t_y^{-1/5} (E_{51}/n_0)^{2/5} \text{ cm}^2 \text{ s}^{-1}. \tag{32}$$

Even in the 'hot' interstellar medium ($n_0 = 10^{-2}$) we then still find that K is substantially less than the 'canonical' cosmic-ray diffusion coefficient of $\sim 10^{28}$ cm^2 s^{-1}. This implies that local effects are required (e.g. wave production by streaming, Lerche, 1967; Wentzel, 1974) which lower K in the vicinity of the SN shock. The logical consequence is that the process only works reasonably well, provided the magnetosonic Mach number $M \gtrsim 3$, where

$$M \equiv 2V_S/(C_S^2 + V_A^2)^{1/2} \tag{3)}$$

and C_S is the sound speed. In the 'warm' medium ($T = 10^4$ K, $n = 1$ cm^{-3}, $B = 3 \mu$G) this yields $V_S \geqslant 45$ km s$^{-1} \equiv V_S^*(w)$, in the 'hot' medium ($T = 10^{5.7}$ K, $n = 10^{-2}$ cm^{-3}, $B = 1 \mu$G) this gives $V_S \geqslant 300$ km s$^{-1} \equiv V_S^*(H)$.

Translated into a SN age limit, above which shock acceleration ceases, we obtain (again using the Sedov solution)

$$\tau_{\text{SN}} \leqslant t^* \equiv 7 \times 10^{16} \left(\frac{E_{51}}{n_0} \right)^{1/3} (V_S^*)^{-5/3} \text{ yr.} \tag{34}$$

For the 'warm' medium (substituting $V_S^* = V_S^*(w)$, $n_0 = 1$) this yields $t_{SN} \leqslant 5 \times 10^5$ yr, for the hot medium (substituting $V_S^* = V_S^*(H)$, $n_0 = 10^{-2}$) this yields $t_{SN} \leqslant 10^5$ yr. Hence we see that in a relatively free environment, SNR's can be 'active', in the sense of cosmic-ray production, for quite a long time.

Morfill *et al.* (1983) have calculated the energy density E_{CR} (in a SNR) of cosmic rays, produced by diffusive shock acceleration. The assumptions were that $\frac{1}{3}$ of the mechanical energy flux into the SN shock is converted into cosmic rays, that cosmic-ray production only becomes very important when the SNR enters the Sedov phase (i.e. after the free expansion phase), that adiabatic expansion losses must be considered throughout the SNR evolution and that this evolution is described reasonably well by the Sedov (1959) similarity solution. The result for the energy density is then

$$E_{CR}(r) = \frac{3}{20} \frac{E_{SN}}{r_3} \left(1 - \frac{r_1}{r} \right), \tag{35}$$

where r is the SNR radius at time t and r_1 the SNR radius when it enters the Sedov phase. Shock acceleration is assumed to become ineffective again, when $t \geqslant t^*$ (see Equation (34)). We can rewrite (35) in terms of the 'active' SNR radius $r^* \equiv \frac{5}{2} V_S^* t^*$. This yields

$$E_{CR} = 6.5 \times 10^{-13} n_0 (V_S^*)^2 \, 1/y^3 \left(1 - \frac{y_1}{y} \right) \text{(eV cm}^{-3}) \tag{36}$$

with $y_1 \equiv r_1/r^*$, $y \equiv r/r^*$; V_S^* is expressed in cm s^{-1}, n_0 in cm^{-3}. The enhancement of the cosmic-ray energy density inside the SNR relative to the value in the solar neighbourhood E_{CR_0} is easy to calculate, since $E_{CR_0} \approx 1$ eV cm^{-3}. Equation (36) is particularly useful, because the product $n_0 (V_S^*)^2 \approx$ const. $\approx 10^{13}$ cm^{-1} s^{-2}, almost irrespective of the background medium.

Figure 12 shows E_{CR}/E_{CR_0} as a function of $y = r/r^*$. The value of $y_1 = r_1/r^*$ was chosen to be 0.1. As can be seen, large cosmic-ray enhancements are possible inside modestly young SNR's, values in excess of ~ 100 may be reached. A SNR interacting with an interstellar cloud at such a stage will convect its enhanced cosmic-ray intensity into the cloud and illuminate it considerably. This situation seems to us a very promising scenario for GRO's. It is one which can be tested relatively easily from observations.

5.4. COMPARISON WITH OBSERVATIONS

Apart from the SNR-cloud interaction features mentioned previously (e.g. enhanced HCO$^+$, cloud heating giving rise to an anomalously high infrared temperature, 'shocked' CO, etc.), the diagnostic of shocked molecule (e.g. CO) motion should indicate a common 'epicentre' for the systematic velocity component – one that should be located within an external young SNR. Compression of the cloud at its interface with the SNR is another additional indicator.

A survey of available data has yielded 5 regions (excepting Carina) where a SNR-cloud interaction may exist. These are W28, W44, IC443, Cyg Loop and

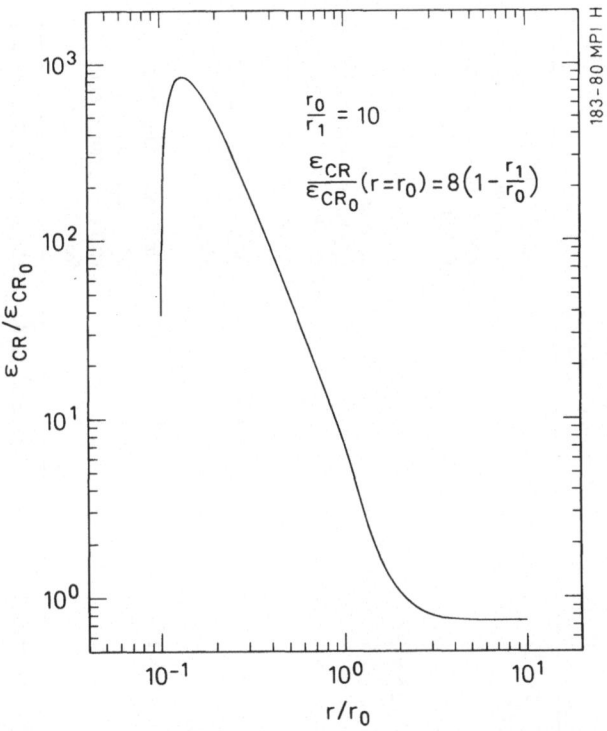

Fig. 12. Cosmic-ray energy density inside a SNR as a function of SNR size. The energy density is normalised to the cosmic-ray level in the solar neighbourhood (= 1 eV cm^{-3}), and the SNR size is normalised to the radius where the Sedov phase begins (or the free expansion phase ends). For more information, see text (from Morfill et al., 1983).

ρ-Ophiuchi. Of these, we can disregard the Cyg Loop immediately, because the cloud masses are too small ($< 500\ M_\odot$ according to Scoville et al., 1977) and the distance from the Sun is too large (~ 1 kpc). This object would require a SNR cosmic-ray enhancement $f \equiv E_{CR}/E_{CR_0}$ which is larger than 10^3. According to our analysis of the previous section (see Figure 12) this is impossible. A summary of the properties of the remaining objects is given in Table III. As usual, the properties of both SNR's and associated interstellar clouds are very uncertain. We have tried to synthesise the most consistent set of values into the table, but to quote a range of uncertainties (either from arithmetic means of separate estimates or from model fitting, etc.) seems to us too speculative. The best way to use Table III is as a guideline, and to employ future observations to improve on the estimates.

We can combine Table III with Figure 12, in order to obtain a measure of the cosmic-ray enhancement in the SNR's concerned. To do this, we have to know the background medium into which the SNR expands. For the N. Polar Spur, Loop I, a 'hot' medium seems appropriate (e.g. Hyakawa et al., 1977, 1978; Morfill et al., 1981), for the others we take both the 'warm' and the 'hot' medium as examples. Experimental determination for the relevant background medium does not exist, although it must be

G. E. MORFILL AND G. TENORIO-TAGLE

TABLE III

Summary of SNR-cloud interaction regions

Object	SNR size (pc)	Energy (ergs)	Age (yr)	Distance (kpc)	Assoc. cloud mass ($M_\odot$)	Cosmic ray enhancement factor	Relevant observations
W28	20	5×10^{50}	$< 6 \times 10^4$	1.8	2×10^4	>30	– HCO^+ enhanced – T enhanced – no IR object detected – CO lines ~ 10 km s^{-1}
W44	25	2×10^{50}	2×10^4	3	2.5×10^4	50	– HCO^+ enhanced – T enhanced – CO lines ~ 10 km s^{-1} – nonth. radio emission – no IR object detected
IC443	30	5×10^{50}	6.5×10^4	3	10^4	20	– Pulsar PSR0611 + 12? – Shocked CO
N. Polar Spur Loop I	100	3×10^{51}	10^5	0.16	4×10^3	4	– No direct evidence of SN-cloud interaction – HCO^+, T enhancement from IR sources?

References: Ilovaisky and Lequeux (1972), van den Bergh (1978), Scoville *et al.* (1977), Wootten (1977, 1981), Lozinskaya (1974), De Noyer (1979), Scoville and Solomon (1974), Morfill *et al.* (1981).

assumed that the 'warm' medium may be somewhat favoured in the vicinity of large interstellar clouds. The results are: W 28 (> 70, > 12), W 44 (220, 40), IC 443 (65, 11), ρ Oph (8). The first figure in the brackets is E_{CR}/E_{CR_0} in the case of a 'warm' medium ($T = 10^4$ K, $n = 1$ cm^{-3}) the second figure corresponds to the 'hot' medium ($T = 10^{5.7}$ K, $n = 10^{-2}$ cm^{-3}). Comparison with the COS-B detection threshold (using the experimentally determined gamma-ray production rate $q_\gamma = 2.1 \pm 0.3 \times 10^{-25}$ photons s^{-1} H-atom, see Lebrun et al., 1982; Strong et al., 1982), yields the result that W 28, W 44, and ρ Ophiuchi might just be visible as point (or point-like) gamma-ray sources above 100 MeV, provided the W 28 and W 44 SN explosions took place in the warm medium. IC 443 is too weak to be observable by COS-B. The large uncertainties of the values in Table III, coupled to the uncertainties in the nature of the ambient medium, make it impossible to give a firm conclusion, however. Perhaps the best way to summarise our results is to say that they do not exclude the process (SNR accelerated cosmic rays interacting with a large cloud) in the case of W 28 and W 44, and suggest it more strongly for the better known ρ Ophiuchi system.

The directional coincidence of W 28 with the GRO 2CG 006–00 appears good (it could, of course, be accidental), the directional coincidence of W 44 with the GRO 2CG 036 + 01 is not adequate, given the catalogue error radius of 1° and the positional separation of 2.7° between the two objects. The cloud ρ Ophiuchi has been discussed by Morfill et al. (1981) in connection with the SNR interaction picture. Directional agreement and cosmic-ray enhancement calculations yield a consistent picture with the gamma-ray measurements of the source 2CG 353 + 16. Recent refined analysis of the gamma-ray data, which suggest that this source is extended, supports its identification with the cloud ρ Ophiuchi (Hermsen, 1983). The SNR interaction would, of course, illuminate the whole cloud with an enhanced cosmic-ray intensity, so that we may take the results of this analysis as support for that model.

5.5. STATISTICAL CONSIDERATIONS

Since the occurrence rate of 'free' SN is relatively well known, and the number of giant molecular clouds in the Galaxy has been estimated also, we can make a statistical argument of the likelihood of SNR-cloud interactions. The SNR volume, which contains an enhanced cosmic-ray energy density $E_{CR} \equiv f E_{CR_0}$, where $f > 1$, is given by $\frac{4}{3}\pi r^3(f)$, where $r(f)$ can be determined from Figure 12. We define $r(f) \equiv r(t = t_f)$, and t_f is the age of the SNR above which the cosmic-ray enhancement drops below a value f. For a galactic SN birth rate $\tau_{SN}^{-1} = \frac{1}{50}$ yr^{-1} we obtain the total volume fraction with enhanced cosmic rays (above a factor f)

$$\mu(f) = \frac{1}{V_{SN}} \int_0^{t_f} dt \, \frac{4}{3}\pi r^3(t) \, \frac{1}{\tau_{SN}}. \tag{37}$$

V_{SN} is the volume within which most galactic SN occur (a disc with radius 12 kpc and height 200 pc (Ilovaisky and Lequeux, 1972) giving $V_{SN} \approx 2.6 \times 10^{66}$ cm^3). Substituting the Sedov solution, and assuming that most SN explode in the 'hot' medium, yields approximately

$$\mu(f) \approx 0.175/f. \tag{38}$$

Next we have to determine the likelihood of chance coincidences of such cosmic-ray enhanced regions with interstellar clouds. If we take the total number of large clouds in the Galaxy to be $N_C = 4000$ (Solomon *et al.*, 1978) and if we assume a homogeneous distribution of SNR's and clouds, the total number of such sources in our Galaxy should be of the order

$$N_\gamma \approx 700/f. \tag{39}$$

Of course, the assumption of a homogeneous distribution of both clouds and SNR's is pessimistic: see e.g. the radial variation of gas (Solomon *et al.*, 1979), SNR's (Kodaira, 1974), and pulsars (Hulse and Taylor, 1975). Pessimistic is also the limitation of the argument to giant ($\gtrsim 10^5 \, M_\odot$) molecular clouds. We have seen that both ρ Ophiuchi and W 28 may be gamma-ray sources of the type discussed here, but neither is very massive. Hence (39) should be regarded as a lower limit. The possibility that a few more GRO's may turn out to be of this type thus cannot be excluded on statistical grounds.

6. High Velocity Clouds

So far we have discussed shock acceleration of cosmic rays by an external (or internal) agent, and the interaction of enhanced relativistic particle intensities with an interstellar cloud. One way to circumvent such chance coincidences would be a 'home made' acceleration, e.g. by the peculiar motion of the cloud itself. Cosmic-ray enhancement by convection and acceleration may play a role. For obvious reasons, only high velocity (HV) clouds need be considered.

We shall calculate the enhanced cosmic-ray intensity inside such a cloud under the most optimistic assumptions. These are: no deviation of external plasma flow around the cloud, good magnetic connection, existence of a large turbulent wake.

We consider neutral HI clouds (e.g. Hulsbosch, 1975) with velocity V_1 and column density nl. In such clouds, waves are damped by ion-neutral friction (e.g. Kulsrud and Pearce, 1969) and cosmic-ray traversal occurs essentially at the speed of light. As argued by Morfill (1982, 1983a) in connection with another phenomenon, this implies that the residence time of cosmic rays inside the cloud is much smaller than all the other time scales in the problem. Mathematically this means that the cloud occupies a δ-function in space.

We can solve for the cosmic-ray distribution $f(p)$ in phase space, by considering convection and diffusion outside the cloud, and losses inside. We have:

(1) upstream (region 1)

$$\frac{\partial}{\partial x} \left(-V_1 f_1 - K_1 \frac{\partial f_1}{\partial x} \right) = 0; \tag{40a}$$

(2) downstream

$$\frac{\partial}{\partial x}\left(-V_2 f_2 - K_2 \frac{\partial f_2}{\partial x}\right) = 0; \tag{40b}$$

(3) continuity of flux across the cloud

$$\left[-V_1 C_g f_1 - K_1 \frac{\partial f_1}{\partial x}\right]_{x=+\varepsilon} + \left[V_2 C_g f_2 + K_2 \frac{\partial f_2}{\partial x}\right]_{x=-\varepsilon} = \frac{l}{\tau} f_c, \tag{41}$$

where $f_c = f_1(\varepsilon \to 0) = f_2(\varepsilon \to 0)$, C_g is the Compton–Getting operator

$$C_g \equiv -\frac{p}{3}\frac{\partial}{\partial p} \tag{42}$$

and $1/\tau$ is the cosmic-ray loss rate. In the case of electrons the main loss process is bremsstrahlung, in the case of relativistic nucleons it is π-production. We shall use

$$1/\tau \equiv 8 \times 10^{-16}\, n, \tag{43}$$

where n is the gas density of the neutral cloud. As a boundary condition far ahead of the cloud, we use

$$f_1(x \to \infty) \equiv f_0 = \begin{cases} F_0 & (p < p_0), \\[2mm] F_0 \left(\dfrac{p}{p_0}\right)^{-3C} & (p \geqslant p_0), \end{cases} \tag{44}$$

where f_0 is the cosmic-ray distribution function in the vicinity of the Sun, where it has been measured. This implies in the case of the nucleonic component $3C \equiv 4.5$, resulting in a cosmic-ray intensity spectrum $I(E) \approx f(p)p^2 \propto E^{-2.5}$.

The value of p_0, where the cosmic-ray spectrum bends over, is not known experimentally, it probably occurs at or somewhat below ~ 0.4 GV.

The solution of Equations (40a) to (44) inside the cloud, at $x = 0$, has been given by Morfill $et\ al.$ (1983). It is independent of the diffusion coefficients K_1 and K_2:

$$\frac{f_c}{f_0} = \frac{L}{Q - C + L}\left\{1 - \frac{C}{Q + L}\left(\frac{p}{p_0}\right)^{-3(Q-C+L)}\right\} \tag{45}$$

where

$$L \equiv V_1/(V_1 - V_2), \tag{46a}$$

$$Q \equiv l/\tau(V_1 - V_2). \tag{46b}$$

It is obvious that cosmic-ray enhancements are only possible if $Q \ll 1$, i.e. if the absorption in the cloud is modest. The largest cosmic-ray enhancement in the cloud is obtained for $V_1 \gg V_2$ (which corresponds to the existence of a large turbulent wake). In keeping with our aim to produce the most optimistic estimate, we examine this case. We can then make a substantial simplification. From measurements of HV clouds (e.g. Hulsbosch, 1975) we know that they have column densities $nl \lesssim 10^{20}$ cm^{-2} and velocities $V_1 \lesssim 2 \times 10^7$ cm s^{-1}. This means values for $Q \approx 10^{-2}$, and Equation (45) reduces to

$$\frac{f_c}{f_0} \approx 3 \left(\frac{p}{p_0}\right)^{3/2}. \tag{46}$$

The maximum in $\pi°$ production occurs at $p/p_0 \approx 10$ (i.e. a few GV for the cosmic-ray spectrum, Stecker, 1971) so that a cosmic-ray enhancement of up to a factor ~ 95 appears possible in HV clouds. We should stress again that this is the most optimistic estimate.

The gamma-ray luminosity of a HV cloud at distance r from the observer is then

$$F_\gamma = q_\gamma \frac{f_c}{f_0} \frac{nlA}{4\pi r^2}. \tag{47}$$

Using a typical angular diameter of $\sim 2°$, which corresponds to $A/R^2 \approx 1.2 \times 10^{-3}$, and a cosmic-ray enhancement factor of 95, as calculated above, yields the upper limit

$$F_\gamma \lesssim 1.8 \times 10^{-7} \text{ ph cm}^{-2} \text{ s}^{-1}. \tag{48}$$

This is too small to be identifiable by COS-B, unless the local cosmic-ray intensity, f_0, is substantially greater than the solar value.

In Figure 13 we have compared the observed density contours of some HV clouds with the gamma-ray sky map (Mayer–Hasselwander et al., 1981). There are also two GRO sources in this region (2CG121 + 04 and 2CG135 + 01). We see that there is no obvious correlation between HV clouds and gamma ray intensity, in agreement with the negative result from our calculations.

However, for a more sensitive gamma-ray detector, or for HV clouds which happen to pass through a young SNR, these objects may be important tracers for cosmic rays. Since HV clouds are observed at high galactic latitudes, they may perhaps one day be used to probe those regions of the Galaxy for cosmic rays and energetic processes.

During their passage through the galactic disc, HV clouds may encounter giant molecular clouds, or at least, denser matter. Such interactions are potential GRO's for two reasons: (1) the enhanced cosmic-ray intensity 'convected' by such clouds is transported to a more massive absorber which it can then illuminate, and (2) the kinetic energy set free in the form of an 'impact shock' is $> 10^{51}$ ergs. The size of the system is ~ 30 pc to start with and adiabatic expansion losses are negligible. Cosmic-ray enhancement at a (practically) parallel shock can take place and local enhancement of the gamma-ray luminosity results.

Tenorio–Tagle (1980, 1981) has studied the effects of collisions between a HV cloud and the galactic disc, both analytically and numerically. From an examination of the

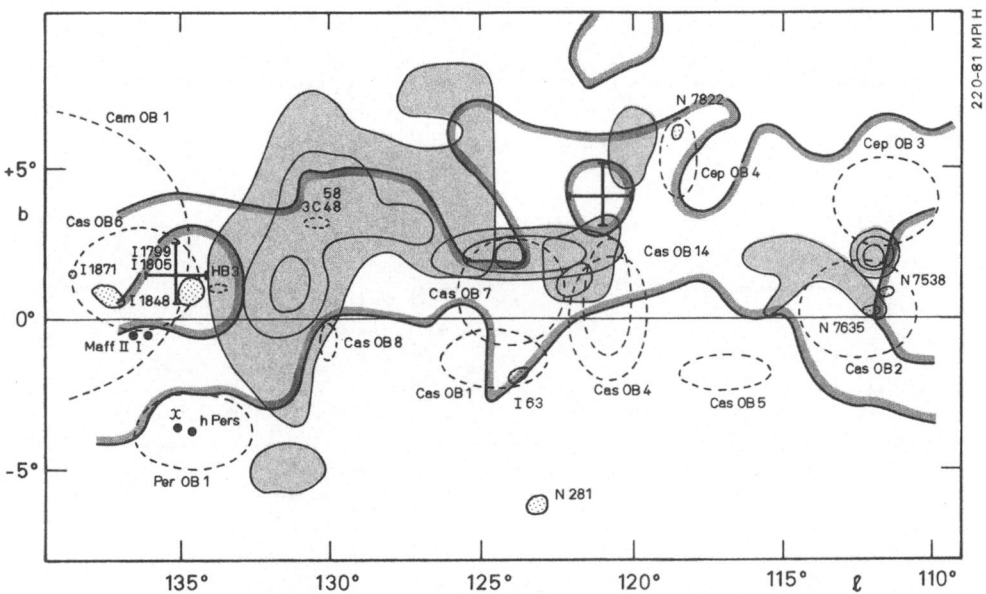

Fig. 13.   Comparison of observed column density contours of high velocity clouds with the gamma-ray sky map I ($E_\gamma > 100$ MeV) (from Hulsbosch, 1975; Morfill *et al.*, 1983). Low latitude HV clouds are shaded, gamma-ray contours are given by the heavy solid lines. H II regions are shown (identifiable by NGC and IC numbers), OB associations (dashed lines) and two SNR's (3C58 and HB3).

allowed parameter space (cloud velocities, sizes, densities and galactic disc local thickness, gas density) we obtain the result that in most cases HV clouds may either punch a hole through the disc, transferring only a small amount of energy and momentum, or they may coalesce with the disc, provided it is locally sufficiently massive and can cool rapidly. Then they transfer all their energy and momentum (see also Tenorio–Tagle *et al.*, 1983).

In Figures 14 and 15 we see two such calculations (Tenorio–Tagle *et al.*, 1983). A meridional cross section of a cloud-disc collision is shown at various stages. The 'age' is shown in units of 10^6 yr in each 'snapshot', and we see in Figure 14 how the cloud punches right through the disc. Cooling was taken into account using the Cox and Daltabuit (1971) relationship. The galactic disc thickness was 100 pc, and cloud and disc densities were taken equal at 1 particle cm^{-3}. Cloud velocity was chosen as 200 km s^{-1}.

In Figure 15 another collision sequence is shown. The essential difference w.r.t. Figure 14 is the cloud density, which was taken here to be 1/10th of the galactic disc value, and the velocity was only 100 km s^{-1}. As can be seen, the cloud is absorbed by the galactic disc. A dense shell is pushed into the disc, and at some stage a reverse shock is set up leading to some mass flow out of the 'crater'. This is different to the case discussed in Figure 14.

Nevertheless, we see in both cases that the enhanced cosmic-ray intensity, which is possibly associated with high velocity clouds, can be swept into the 'target'. If the target

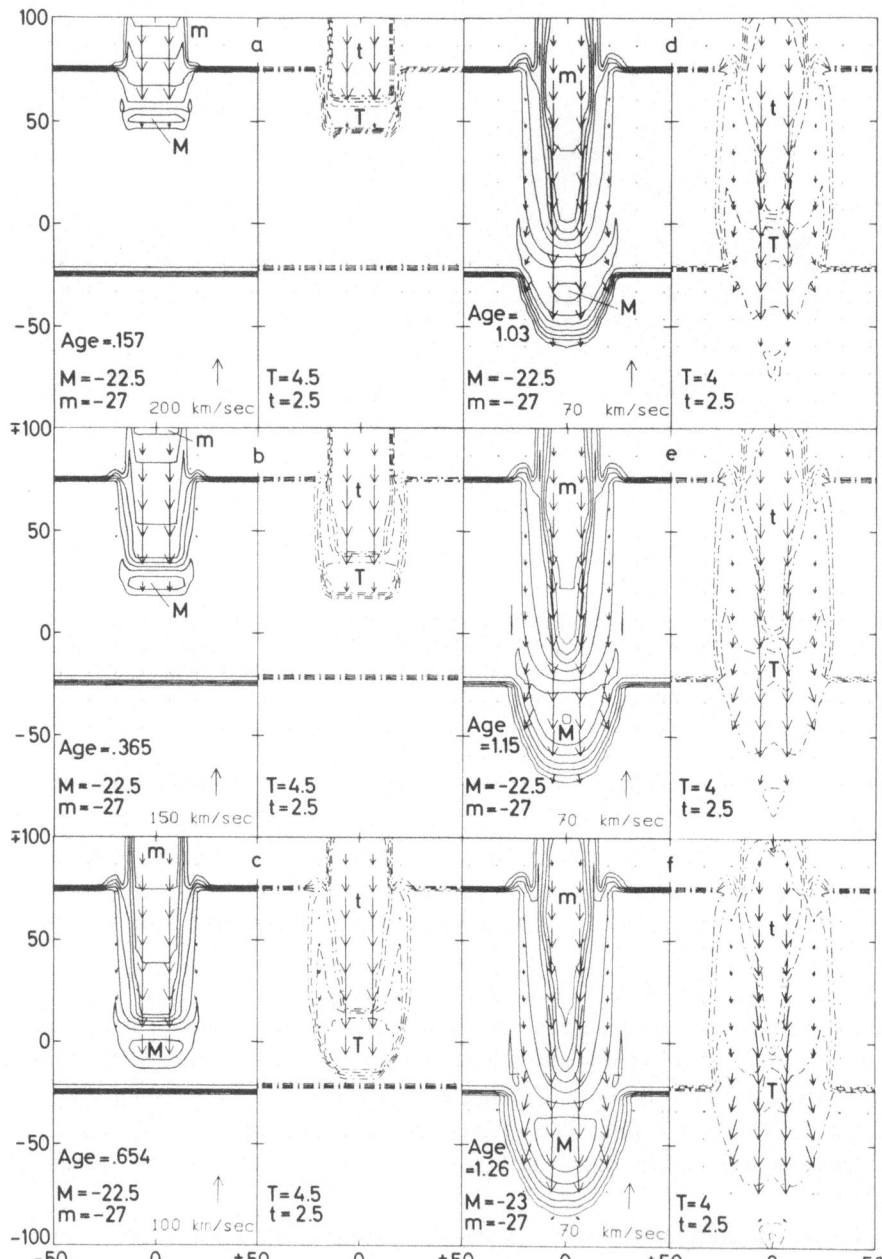

Fig. 14. Cloud-Galaxy Interaction. Meridional cross-section of a cloud–galatic disk collision. The coordinate axis give cylindrical (R, Z) distances in parsecs, and the Z-axis $(R = 0)$ is the axis of symmetry. Isodensity contours and velocity vectors (length proportional to speed) are plotted for six different evolutionary times in one column. In the second column we plot constant temperature contours. Numerical values for the minimum (density, m, temperature, t) and maximum (M, T) contour levels (log ρ_{min}, log T_{min}, log ρ_{max}, log T_{max}, respectively) are also given in each frame. The evolutionary time (Age) is expressed in units of 10^6 yr. In the lower right corner the velocity scale is given for the standard length arrow. Initial conditions: $\rho_{cloud} = 1.67 \times 10^{-24}$ g, $v_{cloud} = 200$ km s^{-1}, $\rho_{galaxy} = 1.67 \times 10^{-24}$ g, disk thickness = 100 pc.

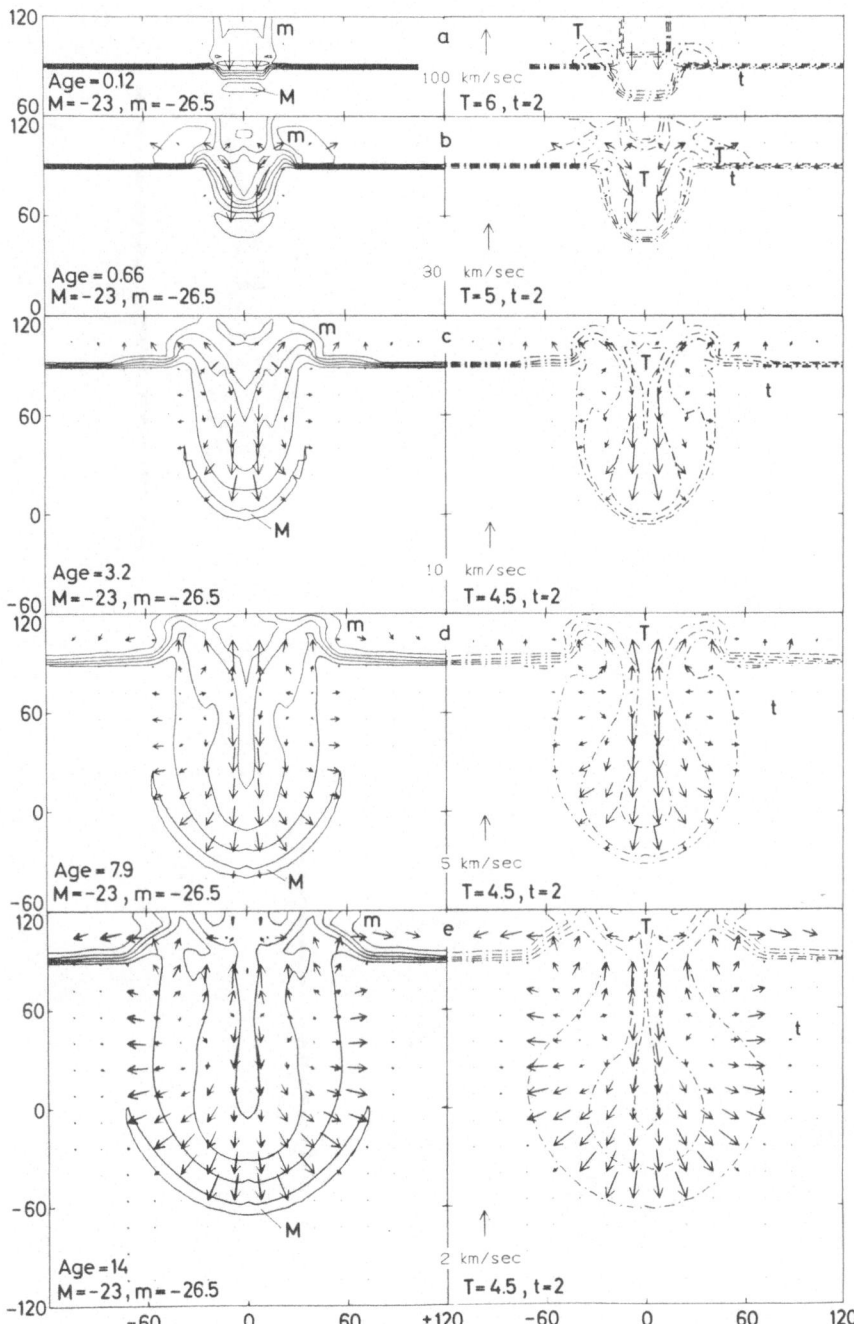

Fig. 15. Cloud-Galaxy Interaction. Initial conditions: $\rho_{\text{cloud}} = 1.67 \times 10^{-25}$ g, $v_{\text{cloud}} = 100$ km s^{-1}, $\rho_{\text{galaxy}} = 1.67 \times 10^{-24}$ g, disk thickness = 100 pc. All symbols and lines have the same meaning as in Figure 14.

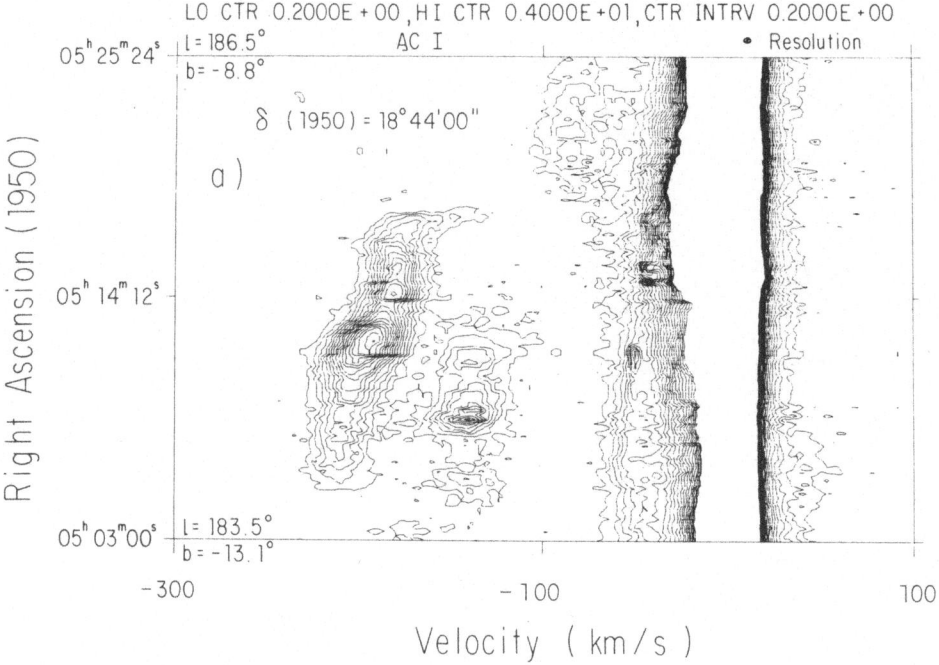

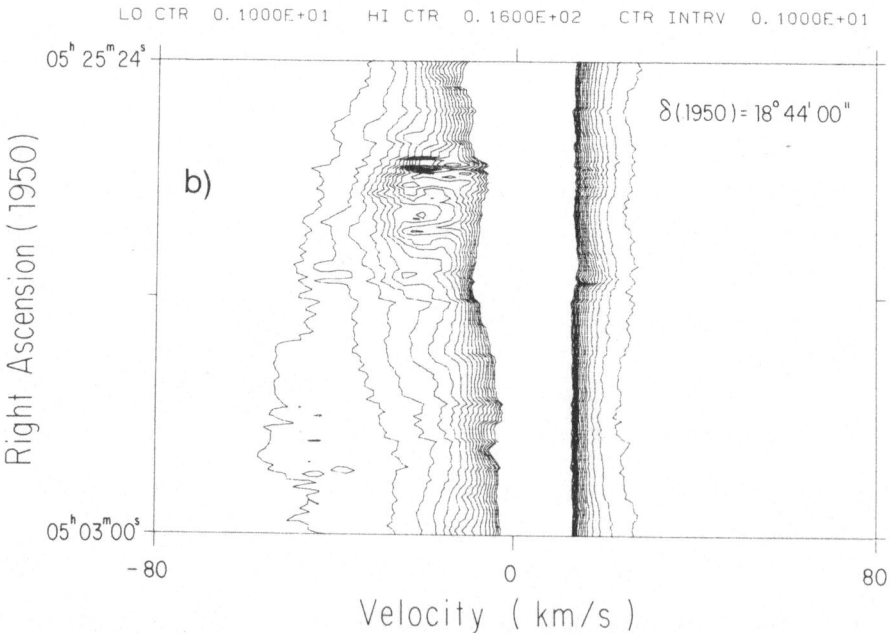

Fig. 16. (a) Velocity contours of neutral hydrogen observed in the direction of the galactic anticentre. (b) Velocity contours of perturbed galactic disc material, presenting evidence that the cloud is colliding with the disc (from Mirabel, 1982).

density is higher than the cloud density (e.g. as in Figure 15) then a local point-like gamma-ray source may result.

Recently, direct evidence for such a cloud-disc collision has been found with the Arecibo telescope (Mirabel, 1982). This is shown in Figures 16a, b. The location of this interaction region is towards the galactic anticentre ($l \approx 185°$, $b \approx -11°$). The angular size of the object is $\sim 2°$, its velocity is ~ 210 km s^{-1}. It contains a total mass of neutral hydrogen of a few $10^4 M_\odot$, which gives it a kinetic energy of the order 10^{52} ergs. It is suggested that this cloud is a segment of the 'Magellanic stream', i.e. matter drawn out from the Magellanic cloud, and that this matter is falling onto our Galaxy. At the interaction point with the galactic disc this extragalactic material produces an observable shock, and there is an abrupt deceleration of the infalling cloud.

From the point of view of comparing this exciting feature with the gamma-ray sky map, we are very unlucky. The area where the cloud-disc interaction takes place is masked by the gamma-ray emission from the Crab pulsar ($l = 184.5$, $b = -5.8$). Thus we are unable to determine, at the present time, whether there is an enhanced gamma-ray emission. A careful analysis of the COS-B data, made with the aim to define point like source emission limits in the interaction region, may be very useful. To do this, some technique has to be developed, which can remove the pulsed component of the nearby Crab.

7. Neutron Stars in Dense Clouds

In this paper we do not discuss 'active' neutron stars (see Salvati, 1983), but confine ourselves to the more 'inert' older objects (e.g. Kundt, 1981). The possibility that *old* neutron stars, whose spin and magnetic field have decayed away, may under special circumstances become GRO's was first discussed by Morfill and Zimmerman (1983).

There may be some 10^9 to 10^{10} old neutron stars in our Galaxy (see e.g. Bonazzola *et al.*, 1981) which are virtually undetectable except by their interaction with the ambient medium. One such signature may be gamma-ray bursts, and as we will show, another signature could be continuous enhanced gamma-ray emission.

Gamma-ray bursts may be caused by accretion of interstellar matter, leading to thermoculear runaway. Hydrogen and helium flashes produced in a system with low accretion rates ($M < 10^{-12} \dot{M}_\odot$ y^{-1}) were studied by e.g. Hameury *et al.* (1982), who found that instabilities may indeed occur when the accreted envelope reaches a mass of 3×10^{23} g.

Continuous gamma-ray emission may be caused by acceleration of cosmic rays at the accretion shock, coupled with adiabatic compression, and subsequent interaction of the enhanced cosmic rays with the accreting matter. For a discussion of the acceleration process see Cowsik and Lee (1981), who did not consider simultaneous losses, however.

7.1. ACCRETION ONTO NEUTRON STARS

In order to make the problem tractable, we assume spherical accretion (see e.g. Shakura, 1974; Börner, 1980, and references therein). We further assume that the neutron star is embedded in a sufficiently dense medium so that its accretion rate may become affected by the radiation pressure produced by the accretion near the Eddington limit, where gravitational and radiation forces balance. We have the Eddington luminosity

$$L_E \equiv 4\pi GMcm\sigma_T^{-1} = 1.2 \times 10^{38} \left(\frac{M}{M_\odot}\right) \text{ erg s}^{-1},\tag{49}$$

where G is the gravitational constant, M the neutron star mass, c the speed of light, m the proton mass, and σ_T the Thomson cross section. The corresponding mass accretion rate is

$$\dot{M}_E \equiv 4\pi cr_s m\sigma_T^{-1} = 1.0 \times 10^{18} R_6 \text{ g s}^{-1},\tag{50}$$

where R_6 is the shock radius in units of the neutron-star radius $R_* \equiv 10^6$ cm. The location of the magnetopause (and the accretion shock) has been calculated by e.g. Arons and Lea (1976), Michel (1977), and Elsner and Lamb (1977). It is given approximately by pressure balance

$$\frac{B^2(r_s)}{8\pi} \approx \frac{\dot{M}}{4\pi r_s^2} \sqrt{\frac{GM}{r_s}},\tag{51}$$

where $B \equiv B_0(R_*/r_s)^3$. Substituting $\dot{M} = \dot{M}_E$ yields

$$\left(\frac{r_s}{R_*}\right)^{2.25} \approx \frac{B_0}{1.4 \times 10^8}.\tag{52}$$

In other words, if the pulsar surface magnetic field, B_0, has decayed to values below $\sim 10^8$ G, the magnetosphere is compressed right down to the stellar radius. Whether such small magnetic field strengths are possible must be doubted (Kundt, 1981). On the other hand, a more reasonable value, $B_0 = 10^{10}$ G, corresponds to $r_s/R_* \approx 10$.

The radiation generated, when the accretion flow is thermalised, has two components, one coming from the shock at r_s the other from the infall onto the neutron star itself. (The problem of magnetospheric penetration of the accreting matter will not be discussed here (for references see e.g. Vasyliunas, 1979; Börner, 1980).). The radiation produced is 'Comptonised' leading to electron temperatures $\sim 10^8$ K (Zel'dovich and Shakura, 1969) or possibly even higher (1–10 MeV) according to Shapiro and Salpeter (1975).

Maraschi *et al.* (1978) considered the effect of radiation pressure on the accreting gas. In the case of steady accretion (steady in the sense that the boundary conditions of the external medium vary slowly over the free-fall time scale) they obtain the result that for an accretion rate $\dot{M} > 0.6 \dot{M}_E$, no shock can occur at any radius, because the flow is slowed down to be subsonic everywhere.

Since the acceleration process, and hence the efficient production of gamma rays, relies on the existence of an accretion shock (Cowsik and Lee, 1981) this represents a constraint to the model.

7.2. Environmental constraints

In order that accreting neutron stars can produce gamma rays and qualify as GRO's, several constraints have to be fulfilled. These are (see Morfill and Zimmermann, 1983):
   (i) There must be sufficient mechanical power in the accretion flow, i.e.

$$\dot{E}_{\text{mech}} > 10^{36} \text{ ergs s}^{-1}. \tag{53}$$

(ii) There must be an accretion shock. Consideration of flow modification by radiation pressure then leads to

$$\dot{E}_{\text{mech}} < 0.6 L_{\text{E}}. \tag{54}$$

(iii) The cosmic-ray acceleration time scale must be less than the loss time due to nuclear collisions with the ambient plasma, i.e.

$$\frac{4K}{V_S^2} < \tau_N(r_s), \tag{55}$$

where V_S, r_s are the accretion shock velocity and radius, respectively.
   Substituting $1/\tau_N \equiv 8 \times 10^{-16} n(r_s)$, and using the Eddington accretion rate, yields

$$n(r_s) = c/\sigma_T r_s V_S \approx 4.5 \times 10^{18} \left(\frac{R_*}{r_s}\right)^{1/2} \text{cm}^{-3} \tag{56}$$

and

$$K < 7 \times 10^{15} \left(\frac{R_*}{r_s}\right)^{1/2} \text{cm}^2 \text{ s}^{-1}. \tag{56a}$$

Writing this in the form of the mean free path ($K = \frac{1}{3}\lambda c$) we obtain from (56)

$$\lambda < 7 \times 10^5 \left(\frac{R_*}{r_s}\right)^{1/2} \text{cm}. \tag{57}$$

We can make a plausibility check here. We know that $\lambda > r_g$, the particle gyroradius. Also, for a particle of rigidity R (volts) we have $R = 300 B r_g$, where B is the ambient magnetic field in gauss and r_g is in cm. Hence inequality (57) implies

$$B > 4.8 R(\text{GV}) \left(\frac{r_s}{R_*}\right)^{1/2} \text{G}. \tag{58}$$

If we assume that flux freezing is not complete (e.g. Mouschovias, 1976; Stenholm et al., 1981) we may take $B \propto n^{1/2}$. Taking as reference values $B = 1\ \mu\text{G}$, $n = 1\ \text{cm}^{-3}$ yields $B(r_s) \approx 2 \times 10^3 (R_*/r_s)^{1/4}$ G. Hence we see that requirement (58) and thus (55) may be easily fulfilled, and thus there is no principal problem with the shock acceleration model in the accretion shocks considered here.

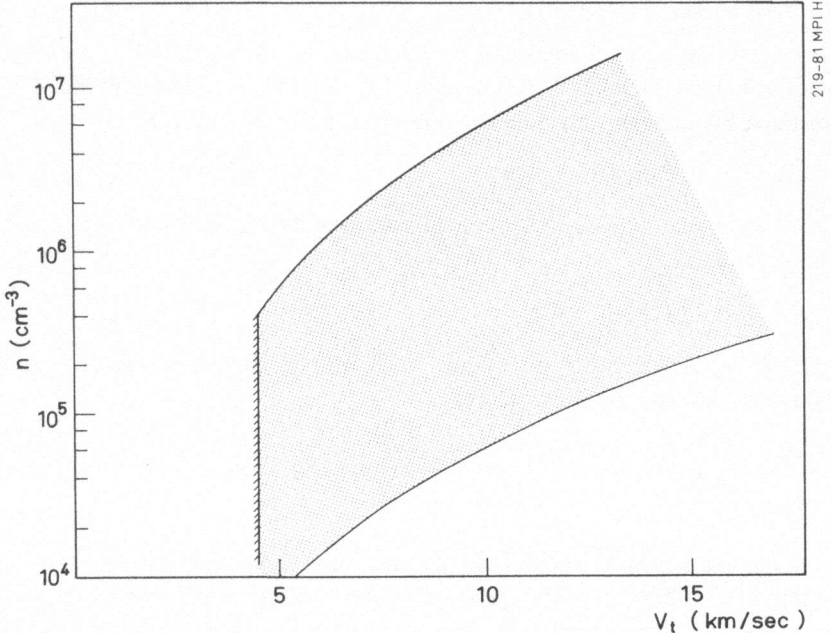

Fig. 17.   Conditions under which neutron stars could become gamma-ray emitters. The lower boundary corresponds to the minimum energy requirement, the upper boundary corresponds to a luminosity = $0.6L_E$ (Eddington limit). The ambient gas density is n, the velocity V_t is a combination of thermal gas speed and neutron star translational velocity (from Morfill and Zimmermann, 1983).

Figure 17 shows the range of ambient gas densities which are acceptable within these two constraints. We have chosen a neutron star mass $M = 1\,M_\odot$ and a shock radius $r_s = R_* = 10^6$ cm. The inequalities (53) and (54) translate into

$$4.8 \times 10^3 \left(\frac{r_s}{R_*}\right)\left(\frac{M}{M_\odot}\right)^{-2} > \frac{n}{V_t^3} > 62.5 \left(\frac{r_s}{R_*}\right)\left(\frac{M}{M_\odot}\right)^{-3}, \qquad (59)$$

where n is in cm^{-3} and V_t in km s^{-1}. We have used the Bondi (1952) accretion rate, which yields

$$n(r_s) = \frac{n}{\pi}\left(\frac{V_S}{V_t}\right)^3 \qquad (60)$$

in the expressions for $\dot{E}_{\text{mech}}$ to arive at (59). Equation (60) relates the accreting gas density at the shock, $n(r_s)$, to the density of the ambient medium, i.e. the cloud fragment, n. The velocity v_t is defined by

$$V_t^2 \equiv C_G^2 + V_{\text{NS}}^2, \qquad (61)$$

i.e., it is the mean of the thermal gas speed, C_G, and the translational neutron star velocity, V_{NS}, relative to the gas. For neutron stars embedded in dense clouds, we have to take into account that ionisation and heating gives an effective thermal gas speed of ~ 10 km s^{-1} (corresponding to $\sim 10^4$ K). Gravitational attraction of the cloud will impart a minimum translational velocity to the neutron star of ~ 5 km s^{-1}. Hence we must consider the range of densities corresponding to values of V_t above $V_t \approx 10$ km s^{-1} in Figure 17. These densities are in excess of $\sim 10^5$ cm^{-3}. This may be considered as protostellar cloud values.

7.3. Occurrence frequency

It is, of course, difficult to obtain a reliable estimate of the number of interstellar cloud fragments with densities above 10^5 cm^{-3}, and thus their volume filling factor. At the same time the spatial density of neutron stars, n_{NS}, is also highly uncertain. We shall choose $n_{NS} = 10^{-2}$ pc^{-3} (see Hameury *et al.*, 1982), corresponding to $\sim 10^9$ neutron stars.

For the dense cloud fragments we proceed in the following way: if we identify these fragments as protostellar clouds at a very early stage, perhaps the best way to estimate the volume filling factor is via the current rate of star formation in our Galaxy. This is estimated to be

$$G = (2–10) \times 10^{-11} \text{ stars pc}^{-3} \text{ yr}^{-1}, \tag{62}$$

(e.g. Schmidt, 1963; Tinsley, 1980). The protostellar evolution life time is a few free-fall times, governed by viscous accretion from a disc (e.g. Tscharnuter, 1980; Lin, 1982; Morfill, 1983b); we take

$$\tau_* \approx 10^6 \text{ yr} \tag{63}$$

Hence the density of protostellar clouds in our Galaxy should be

$$n_* = G\tau_* \approx (2–10) \times 10^{-5} \text{ protostars pc}^{-3}. \tag{64}$$

We assign a typical radius $r_* \approx 0.1$ pc and a mean gas density $n \approx 10^5$ cm^{-3} to the protostellar cloud.

The filling factor for protostellar clouds is then

$$f_* = n_* \tfrac{4}{3} \pi r_*^3 \approx (0.8–4) \times 10^{-7}. \tag{65}$$

Bearing in mind that the scale height of old neutron stars is expected to be larger (~ 200 pc) than that of interstellar clouds (~ 50 pc), we can calculate the number of chance coincidences. For 10^9 neutron stars, and an overlapping volume of $\tfrac{1}{4}$ (= ratio of scale heights) we obtain the current number of GRO's in the galaxy $N_{GRO} = 20–100$. Clearly this number, however large the uncertainties may be, is not intrinsically negligible. Statistically, therefore, this type of GRO cannot be excluded.

The life time of an individual source is of the order $r_*/V_{NS} \lesssim 10^4$ yr. In addition, the sources have to be regarded as compact, having a dimension of the order a few r_s. Finally, it is reasonable to expect these sources to be time variable on time scales of the order months or so, reflecting perhaps density inhomogeneities over distance scales of astronomical units, or variations in gas velocities, or the selfregulating processes for the accretion flow (radiation pressure) may work intermittently.

7.4. CORRELATIONS WITH OTHER OBSERVATIONS

The buried neutron star represents an extremely compact H II source. Figure 18 shows the emission spectrum from such an object which is accreting matter at the rate of $0.1\ \dot{M}_E$. The figure is taken from Shapiro and Salpeter (1975).

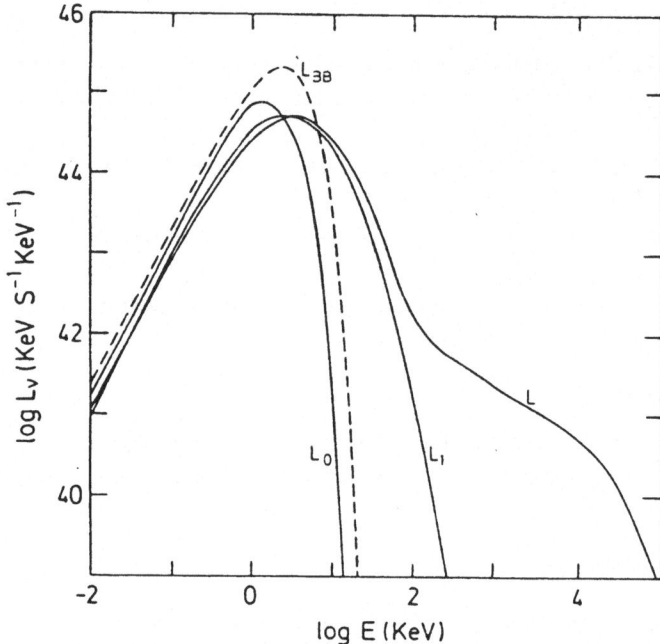

Fig. 18.   Emission spectrum from an accreting neutron star, including Comptonisation effects. The mass accretion rate is $0.1\ \dot{M}_E$. In our case this object is buried inside a dense cloud, and absorption effects are important (from Shapiro and Salpeter, 1975).

In our case we have to consider this object buried deeply inside an interstellar cloud, with perhaps a hydrogen column density of some $10^{22}\ cm^{-2}$ separating it from the outside world. The radiation emitted by the neutron star is largely absorbed by the dust and gas of the surrounding cloud (see Morfill and Zimmermann, 1983) and then re-emitted in the infrared at wavelengths of ~ 10 to $100\ \mu m$. Some hard X-rays may escape and provide an observable counterpart.

Gamma-ray lines (0.5 to 10 MeV) may be observable also, because the kinetic energy of the accreting nucleons is of the order 100 MeV and collisional excitation may occur (Morfill and Zimmermann, 1983).

Finally, it was mentioned already that these sources may be time variable over time scales of months. Statistical evidence for time variable GRO's exists (Bloemen *et al.*, 1981), although direct observations are less obvious. Hermsen (1980) concluded that there was evidence for time variability in the Cygnus region. This (possibly) variable source has been localised at the position $l = 79°$, $b = -1°$ (Bloemen et al., 1981). It could be located inside the giant molecular cloud Cyg OB9 which has an estimated mass of $\sim 10^5 M_\odot$ and is located at a distance of ~ 1.2 kpc away from the Sun (Stark and Blitz, 1978).

8. Cloud Collapse

Chevalier (1977) suggested that compression (by cooling) of the gas behind an old SNR shock could lead to cosmic-ray enhancement by the betatron mechanism. Since in this model cosmic rays are enhanced throughout the collapsing cloud, a similarly enhanced gamma-ray emission should occur. In principle this could be another type of GRO. Whilst it is not clear whether the 'dense shell' stage of SNR's occurs if the interstellar medium is clumpy (McKee and Cowie, 1975; McKee and Ostriker, 1977), there is observational evidence that it happens in some instances (e.g. van den Bergh et al., 1973; Chevalier et al., 1976).

Here we briefly examine the necessary requirements for this process to work. A full discussion is found in Elitzur and Morfill (1983).

8.1. GENERAL REQUIREMENTS

The process will not work for clouds collapsing (or contracting) under the influence of self gravity. This can be seen quite easily by comparing the loss time scale due to nuclear interactions with the adiabatic compression time scale = cloud contraction time scale. We use observations of interstellar clouds (Stenholm et al., 1981) to determine this quantitatively. From radio line emission of a number of molecules, Stenholm et al. (1981) constructed a 'typical' cloud model. The relevant features for us are a mean density $n_c \approx 10^4$ cm^{-3}, mean radius $R_c = l/2 \approx 10^{19}$ cm, mass $\approx 6 \times 10^4 M_\odot$ and a systematic collapse velocity $V_{\mathrm{sys}} \approx 1$ km s^{-1}. The nuclear collision loss rate is

$$1/\tau_N \approx 8 \times 10^{-16} n_c \approx 8 \times 10^{-12} \text{ s}^{-1} \tag{66}$$

and the cloud compression time is

$$\tau_c \approx R_c/V_{\mathrm{sys}} \approx 10^{14} \text{ s}. \tag{67}$$

The ratio $\tau_c/\tau_N \approx 10^3$, which implies that losses dominate by far.

The slow gravitational contraction of massive interstellar clouds does not lead to cosmic-ray enhancement for this reason. Much faster collapse times are required.

The conditions which must be fulfilled for the betatron mechanism to become effective, are described by the following inequality

$$\Omega^{-1} \ll \tau_c \ll l/c. \tag{68}$$

Here $\Omega = eB/mc$ is the gyrofrequency of the cosmic rays, and l/c is (approximately) the time spent by cosmic rays inside a cloud of size l.

However, even then it is not clear whether the system may not relax back rather quickly into a new equilibrium or even become unstable for a while. The reason is the following:

The magnetic field is frozen into the charged component of the interstellar cloud. The ions exchange momentum with the bulk of the cloud (mainly H_2) via ion-neutral collisions and are thus only loosely influenced. This implies that the magnetic field is not perfectly 'tied' to the bulk matter, in disequilibrium it can 'diffuse' out of a compression region or into a rarefaction region – a process known as 'ambipolar diffusion' (e.g. Scholer, 1970). Ionisation is maintained by cosmic-ray interactions, the cosmic rays themselves are also constrained to follow the magnetic field. Hence the various components are internally coupled and a deviation from equilibrium may lead to a nonlinear instability involving all components.

8.2. FORMULATION OF THE COUPLED PROBLEM

The coupled set of equations which has to be solved in order to describe the problem is:

(i) *Magnetic field*

$$\partial \mathbf{B}/\partial t = \operatorname{curl} (\mathbf{v}_d \times \mathbf{B}), \tag{69}$$

where $\mathbf{v}_d$ is the ion drift velocity with respect to the neutral gas

(ii) *Ion equation of motion*

$$n_i m_i \frac{\partial \mathbf{v}_d}{\partial t} = -n_i n_n \sigma_{in} V_{in} \mu_{in} \mathbf{v}_d + \nabla \left(\frac{B^2}{8\pi}\right), \tag{70}$$

subscripts i correspond to ions, n to neutrals; σ_{in}, V_{in}, and μ_{in} are the ion-neutral cross section, relative thermal speeds and reduced mass, respectively. The second term describes ion-neutral friction, the last term magnetic pressure.

(iii) *Ion density*

$$\frac{\partial n_i}{\partial t} = \alpha_p n_n n_c - \alpha_r n_i^2 - \operatorname{div} (n_i \mathbf{v}_d). \tag{71}$$

The second term describes ion production by cosmic-ray collisions with the neutral gas, the third term describes ion losses by recombination and the last term describes 'convection' losses due to ambipolar diffusion of the magnetic field (Equation (69)). α_p is the rate constant for ion production, α_r is the recombination coefficient and n_c is the ionising (i.e. low energy) cosmic-ray particle density.

(iv) *Cosmic-ray density (ionising particles)*

$$\frac{\partial n_c}{\partial t} = \frac{1}{\tau_s} (n_0 - n_c) - n_n n_c \alpha_L - \operatorname{div} (n_c \mathbf{v}_d). \tag{72}$$

The second term describes the ionising cosmic-ray source (see later), the third term is complementary to the second term in Equation (71) – it describes cosmic-ray losses by ionisation (since many ionisations can be produced by a single cosmic-ray particle, $\alpha_L < \alpha_p$) and the fourth term again describes 'convection' losses (or gains) via ambipolar diffusion of the magnetic field.

The cosmic-ray source term is described by the time scale

$$\tau_S \equiv l/V_A, \tag{73}$$

where we have assumed that the ionising particles stream into the cloud, where they are absorbed, with Alfvén speed V_A, (e.g. Skilling and Strong, 1976; Cesarsky and Völk, 1978; Morfill, 1982, 1983a) and that the gradient length scale is typically of the size of the cloud itself (the only external length scale in the problem). The external cosmic-ray intensity, in equilibrium, is n_0.

The second term in Equation (72) ensures the approach to equilibrium for the cosmic rays, with a time scale τ_S. For instance, if we suddenly increase the cosmic-ray density far above the equilibrium value n_0, cosmic rays will stream *out* of the cloud with Alfvén speed – the flow is reversed.

(v) *Relativistic cosmic-ray density*

$$\frac{\partial n_{CR}}{\partial t} - K \frac{\partial^2 n_{CR}}{\partial x^2} = 0, \qquad |x| > 0, \tag{74}$$

$$2K \frac{\partial n_{CR}}{\partial x} = v_d n_{CR}, \qquad |x| = 0. \tag{75}$$

Equation (74) describes the transport in the intercloud medium, and Equation (75) describes the 'losses' by expansion with the ambipolar diffusion speed, v_d, inside the cloud. The boundary conditions are

$$n_{CR}(|x| \to \infty) = 0, \tag{76}$$

$$n_{CR}(x > 0, t = 0) = 0, \tag{77a}$$

$$n_{CR}(x = 0, t = 0) = N_0 - N_{CR}. \tag{77b}$$

As can be seen, we have formulated the problem in such a way that n_{CR} is the excess (or decrease) initially associated with processes inside the cloud. For a fast compression $N_0 \geqslant N_{CR}$, for a fast expansion $N_0 \leqslant N_{CR}$, the intercloud value. Relativistic cosmic rays are dynamically and chemically unimportant in this scenario – they simply follow the magnetic field, which in turn is governed by the ions and their interactions. We may therefore solve the problem in two parts, firstly dealing with the coupled set of Equations (69) through (73). This will yield the ambipolar diffusion speed, $v_d(t)$, of the magnetic field, which can then be used to solve for the relativistic cosmic rays.

8.3. SOLUTION OF THE COUPLED PROBLEM

Elitzur and Morfill (1983) have described a simplified method for solving this coupled system of equations. The geometry was that of a slab, length l, which has collapsed to

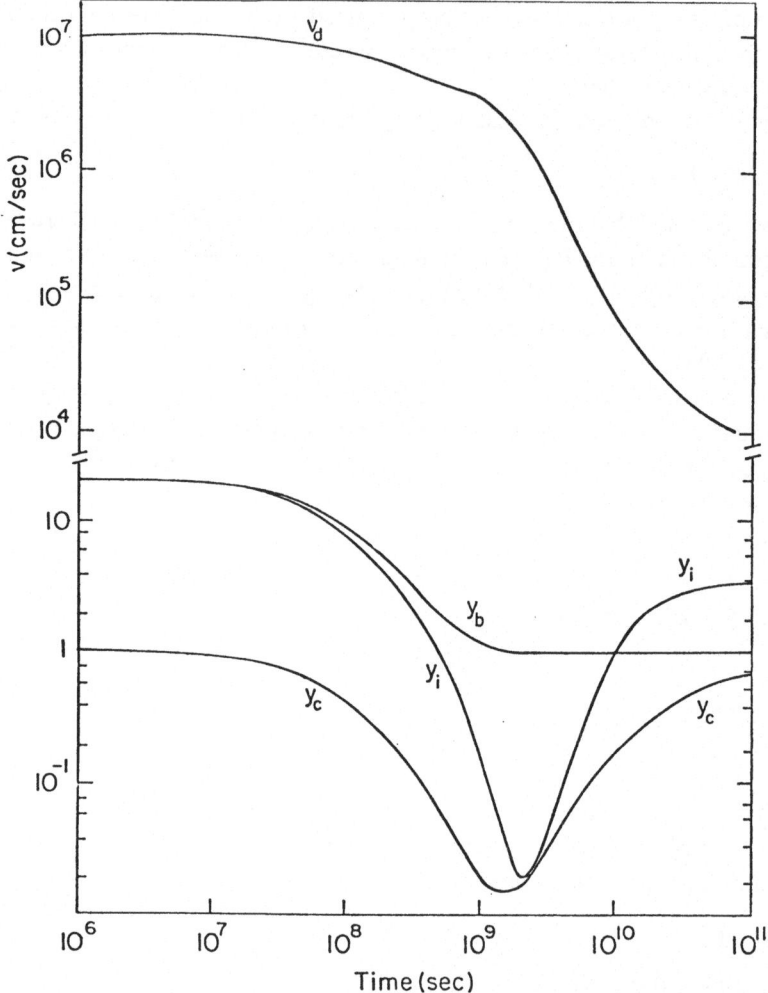

Fig. 19.   Temporal variation of ion density, y_i, cosmic-ray density, y_c, magnetic field strength, y_B, and ambipolar diffusion speed, v_d, in an interstellar cloud which has undergone a compression by a factor 20. For the normalisation see text (from Elitzur and Morfill, 1983).

a thickness, D, perpendicular to the magnetic field. The field lines then 'fan out' from the ends of the slab until the magnetic field strength reaches its average intercloud value. D is thus the scale size of the field compression, and inserting this, yields a set of coupled first order equations which can be solved fairly easily. We will not dwell on the mathematical details here, instead we present a result of the calculations in Figure 19.

In that particular calculation we have assumed that the cloud collapse time $\tau_c > l/V_c$, the time spent by *ionising* cosmic rays in the cloud, so that initially

$$y_c \equiv n_c/n_0 = 1. \tag{78}$$

Parameters used for the purposes of this calculation were:

Compression factor: $X = 20$.

Compression scale: $D = 10^{15}$ cm.

Cloud 'size': $l = XD = 2 \times 10^{16}$ cm.

Cosmic ray ionisation frequency: $\alpha_p n_c = 10^{-17} \, n_c/n_0$ s^{-1}.

Ion-neutral collision rate: $n_n \sigma_{in} V_{in} = 10^{-9} \, n_n$ s^{-1}.

Ion-recombination coefficient: $\alpha_r = 10^{-7}$ cm^3 s^{-1}.

Cosmic-ray supply time scale: $l/V_A = 2 \times 10^9$ s.

Cosmic-ray ionisation loss rate: $\alpha_L n_n = 10^{-15} \, n_n$ s^{-1}.

Representative ion mass: $m_i = 10$ m (proton masses).

The initial cloud density was chosen to be $n_{n0} = 10^4$ cm^{-3}, and the equilibrium magnetic field $B_0 = 3\sqrt{n_{n0}}$ μG. Shown in Figure 19 are the drift velocity, v_d, and the quantities:

$$y_c \equiv n_c/n_0, \tag{79}$$

$$y_i \equiv n_i/n_{i0}, \tag{80}$$

$$y_b \equiv B/B_0. \tag{81}$$

A ratio $l/D = 20$ implies that the magnetic field must be perpendicular to the compression direction to within $3°$. The cosmic-ray residence time in the cloud is then $l/c \approx 2 \times 10^{-2}$ yr, this must be compared with the cooling time for a shocked gas (Cox, 1972) of

$$\tau_{\text{cool}} \approx 10^3/n_{n0} \text{ yr.} \tag{82}$$

This is somewhat larger than the cosmic-ray residence time, and hence our initial condition was $y_c = 1$.

The evolution of the system then follows from Figure 19. At first the magnetic field begins to diffuse out of the cloud, the ambipolar diffusion speed reaching O (100 km s^{-1}). This implies that secondary impact ionisation may occur, which would slow the ambipolar diffusion down again somewhat. (In the calculations performed so far this has not been considered, however.) The magnetic field removes both ions *and* ionising cosmic rays from the cloud. As can be seen, this removal is faster than the replenishment, and the system 'overshoots', before returning to its new equilibrium value.

Such a behaviour appears to be typical, although a steady transition to the new equilibrium (over a time scale $\sim l/V_A$) can also occur. Internal processes, i.e. gas – ions – magnetic field – cosmic-ray interactions, determine the relaxation of the system. In general the relaxation time is short, of the order a few 100 yr. This implies for relativistic cosmic rays (which are simply guided by the magnetic field inside the cloud without having any dynamical importance) that an enhancement, if it occurs, is at least similarly short-lived. Diffusion losses may decrease a relativistic cosmic-ray enhancement even faster.

8.4. RELATIVISTIC COSMIC RAYS

For relativistic cosmic rays we have to differentiate two situations:

(i) If the cloud compression occurs sufficiently fast ($\tau_c < l/c$) so that even relativistic cosmic rays are enhanced, as suggested by Chevalier (1977), there may be cosmic-ray streaming – and as a result cosmic-ray escape from the cloud may be limited to the Alfvén speed (e.g. Lerche, 1967). Then the relativistic cosmic rays behave in exactly the same way as the ionising cosmic rays – with the exception that they are practically not absorbed in the cloud. The time scale for relaxing back to the equilibrium value is again l/V_A. The 'overshoot' would not, however, affect relativistic cosmic rays, because they will simply stop streaming and making waves once the intensity inside and outside the cloud has evened out sufficiently. (The 'pulse' of wave activity produced during the initial streaming out of the cloud may continue to hinder cosmic-ray access for a while, so that some 'overshoot' may occur, but this has not been investigated so far.) Thus we may expect at most a very short lived cosmic-ray increase in a collapsing, shocked, cloud – of a duration O (few 100 yr).

(ii) If the cloud compression is too slow ($\tau_c > l/c$) cosmic-ray transport is diffusive at all times. It is described by Equations (74) and (75) with the boundary conditions (76) to (77b). This system of equations can be solved (see Carslaw and Jaeger, 1959, p. 358). The solution is, for a constant value v_d (see Figure 19):

$$n_{CR}(x = 0, t > 0) = \frac{1}{\sqrt{\pi K t}} - \frac{v_d}{K} e^{v_d^2 t / K} \operatorname{erfc}\left(\sqrt{v_d^2 t / K}\right). \tag{83}$$

We only consider the relativistic cosmic rays inside the cloud ($x = 0$). The first term, which decays as $t^{-1/2}$, is simply due to the diffusive spread of a suddenly released particle population, the second term is due to the convection of particles out of the cloud with the magnetic field. $\operatorname{erf} c(z) = 1 - \operatorname{erf}(z)$, where $\operatorname{erf}(z)$ is the error function with argument z. Using our previously computed value for v_d as a guide (10^7 cm s^{-1}) and taking K to be the canonical cosmic-ray diffusion coefficient (10^{28} cm^2 s^{-1}), since there is no compelling reason why some different value should be preferred, we see that for all time scales of interest $v_d^2 t / K \ll 1$. Hence we can simplify (83)

$$n_{CR}(x = 0, t > 0) \approx \frac{1}{\sqrt{\pi K t}} - \frac{v_d}{K}\left(1 + \frac{v_d^2 t}{K}\right)\left(1 - 1.125 \frac{v_d^2 t}{K}\right)$$

which becomes

$$n_{CR}(x = 0, t > 0) \approx \frac{1}{\sqrt{\pi K t}} - \frac{v_d}{K}\left(1 - 0.125 \frac{v_d^2 t}{K}\right). \tag{84}$$

We see immediately, that the second term (due to ambipolar diffusion losses) dominates if $v_d^2 t / K > 1/\pi$. Since for all reasonable time scales of interest ($t < 10^3$ yr) this condition cannot be reached, we conclude that relativistic cosmic rays are unaffected by the cloud relaxation, at least as long as the local diffusion coefficient is not anomalously low.

The short duration time of such cosmic-ray increases (at best a few 100 yr) makes it very unlikely that they could be important contributors either to the galactic gamma-ray emission, or the GRO's. Taking a supernova rate $1/\tau_{SN} = 1/50 \, \text{yr}^{-1}$, and one collapsing cloud per SNR, yields a total number of such objects in the Galaxy of 5 or so. We have no observational evidence as yet about the frequency of occurrence of sufficiently massive, dense, SN shells – our chosen value of 1 per SNR is, therefore, highly uncertain.

In spite of the expected rarity of such collapsing, cosmic ray producing SN shells, they provide an interesting location for a number of physical processes and interactions, and it seems worthwhile to investigate the few known cases in greater depth. For some additional consequences (e.g. for molecular species) see e.g. Elitzur and Morfill (1983).

9. η Carina

This object was briefly discussed earlier in conjunction with the suggestion by Völk (1981) that it could be a partially buried SNR, lighting up a giant molecular cloud with an enhanced cosmic-ray intensity.

Here we briefly describe an alternative suggestion made by Woosley et al. (1980). We mention this suggestion not because it is in any way well supported by the observations (which are still inadequate – as is the theory, possibly), but because the suggestions have interesting consequences. This was pointed out in a recent paper by Hillebrandt (1982).

The observations of η Carina are apparently consistent with a $\sim 10 \, M_\odot$ star, which has reached its Ne-burning stage and has ejected some $8.5 \, M_\odot$ of envelope with a velocity $V \approx 3{-}5 \times 10^7 \, \text{cm s}^{-1}$. The present size of the envelope is $10^{16} \, \text{cm}$, the luminosity is $\sim 7 \times 10^{38} \, \text{erg s}^{-1}$ (Weaver and Wooseley, 1980). Evolutionary calculations following this stage have shown that the star turns into a supernova, ejecting some $0.06 \, M_\odot$ of matter and releasing some 5×10^{50} ergs in energy. The core mass which remains is $1.44 \, M_\odot$ (Hillebrandt, 1982).

The interesting aspect is that the evolution after the Ne-burning stage takes only a few years !!

The chance that the identification and the theoretical description through the various evolutionary stages are both 'on target', may only be a slim one. Nevertheless, the peculiar object η Carina deserves further study from many different branches of astronomy.

10. Conclusions

In this paper we have discussed astrophysical situations which may give rise to gamma-ray objects (GRO's) with the rough characteristics of the COS-B sources. We have concentrated on point-like objects, which invariably meant situations involving interstellar clouds. Other possibilities, e.g. compact sources, superposition of clouds along the line of sight etc. are discussed elsewhere in this volume (Salvati, 1983; Houston *et al.*, 1983).

We have found a number of possible situations which might yield GRO's, and have tried to assess their luminosity, statistical significance and other observable properties. These properties aid the identification and search for counterparts by giving a set of constraints – sometimes in the infrared emission or in X-rays, sometimes in radio line features or in gamma-ray lines. It is certainly apparent from the survey of point-like sources that GRO's seem to have relatively little difficulty in 'hiding' from us at other wavelengths – especially if energetic processes are going on deeply inside interstellar clouds, the GRO counterpart may appear insignificant at other wavelength ranges. The large COS-B field of view ($\sim 1°$ square) then does not make the task of identification any easier.

The main aims for future gamma-ray missions are then to increase the counting statistics and verify the sources, to improve the determination of GRO positions, to extend the energy range downwards and obtain spectral information, to reduce the limits on angular size and perhaps resolve some of the point-like sources, to obtain light curves showing time variability for individual objects and to improve our statistical picture (e.g. log N – log S curves).

At the same time correlative studies with other branches of astronomy should be pursued (e.g. X-rays, infra-red, nonthermal radio, radio lines, etc.) and tested against special model predictions.

Finally, our theoretical understanding of physical processes and their sometimes complicated interactions has to be improved in order to utilise all the information about GRO's which is at our disposal, and to understand it.

Acknowledgements

We wish to acknowledge fruitful discussions over a prolonged time with Drs G. Bignami, P. Caraveo, L. Drury, M. Forman, T. Hartquist, and H. Völk on many topics related to the subject of this paper.

The editors thank W. Hermsen for assistance in evaluating this paper.

References

Arons, J. and Lea, S.: 1976, *Astrophys. J.* **207**, 914.
Axford, W. I.: 1980, *Proc. IAU/IUPAP Symp.* **94** (Bologna), 339.
Axford, W. I.: 1981a, *Proc. 10th Texas Symp. on Relat. Astrophys. (Baltimore), Ann. N.Y. Acad. Sci.* **375**, 297.
Axford, W. I.: 1981b, European Space Agency SP-161, p. 425.
Axford, W. I., Leer, E., and Skadron, G.: 1977, *Proc. 15th Int. Cosmic Ray Conf. (Plovdiv)* **11**, 132.
Bignami, G. F. and Hermsen, W.: 1982, *Ann. Rev. Astron. Astrophys.*, in press.
Bignami, G. F., Caraveo, P., and Maraschi, L.: 1978, *Astron. Astrophys.* **67**, 149.
Black, D. C. and Bodenheimer, P.: 1975, *Astrophys. J.* **199**, 619.
Bell, A. R.: 1978a, *Monthly Notices Roy. Astron. Soc.* **182**, 147.
Bell, A. R.: 1978b, *Monthly Notices Roy. Astron. Soc.* **182**, 443.
Blandford, R. D.: 1979, *Am. Inst. Phys. Conf. Proc. (La Jolla)* **56**, 335.

Blandford, R. D. and Ostriker, J. P.: 1978, *Astrophys. J.* **221**, L29.

Bloemen, J. B. G. M., Bennett, K., Caraveo, P. A., Hermsen, W., Kanbach, G., Masnou, J. L., Mayer–Hasselwander, H. A., Paul, J. A., Sacco, B., Strong, A. W., and Wills, R. D.: 1981, *17th Int. Cosmic Ray Conf. (Paris)* **9**, 64.

Bondi, H.: 1952, *Monthly Notices Roy. Astron. Soc.* **112**, 195.

Bonnazzola, S., Hameury, J. M., Heyvaerts, J., and Ventura, J.: 1981, *17th Int. Cosmic Ray Conf. (Paris)*, preprint.

Börner, G.: 1980, *Phys. Reports* **60**, 151.

Carslaw, H. S. and Jaeger, J. C.: 1959, *Conduction of Heat Solids*, 2nd Ed., Clarendon Press, Oxford.

Chevalier, R. A.: 1977, *Astrophys. J.* **213**, 52.

Chevalier, R. A., Robertson, J. W., and Scott, J. S.: 1976, *Astrophys. J.* **207**, 450.

Cesarsky, C. J. and Völk, H. J.: 1978, *Astron. Astrophys.* **70**, 367.

Cowsik, R. and Lee, M. A.: 1981, *17th Int. Cosmic Ray Conf. (Paris)* **2**, 318.

Cox, D. P.: 1972, *Astrophys. J.* **178**, 159.

Cox, D. P. and Daltabuit, E.: 1971, *Astrophys. J.* **167**, 113.

De Noyer, L. K.: 1979, *Astrophys. J.* **232**, L165.

Drury, L. O'C.: 1983, *Rep. Progr. Phys.*, in press.

Drury, L. O'C., Axford, W. I., and Summers, D.: 1982, *Monthly Notices Roy. Astron. Soc.* **198**, 833.

Eichler, D.: 1979, *Astrophys. J.* **229**, 419.

Elliott, K. H.: 1979, *Monthly Notices Roy. Astron. Soc.* **186**, 9P.

Elitzur, M. and Morfill, G. E.: 1983, in preparation.

Elsner, R. F. and Lamb, F. K.: 1977, *Astrophys. J.* **215**, 897.

Falle, S. A. E. G. and Garlick, A. R.: 1982, *Monthly Notices Roy. Astron. Soc.* **201**, 635.

Forman, M. and Morfill, G. E.: 1979, *Proc. 16th Int. Cosmic Ray Conf. (Kyoto)* **5**, 328.

Hameury, J. M., Bonnazzola, S., Heyvaerts, J., and Ventura, J.: 1982, submitted to *Astron. Astrophys.*

Hayakawa, S., Kato, T., Nagase, F., Yamashita, K., and Tanaka, Y.: 1978, *Astron. Astrophys.* **62**, 21.

Houston, B., Riley, P. A., and Wolfendale, A. W.: 1983, *Space Sci. Rev.* **36**, 155 (this volume).

Hermsen, W.: 1980, PhD. Thesis Univ. Leiden.

Hermsen, W.: 1983, *Space Sci Rev.* **36**, 61 (this volume).

Hillebrandt, W.: 1982, *Astron. Astrophys.* **100**, L3.

Hulse, R. A. and Taylor, H. J.: 1975, *Astrophys. J.* **201**, L00.

Hulsbosch, A. N. M.: 1975, *Astron. Astrophys.* **40**, 1.

Hovaisky, S. and Lequeux, J.: 1972, *Astron. Astrophys.* **18**, 169.

Kodaira, K.: 1974, *Publ. Astron. Soc. Japan* **26**, 255.

Jokipii, J. R.: 1983, *Space Sci. Rev.* **36**, 27 (this volume).

Kirschner, R. P. and Taylor, K.: 1976, *Astrophys. J.* **208**, L83.

Krymsky, G. F., Kuzmin, A. I., Petukhov, S. I., and Turpanov, A. A.: 1979, *Proc. 16th Int. Cosmic Ray Conf. (Kyoto)* **2**, 39.

Kulsrud, R. M. and Cesarsky, C. J.: 1971, *Astrophys. Letters* **8**, 189.

Kulsrud, R. M. and Pearce, W.: 1969, *Astrophys. J.* **156**, 445.

Kundt, W.: 1981, *Astron. Astrophys.* **98**, 207.

Lebrun, F., Bignami, G. F., Buccheri, R., Caraveo, P. A., Hermsen, W., Kanbach, G., Mayer–Hasselwander, H. A., Paul, J. A., Strong, A. W., and Wills, R. D.: 1982, *Astron. Astrophys.* **107**, 390.

Lerche, I.: 1967, *Astrophys. J.* **147**, 689.

Lin, D. N. C.: 1982, preprint.

Lozinskaya, T. A.: 1974, *Soviet Astron.* **17**, 603.

Maeder, A.: 1981, *Astron. Astrophys.* **99**, 97.

Maraschi, L., Reina, C., and Treves, A.: 1978, *Astron. Astrophys.* **66**, 99.

Mayer–Hasselwander, H. A., Bennett, K., Bignami, G. F., Buccheri, R., Caraveo, P. A., Hermsen, W., Kanbach, G., Lebrun, F., Lichti, G. G., Masnou, J. L., Paul, J. A., Pinkau, K., Sacco, B., Scarsi, L., Swanenburg, B. N., and Wills, R. D.: *Astron. Astrophys.* **105**, 164.

McKee, C. F. and Cowie, L. L.: 1975, *Astrophys. J.* **195**, 715.

McKee, C. F. and Hollenbach, D. J.: 1980, *Ann. Rev. Astron. Astrophys.* **18**, 219.

McKee, C. F. and Ostriker, J. P.: 1977, *Astrophys. J.* **218**, 148.

McKenzie, J. F. and Westpahl, K. O.: 1969, *Planet. Space Sci.* **17**, 1029.

Mezger, P. G.: 1978, *Astron. Astrophys.* **70**, 565.

Michel, C. F.: 1977, *Astrophys. J.* **216**, 838.

Mirabel, I. F.: 1982, *Astrophys. J.* **256**, 120.

Morfill, G. E. and Drury, L. O'C.: 1981, *Monthly Notices Roy. Astron. Soc.* **197**, 369.

Morfill, G. E.: 1982, *Monthly Notices Roy. Astron. Soc.* **198**, 583.

Morfill, G. E.: 1983a, *Astrophys. J.* **262**, 749.

Morfill, G. E.: 1983b, *Icarus* **53**, 41.

Morfill, G. E. and Meyer, P.: 1981, *Proc. 17th Int. Cosmic Ray Conf. (Paris)* **1**, 116.

Morfill, G. E. and Scholer, M.: 1978, AFGL-TR-77-0309 Special Report No. 209, p. 23 (Proceedings of the L. De Feiter Symposium, Tel Aviv, 1977).

Morfill, G. E. and Scholer, M.: 1979, *Astrophys. J.* **232**, 473.

Morfill, G. E. and Zimmermann, H.-U.: 1983, submitted to *Monthly Notices Roy. Astron. Soc.*

Morfill, G. E., Völk, H. J., Drury, L. O'C., Forman, M. A., Bignami, G. F., and Caraveo, P. P.: 1981, *Astrophys. J.* **246**, 810.

Morfill, G. E., Bignami, G. F., and Forman, M. A.: 1983, in preparation.

Mouschovias, T. Ch.: 1976, *Astrophys. J.* **206**, 753.

Oort, J. H.: 1951, *Problems of Cosmical Aerodynamics*, Dayton, Ohio Central Documents Office, p. 118.

Parker, E. N.: 1963, *Interplanetary Dynamical Processes,* Interscience, New York.

Rappaport, S., Doxsey, R., Solinger, A., and Borken, R.: 1974, *Astrophys. J.* **194**, 329.

Raymond, J. C.: 1979, *Astrophys. J. Suppl.* **39**, 1.

Salvati, M.: 1983, *Space Sci. Rev.* **36**, 145 (this volume).

Schmidt, M.: 1963, *Astrophys. J.* **137**, 758.

Scholer, M.: 1970, *Planet. Sci.* **18**, 977.

Scholer, M. and Morfill, G. E.: 1975, *Solar Phys.* **45**, 227.

Scoville, N. Z. and Solomon, P. M.: 1974, *Astrophys. J.* **187**, L67.

Scoville, N. Z., Irvine, W. M., Wannier, P. G., and Predmore, G. R.: 1977, *Astrophys. J.* **216**, 320.

Sedov, L. I.: 1959, *Similarity and Dimensional Methods in Mechanics*, Academic Press, London.

Shakura, N. I.: 1974, *Astron. Zh.* **51**, 441.

Shapiro, S. L. and Salpeter, E. E.: 1975, *Astrophys. J.* **198**, 678.

Shull, J. M.: 1980a, *Astrophys. J.* **237**, 769.

Shull, J. M.: 1980b, *Astrophys. J.* **238**, 860.

Shull, J. M. and McKee, C. F.: 1979, *Astrophys. J.* **227**, 131.

Solomon, P. M.: 1978, Communication quoted by Stark and Blitz (1978) *loc. cit.*

Solomon, P. M., Scoville, N. Z., and Sanders, D. B.: 1979, *Astrophys. J.* **232**, L89.

Stark, A. A. and Blitz, L.: 1978, *Astrophys. J.* **225**, L15.

Stecker, F. W.: 1971, *Cosmic Gamma Rays*, NASA SP–249.

Stenholm, L. G., Hartquist, T. W., and Morfill, G. E.: 1981, *Astrophys. J.* **249**, 152.

Skilling, J. and Strong, A. W.: 1976, *Astron. Astrophys.* **53**, 253.

Strong, A. W., Bignami, G. F., Bloemen, J. B. G. M., Buccheri, R., Caraveo, P. A., Hermsen, W., Kanbach, G., Lebrun, F., Mayer–Hasselwander, H. A., Paul, J. A., and Wills, R. D.: 1982, *Astron. Astrophys.* (in press).

Swanenburg, B. N., Bennett, K., Bignami, G. F., Buccheri, R., Caraveo, P., Hermsen, W., Kanbach, G., Lichti, G. G., Masnou, J. L., Mayer–Hasselwander, H. A., Paul, J. A., Sacco, B., Scarsi, L., and Wills, R. D.: 1981, *Astrophys. J.* **243**, L69.

Tenorio–Tagle, G.: 1980, *Astron. Astrophys.* **88**, 61.

Tenorio–Tagle, G.: 1981, *Astron. Astrophys.* **94**, 338.

Tenorio–Tagle, G., Yorke, H. W., and Bodenheimer, P.: 1981, in J. P. Sivan (ed.), *Mecanismes de Production d'Energie Dans Le Milieu Interstellaire.*

The, P. S. and Vleming, G.: 1971, *Astron. Astrophys.* **14**, 120.

Tinsley, B. M.: 1980, *Fund. Cosmic Phys.* **5**, 287.

Tscharnuter, W.: 1980, in P. Stobie (ed.), 'Stellar Hydrodynamics', *IAU Colloq.* **10**.

Van den Bergh, S., Marscher, A. P., and Terzian, Y.: 1973, *Astrophys. J. Suppl.* (227), **26**, 19.

Vasyliunas, V. M.: 1979, *Space Sci. Rev.* **24**, 609.

Völk, H. J.: 1981, *Izv. AN SSSR, Ser. Fiz.* **45**, No. 7, p. 1125 (in Russian-English translation: *Bull. Acad. Sci. USSR, Physical Series* **45**, No. 7, p. 1, Allerton Press Inc.).

Völk, H. J.: 1983, *Space Sci. Rev.* **36**, 3 (this volume).

Völk, H. J., Morfill, G. E., and Forman, M. A.: 1981, *Astrophys. J.* **249**, 161.

Weaver, T. A. and Woosley, S. E.: 1980, *Bull. Am. Astron. Soc.* **4**, 724.

Wentzel, D. G.: 1974, *Ann. Rev. Astron. Astrophys.* **12**, 71.

Wheeler, J. C., Mazurek, T. J., and Sivaramakrishnan, A.: 1980, *Astrophys. J.* **237**, 781.

Wooten, A.: 1977, *Astrophys. J.* **216**, 440.

Wooten, A.: 1981, *Astrophys. J.* **245**, 105.

Woosley, S. E., Weaver, T. A., and Taam, R. E.: 1980, in J. C. Wheeler (ed.), *Type I Supernovae*, Univ. of Texas Press, p. 96.

Yorke, H. W., Tenorio–Tagle, G., and Bodenheimer, P.: 1982, in P. Gorenstein and J. Dizinger (eds.), 'Supernovae Remnants and Their X-Ray Emission', *IAU Symp.* **101**.

Zel'dovich, Ya. B. and Shakura, N. I.: 1969, *Soviet Astron. AJ* **13**, 175.

GAMMA-RAY PULSARS IN THE MeV REGION*

MARCO SALVATI

Istituto di Astrofisica Spaziale, C.P. 67, 00044 Frascati, Italy

Abstract. Most models for the magnetosphere of pulsars assume ultrarelativistic primary particles streaming along the field lines; the resulting curvature photons can initiate electro-optical showers, which provide a secondary particle population at a lower energy. Both gamma and radio radiation are thought to be explicable within this framework.

We show that – because of the properties of curvature radiation – there is an intrinsic lower limit to the gamma-ray frequency which a pulsar can emit, if the power level is comparable to the rotational energy loss. Furthermore, independently of the radiative efficiency, a short duty cycle cannot be obtained at low gamma-ray frequencies, without producing at the same time certain observable spectral features.

Low-energy gamma-ray experiments on suitable candidates are expected to provide far-reaching information on the pulsar mechanisms.

1. Introduction

The class of gamma-ray emitting pulsars is not very crowded: only PSR 0531 and 0833 – the pulsars associated with the Crab and Vela SNR's – have been detected with absolute confidence. They have been observed several times by different experimental groups, and considerable information has been accumulated on pulse shapes, spectra, and phase properties – see, for instance, the review by Buccheri (1981). Although PSR 0531 was also seen in the MeV range (Kurfess, 1971; Penningsfeld *et al.*, 1979; Knight, 1982), most of our knowledge about these objects is due to photons above $\simeq 50$ MeV.

Detection of other pulsars has been claimed in several instances. The SAS-II team reported significant signals from PSR 1747 and 1818 (Thompson *et al.*, 1976; Ogelman *et al.*, 1976), but a conflicting upper limit to the emission from the latter was obtained by COS-B (Kanbach *et al.*, 1977). On the other hand, a preliminary report on the COS-B discrete source program contained PSR 0740 and 1822 (Pinkau, 1979), which had to be rejected later because of a discrepancy between the radio and gamma-ray periods.

Note however that PSR 1822 was again detected somewhat above the 3σ level by the OPALE NaI (Tl) telescope (Mandrou *et al.*, 1980); and that two observations of PSR 0950 with the MPI Compton telescope resulted in a signal somewhat below 3σ's (Diehl *et al.*, 1981). As pointed out by the authors, these results are certainly far from conclusive; yet, they are particularly interesting since they have been obtained at photon energies around 1 MeV, where crucial information about the emission processes is contained, and where the following discussion will be focussed.

Finally, many gamma-ray source error boxes from the COS-B lists have been searched for pulsations in different frequency bands. The most important findings are

* Proceedings of the XVIII General Assembly of the IAU: *Galactic Astrophysics and Gamma-Ray Astronomy*, held at Patras, Greece, 19 August 1982.

Space Science Reviews **36** (1983) 145–153.   0038–6308/83/0362–0145$01.35.

the radio and X-ray pulsations from MSH 15–52 (Seward and Harnden, 1982; Manchester *et al.*, 1982): they do not necessarily imply that the gamma-ray emission is also due to the pulsar, since the positional coincidence is very loose, and the timing is not accurate enough for the construction of a gamma-ray light curve; the X-ray data, though, have a relevance of their own to what will be said in the following.

In summary, the gamma-ray pulsar class is not empty, but the objects which belong to it are certainly not very numerous. Our point here is that these few objects are nevertheless a precious source of information about the radio pulsar phenomenon, which characterizes a much wider and much more studied astronomical class.

In Section 2 we give experimental and theoretical evidence that the gamma-ray and radio luminosities of pulsars are physically related, insofar as they share the basic electrodynamical mechanisms. In Section 3 we discuss the commonly accepted sparking gap approach to the pulsar magnetosphere, and in Section 4 we point out certain general constraints which are intrinsic to it; the constraints could be tested by observing selected objects at low gamma-ray energies, and – because of the conclusions of Section 2 – the outcome would directly affect our understanding of pulsars in the broadest sense.

2. Gamma Rays vs Radio

If the spectra of PSR 0531 and 0833 are scaled to the same distance, and compared frequency band by frequency band, a well known feature becomes apparent. The two objects are not very different in the radio and gamma-ray domains, at least order-of-magnitude wise; whereas at optical and X-ray wavelengths Vela is exceedingly fainter than the Crab, as if the emission there was a very steep function of the period.

Further evidence along the same line is the absence of a strong correlation between radio luminosity and period among all known pulsars, on the one side, and, on the other, the lack of optical detection to very low levels in the few studied objects. Also, the optical luminosity of the Crab pulsar has been observed to decrease, and the rate is about the one which would be deduced from the Crab–Vela scaling.

The X-ray pulsar in MSH 15–52 is admittedly intriguing: it lies orders of magnitude above the line drawn through Crab and Vela, and clearly demonstrates that nice correlations do not survive the third experimental point. The scenario could be maintained if the X-rays from MSH 15–52 were interpreted as the low-frequency tail of the gamma rays, rather than the high-frequency extension of the optical emission. Then this object would become a candidate for the low-frequency spectral signatures discussed in Section 4.

Our conclusion is that there is some empirical basis for recognizing two domains within a pulsar spectrum; they are distinguished by a rather mild, or, respectively, a very strong dependence on the pulsar period, and the radio- and gamma-ray luminosities partake together of the former behavior.

The close physical association between radio and gamma radiation is also supported by general theoretical arguments.

We take it for granted that pulsars are rotating magnetized neutron stars, whose primary energy stockpile is the kinetic energy of rotation. A critical surface is the locus where the corotation velocity would equal the velocity of light – the so-called light cylinder; the magnetic lines of force which close within the critical surface give rise to a kind of van Allen belts, the other ones are distorted by the induced currents – either displacement or convection – and are responsible for the energy outflow. The conventional terminology refers to these regions as to the closed and open magnetosphere, respectively.

A key issue is the amount of plasma which pollutes the open magnetosphere, and along the history of pulsar studies both extreme views have been considered. It was argued that gravitational forces are enormous at a neutron star's surface, and the height of a thermally supported atmosphere must be negligible (Pacini, 1967, 1968). So the pulsar is surrounded by an almost perfect vacuum, and along the open lines one has available for particle acceleration exceedingly large potential differences, of the order of those induced by rotation among the open lines themselves

$$\Delta V \simeq 7 \times 10^{12} P^{-2} B_{12} \, \text{V} \,, \tag{1}$$

here P is the period in seconds, B_{12} the surface magnetic field in units of 10^{12} G (Goldreich and Julian, 1969).

If this were the case, the particle number flux should be rather modest, but the energy per particle would be measured in macroscopic units.

The opposite view contends that the B-parallel E-field is large also at the star surface, and easily overcomes both the gravitational pull and the binding forces of the solid crust (Goldreich and Julian, 1969). Plasma is injected from the star into the magnetosphere until the magnetic lines become very nearly equipotential; the ΔV available for acceleration would drop dramatically with respect to the value of Equation (1), and the steady state would be characterized by a large number of relatively low-energy particles.

Particles which have crossed the ΔV of Equation (1) are natural candidates for the emission of hard gamma rays, even for a radiation process as inefficient as curvature radiation. In a plasma-screened magnetosphere, instead, the behavior of the particles is more difficult to predict, since they could well have energies around 1 GeV, and could produce the entire electromagnetic spectrum if the appropriate emission processes were assumed (Goldreich et al., 1972): their typical curvature radiation frequency falls in the radio range, while a suitable pitch angle would allow them to radiate optical photons at the light cylinder, and gamma rays at the star; the latter could also originate from inverse Compton interactions.

It is fair to say that these low-energy particles are not per se sufficient to explain the radio emission, but are necessary to it, since higher-energy particles would certainly emit at higher frequencies. It would then appear as if a completely empty magnetosphere corresponded to the perfect gamma-ray pulsar, whereas the magnetosphere of the perfect radio pulsar had to be completely screened. Such a conclusion is doubtful even under the assumption that the idealized, extreme cases do occur in real life.

On the one hand, a perfect gamma-ray pulsar would not be a radio pulsar, and could not be identified by time-analyzing the gamma rays because of the absence of a radio period (Buccheri et al., 1977); secondly, acceleration in an empty magnetosphere would occur predominantly at large distances, and would result in a broad modulation instead of the sharp peaks observed, e.g., in Vela. On the other hand, the perfect radio pulsar would appear as such only if the low-energy plasma could be forced into a coherent radiation mode, and the interaction of the high-energy particles with the low-energy ones, plus the resulting instabilities, are a common ingredient of most models (Ruderman and Sutherland, 1975; Arons, 1981; Usov, 1981).

Pulsars as we know them are very likely a cooperative effort of both populations of particles, because of which a certain degree of correlation between radio and gamma rays is not unexpected. The magnetosphere of a working pulsar should be in an intermediate state, rather than at either extreme, and a suitable description should be looked for among the models of the so-called sparking gap class (Ruderman and Sutherland, 1975).

3. The Sparking Gaps

A sparking gap arises in a magnetosphere which tends to become completely screened, but fails to accumulate the right amount of charge in some circumscribed places. The failure may be due to different reasons: the charge carriers, especially the ions, have a finite mass, and they lag for a while before being accelerated to the terminal velocity (Fawley et al., 1977); the binding energy of the ions in a low surface temperature pulsar can be so high as to prevent their extraction (Ruderman, 1980); field geometry and particle number conservation along the stream lines can make the locally available charge smaller than required for a complete screening (Scharlemann et al., 1978), or even of the opposite sign (Holloway, 1973).

The different gaps occur in different places within the magnetosphere, and further differences arise. The emphasis, however, is on the common properties of all the models in the class.

A certain fraction of the maximum possible ΔV, Equation (1), appears in the gaps along the field lines. Highly energetic primary particles are produced, which in turn radiate curvature photons in the hard gamma-ray range; other radiation processes should be irrelevant, since a transverse momentum component would be very difficult to excite, and would die out very quickly. Under most circumstances the gap growth is limited by electro-optical showers – the sparks – which are ignited when a certain fraction of the primary photons is converted into electron positron pairs inside the gap. Conversion outside the gap results in a low-energy secondary plasma; by adding the high-energy primaries which stream out of the acceleration region the two required particle populations are obtained.

From this point onward the picture becomes very model-dependent, and there is much additional branching beyond the geometrical differences described above. For instance, at least two pair-producing mechanisms have been invoked, namely, the interaction with the magnetic field (Sturrock, 1971), and the interaction with soft target photons (Ruderman, 1980). Then the secondary particles might undergo a variety of instabilities (Usov, 1981), and might radiate in a variety of modes.

We will not examine the individual models in any detail; we will instead discuss certain observable features directly relevant to the primary processes – curvature radiation and pair production. Such processes are basic to all the models of the class.

4. Testing the Gap Models

Curvature radiation is a rather inefficient process, especially at low frequencies, so the most severe observational constraints are provided by those pulsars which emit a large fraction of their rotational energy in the form of low-energy gamma rays. MSH 15–52 is not a suitable candidate – despite the possible interpretation of the X-rays as an extension of the γ-rays – because the radiative efficiency is quite small. An ideal target could be PSR 1822, which was marginally detected at $\simeq 0.5$ MeV (Mandrou et al., 1980), and whose nominal efficiency is even larger than 1 – this follows from the canonical values of the neutron star's parameters, and from the distance given by the dispersion measure.

If most of the rotational energy goes into gamma rays, to a large extent these must be primary curvature photons: a substantial production of secondary photons would require frequent sparking and numerous secondary particles, in which case the gap potential drop would be small with respect to Equation (1), and the efficiency would be low.

Now consider an open line with radius of curvature R; the extent to which that radius can remain approximately constant is at most $s = \pi R$, otherwise the line would become closed. If a primary particle is streaming along it with Lorentz factor γ, the radiated power and the typical frequency are (e.g., Salvati and Massaro, 1978)

$$ mc^2 \, \frac{d\gamma}{dt} = - \frac{2}{3} \frac{e^2 c}{R^2} \, \gamma^4 \,, \qquad \nu \simeq \frac{c}{4\pi R} \, \gamma^3 \,. \tag{2} $$

We integrate the energy equation out to s, indicating by γ_1 and γ_2 the values of γ at the beginning and, respectively, at the end of the active region

$$ \gamma_2^3 = \gamma_1^3 \left(1 + \frac{2 s r_e}{R^2} \, \gamma_1^3 \right)^{-1} \,, \qquad r_e = \frac{e^2}{mc^2} \,. \tag{3} $$

We then insert the adopted value of s, and replace γ^3 with the emitted frequency, Equation (2)

$$ \gamma_2^3 = \gamma_1^3 \left(1 + \frac{\nu_1}{\nu_*} \right)^{-1} = \gamma_1^3 \left(1 - \frac{\nu_2}{\nu_*} \right) \,, \tag{4} $$

$$ h\nu_* = \frac{hc}{8 \pi^2 r_e} = 5.5 \text{ MeV} \,, $$

where ν_1 and ν_2 correspond to γ_1^3 and γ_2^3, respectively. Finally, the radiative efficiency

ε can be written

$$\varepsilon = 1 - \frac{\gamma_2}{\gamma_1} = 1 - \left(1 + \frac{\nu_1}{\nu_*}\right)^{-1/3} = \tag{5}$$

$$= 1 - \left(1 - \frac{\nu_2}{\nu_*}\right)^{1/3}.$$

We stress that Equation (5) gives the efficiency with which the energy of a single particle is converted into gamma rays of frequency ν_1 to ν_2; hence it only gives an upper limit to the efficiency with which the pulsar rotational energy is converted into the required radiation. The real efficiency is expected to be lower, even much lower, because the particle acceleration mechanism can be inefficient, and also because particles can be produced which emit in a different spectral band. If follows that Equation (5) is useless in connection with low-efficiency objects – such as PSR 0531 – or with objects, whose observation has produced only upper limits.

An interesting candidate is instead PSR 1822. The photons apparently detected from it have energies $\simeq 0.5$ MeV, and it makes no difference to set them equal to ν_1 or ν_2; in both interpretations $\varepsilon \simeq 0.04$, largely below the observed value. We feel that the low-energy gamma-ray emission from PSR 1822 – if confirmed at the claimed level – would severely strain the current approach to pulsar electrodynamics, and would require some ingenious modification in order to maintain its basic principles.

A similar constraint applies to low-energy gamma-ray pulsars which are not very efficient radiators, but emit very narrow pulses. Vela would be an excellent example, if future 1-MeV experiments showed the same light curve which has already been found at higher photon energies. We must stress that the following argument is not as general as the one which led to Equation (5): our considerations will assume that the field geometry is dipolar, that the primary photons are absorbed because of the interaction with the magnetic field, and that the secondary particles emit gamma rays via the synchrotron mechanism, and not, say, via inverse Compton scattering. Some of the assumptions could be relaxed: it would be easy, for instance, to generalize the following formulae to an arbitrary multipolar geometry.

In our hypothetical object the narrow low-frequency pulses could not be due to primary curvature photons: the radiating particles would have a relatively low Lorentz factor, and their radiation length would easily be larger than the whole magnetosphere, so that the photons would not be beamed at all. One is forced to conclude that secondary emission dominates; this again entails a low radiative efficiency, which however is consistent with a Vela-like object.

The magnetic production of secondaries only occur if the primary photons have a large enough frequency, and see a large enough transverse component of the field; the transverse component increases at first, as the field line bends away from the photon path, then decreases because of the general dipolar behavior. Thus, for any given birth point of the primary photons, there is a minimum frequency which can give rise to secondaries. Higher frequencies are absorbed closer to the birth point; but the absorption cross-section is such, that the quantity $v^2 B_\perp$ after one mean free path increases with increasing v. If the energy of the secondary particles is approximated as half the energy of the parent photon, one finds the minimum secondary synchrotron frequency in correspondence to the minimum absorbed primary frequency.

Let

$$P = \text{pulsar period in seconds}, \qquad r_* = \text{star radius}, \qquad \psi = \text{pitch angle};$$

$$r = \text{distance of the birth point of the primary photons};$$

$$\vartheta_* = \text{magnetic colatitude at the star of the line which goes through the birth point};$$

$$\vartheta_c = \text{the same as } \vartheta_* \text{ for the last open line}.$$

(6)

Then we find

$$\gamma_{\text{sec}} \simeq 500 \, \frac{\vartheta_c}{\vartheta_*} \left(\frac{r}{r_*}\right)^{5/2} P^{1/2},$$

$$\psi_{\text{sec}} \simeq \frac{1}{\gamma_{\text{sec}}} \left(\frac{r}{r_*}\right)^3,$$

(7)

$$h\nu_{\text{sec}} \simeq 4.6 \, \frac{\vartheta_c}{\vartheta_*} \left(\frac{r}{r_*}\right)^{5/2} P^{1/2} \, \text{MeV}.$$

Frequencies lower than ν_{sec} are emitted only by secondary particles which are freely cooling via synchrotron radiation, hence one expects a power law photon number spectrum with slope -1.5. The Vela spectrum at higher frequencies has a slope of about -1.9 (Kanbach et al., 1980), and it is not clear whether present-technology experiments could detect such a mild break. If they could, however, Equation (7) would provide a powerful diagnostic, and possibly an interesting inconsistency, if $r < r_*$ or $\vartheta_* > \vartheta_c$ were implied.

Note that the extrapolation with slope -1.5 from the lowest COS-B point is in conflict with the X-ray upper limits and with the optical measurement, so that beyond the hypothetical break one has to predict a major spectral cutoff. In the present approach the cutoff is expected at the frequency ν_c, where the secondary particles are so cold, that they stop emitting ordinary synchrotron, and switch to small pitch angle synchrotron; this is not because the pitch angle has changed during the cooling, but because $1/\gamma$ has grown large enough

$$\nu_c \simeq \nu_{\text{sec}}(\gamma_{\text{sec}} \psi_{\text{sec}})^{-2}.$$

(8)

Measuring ν_c should not require a very sophisticated experiment, and would determine the position of the gap inside the magnetosphere.

At present, the only object which is known to emit narrow pulses at MeV energies is the Crab pulsar (Graser and Schönfelder, 1982). Here the picture is complicated by the X-rays, which, as discussed in the introduction, are likely not due to the same mechanism as the gamma rays. One then expects some feature at the transition between the two spectral ranges, which could prevent the identification of ν_{sec} and/or ν_c. On the experimental side the situation is still unsettled: there have been suggestions of a spectral break around 1 MeV (Knight, 1982), but a straight power-law spectrum has also been proposed (Graser and Schönfelder, 1982).

A safe way of using Equation (7) is to replace ν_{sec} with the lowest gamma-ray frequency, ν_m, down to which the high-energy gamma-ray spectrum has been firmly established not to break or bend: this produces a firm upper limit on the distance of the active region from the star

$$r \lesssim r_* \left(\frac{h\nu_m}{840 \text{ keV}} \right)^{2/5}. \tag{9}$$

5. Conclusions

We have shown that the sparking gap approach to the pulsar magnetosphere is constrained in two different ways. First, it is impossible to emit MeV photons with very high efficiency; second, narrow pulses in the MeV region must be accompanied by distinctive spectral features.

The pulsars which have been recognized as gamma-ray emitters (or possible gamma-ray emitters) should be observed at MeV energies; for the understanding of the pulsar phenomenon, such observations may well prove more useful than the addition of a few more objects to the class.

Acknowledgement

The editors thank V. Schönfelder for assistance in evaluating this paper.

References

Arons, J.: 1981, in W. Sieber and R. Wielebinsky (eds.), 'Pulsars', *IAU Symp.* **95**, 69.
Buccheri, R.: 1981, in W. Sieber and R. Wielebinsky (eds.), 'Pulsars', *IAU Symp.* **95**, 241.
Buccheri, R., D'Amico, N., Scarsi, L., Kanbach, G., and Masnou, J. L.: 1977, in R. D. Wills and B. Battrick (eds.), *Recent Advances in Gamma-Ray Astronomy*, Proc. 12th ESLAB Symp., ESA SP-124, p. 309.
Diehl, R., Graser, U., and Schönfelder, V.: 1981, *MPI-PAE/Extraterr.* **172**, 71.
Fawley, W. M., Arons, J., and Scharlemann, E. T.: 1977, *Astrophys. J.* **217**, 227.
Goldreich, P. and Julian, W. H.: 1969, *Astrophys. J.* **157**, 869.
Goldreich, P., Pacini, F., and Rees, M. J.: 1972, *Comm. Astrophys. Space Sci.* **3**, 185.
Graser, U. and Schönfelder, V.: 1982, *Astrophys. J.* **263**, 677.
Holloway, N.: 1973, *Nature Phys. Sci.* **246**, 6.
Kanbach, G., on behalf of The Caravane Collaboration: 1977, in R. D. Wills and B. Battrick (eds.), *Recent Advances in Gamma-Ray Astronomy*, Proc. 12th ESLAB Symp., ESA SP-124, p. 21.
Kanbach, G., Bennet, K., Bignami, G. F., Buccheri, R., Caraveo, P., D'Amico, N., Hermsen, W., Lichti, G. G., Masnou, J. L., Mayer-Hasselwander, H. H., Paul, J. A., Sacco, B., Swanenburg, B. N., and Wills, R. D.: 1980, *Astron. Astrophys.* **90**, 163.

Knight, F. K.: 1982, *Astrophys. J.* **260**, 538.

Kurfess, J. D.: 1971, *Astrophys. J.* **168**, L39.

Manchester, R. N., Tuohy, I. R., and D'Amico, N.: 1982, *Astrophys. J.* **262**, L31.

Mandrou, P., Vedrenne, G., and Masnou, J. L.: 1980, *Nature* **287**, 124.

Ogelman, H. B., Fichtel, C. E., Kniffen, D. A., and Thompson, D. J.: 1976, *Astrophys. J.* **209**, 584.

Pacini, F.: 1967, *Nature* **216**, 567.

Pacini, F.: 1968, *Nature* **219**, 145.

Penningsfeld, F., Graser, U., and Schönfelder, V.: 1979, *Proc. XVI ICRC* **1**, 105.

Pinkau, K.: 1979, *Nature* **277**, 17.

Ruderman, M.: 1980, *Ann. N.Y. Acad. Sci.* **336**, 409.

Ruderman, M. and Sutherland, P.: 1975, *Astrophys. J.* **196**, 51.

Salvati, M. and Massaro, E.: 1978, *Astron. Astrophys.* **67**, 55.

Scharlemann, E. T., Arons, J., and Fawley, W. M.: 1978, *Astrophys. J.* **222**, 297.

Seward, F. D. and Harnden, F. R.: 1982, *Astrophys. J.* **256**, L45.

Sturrock, P. A.: 1971, *Astrophys. J.* **164**, 529.

Thompson, D. J., Fichtel, C. E., Kniffen, D. A., Lamb, R. C., and Ogelman, H. B.: 1976, *Astrophys. Letters* **17**, 173.

Usov, V.: 1981, *Adv. Space Res.* **1**, 125.

COSMIC GAMMA RAYS AND MOLECULAR GAS
IN THE GALAXY*

B. HOUSTON, P. A. RILEY, and A. W. WOLFENDALE

Physics Department, University of Durham, U.K.

Abstract. A brief analysis is made of the interrelation of the intensity of cosmic-ray particles, the column density of gas and the intensity of cosmic γ-rays. It is shown that, locally, γ-ray data enable the calibration of H_2 densities to be inferred from CO data and elsewhere the variation of cosmic-ray intensity with position to be assessed. Finally, the importance of cosmic-ray irradiated molecular clouds in simulating γ-ray 'sources' is reiterated.

1. Introduction

There are three components to the cosmic γ-ray flux, one from known discrete sources, another from CR interactions and a third from apparent sources, the origin of which is arguable and is the subject of later discussion. It is commonly held that the majority of the flux comes from cosmic-ray interactions, electrons and nuclei interacting with gas in the ISM, and electrons interacting by the Inverse Compton process with the various photon fields, and this view will be followed here. Of the two main 'targets' for CR in the ISM – gas and photons – the former appear to be more important.

The gas under consideration is both atomic and molecular but whereas the distribution of atomic gas is known rather well, that of the molecular gas is known only poorly. Nevertheless, the influence of the molecular gas, which is condensed largely in rather dense clouds, is considerable and this is the main topic of consideration in the present paper. Two strands can be discerned: the use of gamma rays to determine cosmic-ray intensities where the mass is known and their use in the interpretation of some of the so-called γ-ray sources in terms of cosmic-ray interactions (CRI) with gas clouds. Inevitably there is some entanglement of these threads but in what follows we endeavour to keep them separate.

2. Determination of Cloud Masses

In recent work (Issa and Wolfendale, 1981) we assumed that the cosmic-ray intensity within about 2 kpc of the Sun is the same as that locally and compared the observed and expected γ-ray fluxes from the known molecular clouds in this region. There is no doubt that the procedure is rather uncertain for many reasons (not least the difficulties in deriving γ-ray fluxes) – and is strictly premature – but the principle is sound and the results have some value. The comparison shows that the masses assigned to the majority of the clouds (which came largely from a CO → H_2 conversion advocated by Blitz, 1978) were 'correct'. Such a demonstration is of value because of the notorious difficulty of

* Proceedings of the XVIII General Assembly of the IAU: *Galactic Astrophysics and Gamma-Ray Astronomy*, held at Patras, Greece, 19 August 1982.

Space Science Reviews **36** (1983) 155–159.   0038–6308/83/0362–0155$00.75.

estimating the conversion factor for CO → H$_2$ applicable to the bulk of gas in a cloud.

Analyses using the same philosophy, at least in part, have been made by a number of workers recently (Lebrun and Paul, 1979; Strong and Lebrun, 1982; Strong and Wolfendale, 1981; Strong et al., 1982) by examining γ-ray intensities at high galactic latitudes ($|b| \gtrsim 10°$). It is claimed that there is consistency between the column densities of H$_2$ derived from the analysis of galaxy counts, N_g (N_g gives the column density of all gas from which must be subtracted the CD of H I) and from γ-rays, assuming constancy of the CR intensity. Thus, (Strong, 1982), γ-rays alone can be used to estimate the CD of H$_2$ in regions of the sky where N_g values are not available. In fact, inevitably, there are problems; uncertainty in the number of galaxies which would be visible in the absence of interstellar absorption, in the galaxy counts themselves and in the contribution of the γ-ray intensity from extragalactic processes. Nevertheless, the results for H$_2$ column densities are probably accurate to a factor of two.

3. Cosmic-Ray Intensity Variations

The search for variations in cosmic-ray intensity over the Galaxy is considered by us to be a key application of diffuse γ-ray data in view of its relevance to the origin of the cosmic rays. What has become increasingly clear is that although there is undoubtedly a large scale radial gradient of γ-ray emissivity, the peak at $R \sim 6$ kpc being about 3 times the local value, the gradient in cosmic-ray intensity, I_{CR}, is smaller. Indeed, the demonstration of any increase in I_{CR} in the Inner Galaxy is difficult. The problem is, of course, the conversion CO → H$_2$ in this region; whereas the conversion may be known locally (see Section 2) the possible effect of the 'metallicity gradient' in the Inner Galaxy is very serious (Blitz and Shu, 1980; Li et al., 1982).

In the Outer Galaxy our early work (Dodds et al., 1975) using the SAS II observations indicated a gradient of I_{CR} such that for l: 90°–250° (with a number of gaps) the intensity fell by ∼ 50% in going from $R = 10$ to $R = 12$ kpc. It now seems that the gradient is probably not quite as steep (although the discovery of more gas in the Outer Galaxy – a likely occurrence – will increase it again). Still averaging over a wide range of l, the present situation is as shown in Figure 1, viz. that adoption of the averaged radial distribution of CR electrons implied by the studies of Phillipps et al. (1981), $I_{CR} \propto \exp(-R/8 \text{ kpc})$, yields a reasonable fit to the γ-ray emissivity versus R plot. The subtleties are indicated in the caption to the figure.

In fact, the adoption of a simple exponential fall-off in I_{CR} with R is an oversimplification and there is evidence for different gradients in different directions. Such a situation is not unexpected if the cosmic rays are generated in galactic sources (SNR, pulsars, shocked regions of the ISM ...) because these sources have a patchy distribution. It is interesting to note that the lowest CR intensities (inferred from the γ-ray results) are in the 3rd quadrant (Issa et al., 1981a) and it is here that young galactic objects are sparse, too.

The search for cosmic-ray intensity differences in different parts of the Galaxy will be a continuing one. If cosmic rays are galactic in origin, if their sources are most

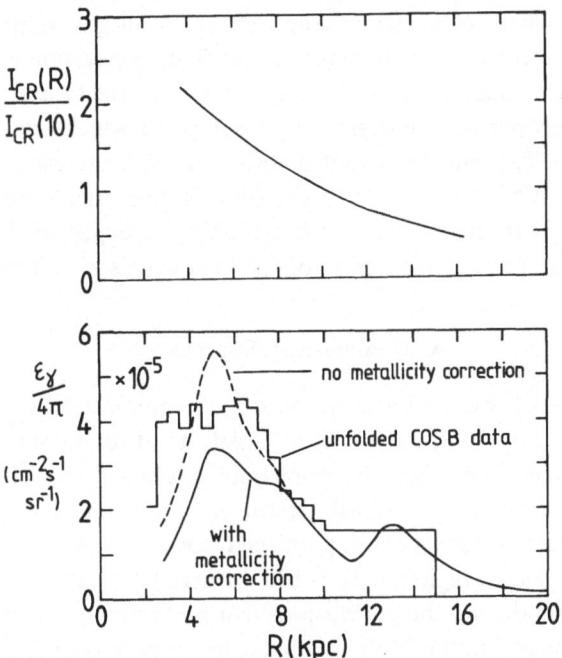

Fig. 1. Dependence of cosmic-ray intensity (normalised to unity at $R = 10$ kpc) and γ-ray emissivity on galactocentric distance, R. The CR versus R dependence of the upper figure has been used to predict the variation of $\varepsilon_\gamma/4\pi$ with R shown in the lower figure both with and without the metallicity correction. The lower dependence ('with metallicity correction') is preferred so that the gradient is probably a little steeper than indicated in the upper figure. An alternative is that the difference between our best estimate of $\varepsilon_\gamma/4\pi$ and the 'unfolded COS-B data' is due to unresolved γ-ray sources; although we believe that such sources do not contribute much of the flux they could just be responsible for the required amount.

frequent in regions of high gas density and if their diffusion distance before escape is only a small fraction of the linear dimension of the Galaxy then there should be a correlation of CR intensity with gas density. Already there have been claims for such a correlation (most recently by Fichtel, 1982) on a large scale, i.e. several kpc scale, but we must regard it as yet unproven, largely because of the problem with the CO → H_2 conversion. It would be interesting to know down to what linear dimension such a correlation exists (if it does exist on a large scale). The apparent observation of some molecular clouds which need an enhanced value of I_{CR} (Morfill, 1981; Morfill and Drury, 1981; Montmerle and Cesarsky, 1980; Issa *et al.*, 1981c) means that CR concentrations on a sub kpc scale may need to be taken seriously. In this connection, some recent COS B observations of good statistical quality by Strong *et al.* (1982) for $E_\gamma > 70$ MeV which cover the whole range of longitude are of considerable interest. Comparison of the γ-ray intensities from these observations with those predicted by us adopting a constant value of I_{CR} and our estimates of column densities does not show good agreement; instead, there is evidence for higher values of I_{CR} where the CD are greatest. An empirical approach has been adopted (Riley and Wolfendale, 1982) by writing $I_\gamma = N_t^n + I_b$, where N_t is the column density of all gas, derived using galaxy counts, and it is found that $n \simeq 1.5$ gives the best fit. This conclusion depends upon the assumption of a linear relationship between gas column density and the logarithm of the number of galaxies per square degree, an assumption which has been questioned recently by Strong (1983) who finds that it results in an underestimation of column densities above 2.5×10^{21} atoms cm^{-2}. Although it is thus premature to be certain, it is interesting to note that such a value of n is roughly as expected from the previous remarks, viz. that there is an enhancement of CR in at least some of the denser clouds.

4. Gamma-Ray Sources

The final topic concerns the nature of the peaks of γ-ray intensity which have been interpreted by some as corresponding very largely to discrete sources (unresolved pulsars ...). Our contention has been for some time that many, probably $\sim 50\%$, of the 2CG catalogue of sources from COS-B (Hermsen, 1980) are due rather to CRI of molecular clouds. The most recent of our published work (Li and Wolfendale, 1982) has concentrated on a restricted region, l: $10° - 180°$, where CO data are available over an adequate range of latitude and the conclusions can be summarised, as follows. Out of the 9 unidentified 'sources' in the 2CG catalogue, 6 can be explained as due to CRI of the clumpy ISM; of the 6, 4 need a negligible increase in I_{CR} or underestimate of H_2 column density (by 30%) but the other 2 need a significant increase.

The four needing only a small increase are $013 + 00$, $036 + 01$, $075 + 00$, $078 + 01$ (and perhaps $095 + 04$); the two needing a significant increase are $121 + 04$ ($F \cong 4$) and $135 + 01$ ($F \cong 10$), where F is the factor by which I_{CR} must be increased (or the H_2 mass increased).

There have been other developments, too. The 2CG source $353 + 16$ associated with the ρOph cloud complex, which had been claimed previously to be discrete, has now been shown to be extended (COS-B group, private communication) and to mirror rather well the distribution of gas in that vicinity. It remains to be seen whether a significantly enhanced value of I_{CR} is needed; previously (Issa et al., 1981b; Issa and Wolfendale, 1981) we needed an enhancement factor $F \sim 1.5$ which cannot be regarded as significantly greater than unity and we know of no reason yet to increase this value.

Another feature of the analysis by Li and Wolfendale was the demonstration that the lumpy ISM coupled with the limited number of γ-rays detected in the COS-B experiment (typically 80 above 100 MeV for a flux of 1.0×10^{-6} cm^{-2} s^{-1}) causes the appearance of many apparently significant γ-ray 'sources'. Specifically, at the threshold of significance $C_s/\sigma_0' = 4.75$ adopted in the 2CG catalogue the probability of a 'spurious' source appearing is increased by a factor of about 500.

Finally, reference can be made to our very recent analysis of the SAS II data (Houston and Wolfendale, 1982) using similar techniques to those adopted by COS-B. As yet the analysis is not complete but the preliminary results are interesting. In summary, they are:

(i) There is support for the existence of peaks at the positions of over half of the 2CG 'sources'. There is no strong evidence against the other 2CG sources in view of the poor statistics.

(ii) The SAS II fluxes of the brighter confirmed 'sources' are similar to those from COS-B.

(iii) The number of 'spurious' sources thrown up by the analysis is again larger than expected on the basis of the derived statistical confidence levels, adding weight to the analysis by Li and Wolfendale referred to earlier.

(iv) Comparison made with the expected fluxes from CRI of gas in the ISM again points to the likelihood of very approximately half of the 2CG sources being due to irradiated molecular clouds.

Acknowledgements

The editors thank C. Fichtel for assistance in evaluating this paper.

References

Blitz, L.: 1978, Ph.D. Thesis, Columbia Univ. (also NASA Tech. Mem. 79708).

Blitz, L. and Shu, F. H.: 1980, *Astrophys. J.* **238**, 148.

Dodds, D., Strong, A. W., and Wolfendale, A. W., 1975, *Monthly Notices Roy. Astron. Soc.* **171**, 569.

Fichtel, C. E.: 1982, *COSPAR Meeting*, Ottawa, (in press).

Hermsen, W.: 1980, Ph.D. Thesis, Leiden Univ.

Houston, B. and Wolfendale, A. W.: 1982, *Irish Astron. J.*, (in press).

Issa, M. R. and Wolfendale, A. W.: 1981, *Nature* **292**, 430.

Issa, M. R., Riley, P. A., Strong, A. W., and Wolfendale, A. W.: 1981a, *J. Phys. G.* **7**, 973.

Issa, M. R., Strong, A. W., and Wolfendale, A. W.: 1981b, *J. Phys. G.* **7**, 565.

Issa, M. R. *et al.*: 1981c, *Proc. 16th I.C.R.C.* **1**, 150.

Lebrun, F. and Paul, J. A.: 1979, *Proc. 16th I.C.R.C.* **12**, 13.

Li, T. P. and Wolfendale, A. W.: 1982, *Astron. Astrophys.* **116**, 95.

Li, T. P., Riley, P. A., and Wolfendale, A. W.: 1982, *J. Phys. G.* **8**, 1141.

Montmerle, T.: 1981, *Phil. Trans. Roy. Soc.* **A301**, 505.

Montmerle, T. and Cesarsky, C. J.: 1980, *Adv. Space Explor.* **7**, 61.

Morfill, G. E.: 1981, *Proc. 16th I.C.R.C.* **1**, 170.

Morfill, G. E. and Drury, L. O'C.: 1981, *Proc. 16th I.C.R.C.* **1**, 171.

Phillipps, S., Kearsey, S., Osborne, J. L., Haslam, C. G. T., and Stoffel, H.: 1981, *Astron. Astrophys.* **103**, 405.

Riley, P. A. and Wolfendale, A. W.: 1982, (in preparation).

Strong, A. W. and Lebrun, F.: 1982, *Astron. Astrophys.* **17**, 73.

Strong, A. W. and Wolfendale, A. W.: 1981, *Phil. Trans. Roy. Soc. London* **A301**, 541.

Strong, A. W.: 1982, *Proc. Leiden Conf. on S. Hemisphere*.

Strong, A. W., *et al.*: 1982, *Astron. Astrophys.* **115**, 404.

Strong, A. W.: 1983, *Monthly Notices Roy. Astron. Soc.* **202**, 1015.

GAMMA RAYS FROM CLOSE BINARY SYSTEMS*

L. MARASCHI and A. TREVES

Istituto di Fisica Cosmica del CNR, Milano, Italy
Dipartimento di Fisica, Università di Milano, Milano, Italy

Abstract. The possibility of producing γ-rays in a close binary system is examined. First the case of systems containing a young pulsar is discussed. It is shown that high energy radiation should be efficiently produced at a shock front, where the pressure of the relativistic wind of the pulsar equals the ram pressure of the primary star wind. The expected number of such systems in the Galaxy is evaluated and the model is applied in some detail to the cases of Cyg X-3 and LSI + 61°303, two binary systems which are possibly sources of γ-rays.

In the second part of the paper, models of accreting black holes are considered with accretion rates as may occur in close binary systems. If electrons and protons are thermally decoupled, the proton temperature, in the vicinity of the hole horizon, is large enough for pion production. However the γ-ray luminosity and spectrum indicate that the process is of little interest for explaining the γ-ray sources discovered by the COS-B satellite. Accretion models where there is a non thermal component in the particle energy distribution may be relevant to γ-ray astronomy.

1. Introduction

Among the 25 sources listed by the COS-B catalogue (Swanenburg *et al.*, 1981), selected with criteria of statistical significance and of pointlike appearance, only two are firmly identified, with the Crab and Vela pulsar respectively, on the basis of the observed periodicity in γ-rays. The proposed identifications of 2CG 289 + 64 with the quasar 3C 273 and of 2CG 353 − 116 with the dense molecular cloud ρ Oph, though not without question, are convincing because both objects are, in their respective categories, unique. The identification of ρ Oph is further strengthened by the detection of a broad extended γ-ray excess from the region of the Orion Nebula (Caraveo *et al.*, 1980).

The case of 2CG 135 + 15 illustrates well the difficulty of identifying γ-ray sources. In this error box fall two exceptional objects, one is a very close by Quasar (0241 + 622) $z = 0.04$) of low luminosity (Apparao *et al.*, 1978), the other a very unusual, periodically variable radio source (GT 0236) identified with a luminous star (LSI + 61°303) in a binary system (Gregory and Taylor, 1978). Whether the γ-ray source should be identified with either object is therefore a matter of discussion. In the following we will discuss the implications of the association with the second object.

Another case which awaits clarification is that of the short period binary and strong X-ray source, Cyg X-3. The SAS-2 team (Lamb *et al.*, 1977) reported a positive detection in γ-rays on the basis of the observation of the 4.8 hr periodicity. The observations took place after a major radio outburst. The COS-B data (Swanenburg *et al.*, 1981; Bennett *et al.*, 1977) do not show periodic modulation, but do not exclude a point-source compatible with the Cyg X-3 position. Positive detections of periodic γ-ray

---

* Proceedings of the XVIII General Assembly of the IAU: *Galactic Astrophysics and Gamma-Ray Astronomy*, held at Patras, Greece, 19 August 1982.

Space Science Reviews **36** (1983) 161–171.   0038–6308/83/0362–0161$01.65.

emission have been claimed at higher energies ($E \simeq 10^2 - 10^3$ GeV) (Vladimirsky *et al.*, 1973).

Confusion problems are even greater in the galactic centre region where the insufficient spatial resolution of the instruments above 10 keV prevents any detailed association of the high energy sources with the structures observed at longer wavelengths. However a clear trend emerges from the data as discussed by Matteson (1982), that is the increasing dominance of the region within 1° of the galactic centre with increasing energy. A variable source is observed from 10 to 100 keV, whose spectrum hardens significantly in the $10^2 - 10^3$ keV range. The unique nature of this source makes it likely that it is the galactic nucleus itself and the 100 MeV source 2CG 359 – 00 could also be associated with it.

Except for the above mentioned cases in which more or less reliable candidates exist, the rest of the COS-B sources, i.e. the great majority, are as yet unidentified. The weakness of the X-ray counterparts, if any, which implies $L_x \lesssim 0.1 \, L_\gamma$, indicates that the sources are 'true' γ-ray sources, that is they emit a large fraction of their luminosity in the γ-ray band.

The average properties of the γ-ray source population are a 'typical' distance of 2–7 kpc with a corresponding luminosity L_γ between 0.4 and 5×10^{36} erg s^{-1} (see Swanenburg *et al.*, 1981, 1983).

This brief review sets the framework for γ-ray source models. From the identified or possibly identified sources it is apparent that at least 5 classes of objects may be relevant to the problem and are therefore prime candidates for explaining the unidentified sources. They are:

(a) young pulsars (2CG 184 + 05; 2CG 263 – 02);
(b) molecular clouds (2CG 253 + 16);
(c) close binary systems (Cyg X-3; 2CG 135 + 01 ?);
(d) active Galactic Nuclei (3C 273);
(e) medium sized accreting black hole (GCX ?).

Classes (a) and (b) are covered by other authors at this workshop. This paper concerns mainly class (c) but touches also (d) and (e). For these three classes the observational evidence is far from conclusive, nevertheless the theoretical problems raised are of great interest.

The most natural model for producing γ-rays in a binary system is that it contains a young, i.e. fast rotating, pulsar, since it is known that young pulsars can emit γ-rays. Systems of this type exist necessarily because they are the progenitors of the pulsing X-ray sources (also called X-ray pulsars), which are powered by accretion rather than by rotation. Therefore two basic questions arise: (a) how many of these systems are expected in the Galaxy and (b) whether γ-rays are a good means for detecting them. The answer is related to the physical processes which affect the pulsar rotation and emission in the presence of a companion, which determines the environment in which the pulsar machine works. These problems are reviewed in the first part of this paper (Section 2).

γ-rays could also be generated in accreting neutron stars, (see e.g. Shapiro and

Salpeter, 1975), however they would necessarily be associated with a copious X-ray flux. Therefore this class of models could perhaps be applied to Cyg X-3 but not to the 'average' γ-ray source, which has only a weak X-ray counterpart if any.

The other possibility which is discussed here in some detail (Section 3) is that of γ-ray production in accreting black holes. Because accretion models follow simple scaling laws the calculations are not very different for high mass and low mass black holes provided that the gas supply scales as the mass. Therefore the results are similar for a black hole of stellar mass ($10\ M_\odot$) accreting gas from a companion and for a black hole of medium or high mass ($10^3 - 10^8\ M_\odot$) accreting from the ambient medium. This class of models may be relevant for black holes in binaries, for the galactic nuclei and for Quasars.

2. Young Pulsars in Close Binaries

The issue of whether the majority of the 2CG sources could be accounted for by other, as yet undiscovered pulsars, has been discussed by several authors with a substantial agreement on the affirmative. However the unfruitful searches for young radio pulsars in the COS-B error boxes (see D'Amico, 1983) lead us to discuss the following scenario.

Consider a system where a young pulsar, of age comparable to that of the Vela pulsar, $t \simeq 2 \times 10^4$ yr, has a close companion with an orbital period of the order of days. Supposing that the pulsar radiates as PSR 0833 – 45 one expects an emission peaked in γ-rays with weak X-ray and optical contribution. The requirements for explaining the unidentified sources are thus satisfied.

Assuming that the unevolved star loses mass through a spherical wind at a rate $\dot{M}$ and with velocity $v_w \simeq 10^8$ cm s^{-1}, the wind will be opaque to radio waves of wavelength greater than λ if (see Illarionov and Syunyaev, 1975),

$$\dot{M} \gtrsim 3.6 \times 10^{-10}\ (D/10^{10}\ \text{cm})^{3/2}\ (\lambda/100\ \text{cm})^{-1}\ M_\odot\ \text{yr}^{-1}. \tag{1}$$

Therefore even a modest mass loss, $\dot{M} \simeq 10^{-9}\ M_\odot\ \text{yr}^{-1}$, appropriate for a moderately luminous Main Sequence star is sufficient to screen-off the pulsed radio emission, which is usually observable at long wavelengths.

For winds of this order, the pulsar rotation and the associated pulsar mechanism and γ-ray emission should be unaffected.

A. STATISTICAL CONSIDERATIONS

We give here a simplified version of the discussion published earlier (Maraschi and Treves, 1979). According to the evolutionary scenario for massive X-ray binaries (e.g. Van den Heuvel, 1978) the X-ray active phase occurs when the companion starts its evolution towards the giant phase and ends before this transition is completed. This implies a life-time of the X-ray phase of $10^4 - 10^5$ yr, comparable to the age of a young pulsar. Since the number of massive X-ray binaries in our Galaxy is ~ 20 (e.g. Bradt and McClintock, 1983), the number of close binaries containing young pulsars should be roughly equal. However to a limiting sensitivity of 10^{36} erg s^{-1}, the COS-B survey covers about 10% of the galactic plane which leads to an expectation of $\sim 2\ \gamma$-ray sources of this type. Since in this phase the X-ray production may be very low the possibility of identifying such systems through their γ-ray emission is of great interest.

B. PHYSICAL PROCESSES NEAR A YOUNG PULSAR IN A BINARY SYSTEM

The number estimated above is based on the assumption that the efficiency of γ-ray production in isolated and binary pulsars is the same, but it may be increased if the environment provided by the binary system enhances the γ-ray yield.

In fact the largest fraction of the energy loss associated with the pulsar spin down (much larger than that emitted in the form of pulsed radiation) is supposed to be emitted in the form of a relativistic wind, that is low frequency electromagnetic waves, relativistic particles, magnetic field, etc. (Rees and Gunn, 1974). The interaction of this relativistic wind with the stellar wind or envelope of the companion generates a shock which can roughly be described balancing the two main pressures (Illarionov and Syunyaev, 1975; Bignami et al., 1977; Maraschi and Treves, 1981a, 1982). The location of the shock, i.e. its distance, r_b, from the neutron star, in the region between the two stars can therefore be estimated from the condition

$$\frac{\dot{M}v_w}{(D - r_b)^2} = \frac{L_{\text{pulsar}}}{cr_b^2} , \qquad (2)$$

where D is the separation of the binary system.

At the shock boundary one expects heating of the surrounding matter, local particle acceleration, magnetic field compression etc. These phenomena will produce radiation through different processes (thermal bremsstrahlung, synchrotron radiation, Compton scattering and π° decay) whose relative importance depends on the physical parameters of the shock.

C. THE CASE OF Cyg X-3 AND LSI61°303

The same framework enables us to construct models for two systems as different as Cyg X-3 and LSI + 61°303 (Bignami et al., 1977; Maraschi and Treves, 1981a, 1982).

The observational characteristics of these two sources are summarized in Table I. The main features that they have in common is that both may be γ-ray sources (for Cyg X-3 up to 10^{14} eV) and both have strong radio activity, in the form of large flares for Cyg X-3 and of periodic outbursts for LSI + 61°303. But the binary periods are very different, implying different separations and the non collapsed component is, in the first case, probably a low mass Main Sequence star and, in the second, a luminous early type Main Sequence or giant.

Accordingly the parameters of the models derived for the two systems are different and are compared in Table I. Cyg X-3 requires a very active young pulsar ($L_{\text{rotation}} \simeq 10^{38}$ erg s^{-1}) in a very dense environment, where free-free radiation dominates the X-ray emission. Due to the high density the shock is close to the neutron star, $r_b = 3 \times 10^{10}$, and the associated magnetic field is high, $B \simeq 10^3$ G. Synchrotron radiation can therefore extend from the radio to the X-ray band. In particular the

TABLE I

Two systems possibly consisting of a young pulsar in a close binary system

Phenomenology of the systems	
Cyg-3	LSI 61°303
(Boldt and Kondo, 1976, and references therein)	(Taylor and Gregory, 1982)
Strong X-ray source	Weak X-ray source
$L_{comp}\ 10^{33}$ erg s^{-1}	$L_{comp}\ 10^{38}$ erg s^{-1}
100 MeV source (Lamb *et al.*, 1977)	= 2CG 135 + 1 (?)
10^{12}–10^{14} eV source	1 MeV source (?)
Orbital period P = 4.8 hr	$P = 26.6^d$
Separation $D \simeq 10^{11}$ cm	$D \simeq 3 \times 10^{12}$ cm
Radio flares	Periodic radio emission

Models		
	Cyg X-3 (Bignami *et al.*, 1977)	LSI 61°303 (Maraschi and Treves, 1981a)
Density at the shock	$n \simeq 10^{14}$ cm^{-3}	$n \simeq 10^8$ cm^{-3}
Magnetic field at the shock	$B \simeq 10^3$ G	$B \simeq 6$ G
Thermal bremsstrahlung	X-rays	–
Synchrotron radiation	Radio X-rays	Radio-infrared
Inverse Compton scattering	10 MeV – E_{max}	X-ray – γ-rays
γ-rays from π° decay	Important	–

observed infrared emission should be explained in this way. Compton scattering of the synchrotron photons should contribute to the emission from 10 MeV up to the maximum energy at which electrons are accelerated. If the protons propagate diffusively strong interactions leading to γ-ray production via π° decay can be important.

For LSI + 61°303 a moderately active pulsar is required, with $L_{rot} = 10^{36}$ erg s^{-1}. The shock boundary r_b is at 3×10^{12} cm with a magnetic field $B \simeq 6$ G. The energy density of photons emitted by the primary star is larger than the magnetic energy density, therefore relativistic electrons should radiate the bulk of their energy through inverse Compton scattering producing X-ray and γ-rays. The radio emission derives from the synchrotron process and the periodicity could be understood as a consequence of an eccentric orbit.

D. OTHER SYSTEMS WHICH MAY CONTAIN A YOUNG PULSAR

Up to now, only one binary system with mass transfer is known to contain a rapidly spinning neutron star. This is the recurrent X-ray transient AO 538 – 66 in the LMC, where a pulsation period of 69 ms has been observed in X-rays during outburst (Skinner *et al.*, 1982). The inferred high orbital eccentricity is indicative of the young age of the pulsar, whose rotation period is very close to that of the Vela pulsar. The model here involves a transition from the pulsar phase at aphastron, to an accretion phase at periastron, when the neutron star crosses the densest region of the envelope of the companion which is a Be star (see Maraschi *et al.*, 1983, and references therein).

The period, the strongly asymmetric X-ray light curve and high eccentricity of AO 538 – 66 are very reminiscent of another enigmatic X-ray source, Cir X-1, in which the presence of a young pulsar has also been suggested (Maraschi and Treves, 1981b, 1982). The latter source is a periodic radio source and in this respect similar to LSI + 61°303. No radio emission has been as yet discovered from AO 538 – 66. The model would predict radio emission at least at high frequencies (~ 10 GHz), in anticorrelation with the X-ray emission. Neither of these sources appears in the COS-B catalogue. However AO 538 – 66 is at very large distance and Cir X-1 is in a region of high γ-ray background, so that in both cases the γ-ray flux could be below the detection threshold.

3. γ-Rays from Accreting Black Holes

Within the class of black hole accretion models, involving only thermal processes, i.e. with Maxwellian particle distribution functions and without suprathermal particle acceleration, the most important mechanism for producing high energy γ-rays is via nuclear interaction of high temperature protons and subsequent π° decay. Therefore the question is, under what conditions can the protons reach a high enough temperature and what is the emergent γ-ray and X-ray luminosity, taking into account radiative losses and general relativistic effects, which are dominant in the vicinity of black holes.

It was first noted by Dahlbacka *et al.* (1974) that the time scale for collisional energy exchange between protons and electrons in a spherically accreting plasma may be larger than the free-fall time-scale. Therefore protons may fall adiabatically, while radiative losses will limit the electron temperature to lower values. In the Bondi adiabatic theory with $\gamma = \frac{5}{3}$ one has $T_p = (\frac{3}{10})mGM/(kr)$, which for $r = r_S = 2GM/c^2$ gives $T_p = (\frac{3}{20})mc^2/k = 1.7 \times 10^{12}$ K. Because this value is very close to the threshold temperature for the process to become efficient (Dahlbacka *et al.*, 1974) the calculated γ-ray luminosity in specific models is extremely sensitive to the precise value of the final temperature obtained.

Dahlbacka *et al.* (1974), Kolykhalov and Syunyaev (1979), and Giovannelli *et al.* (1982) evaluated the γ-ray luminosity and emergent spectrum due to this process assuming complete decoupling of protons and electrons. With this hypothesis the γ-ray luminosity L_γ scales as $M^{-1}\dot{M}^2$.

The γ-ray spectrum is found to be peaked at 20 MeV due to the gravitational and Doppler redshifts and, for $M = 10 M_\odot$ and $\dot{M} = 10^{18}$ G s^{-1} as plausible in a close binary system, Dahlbacka *et al.* (1974) and Kolykhalov and Syunyaev (1979) estimate $L_\gamma \simeq 10^{36}$ erg s^{-1} while Giovannelli *et al.* (1982) give $L_\gamma = 10^{34}$ erg s^{-1} (Giovannelli *et al.*, 1982). Although these results differ significantly they are in the range of interest for comparison with observations.

However for large accretion rates the energy transfer from protons to electrons becomes important and consequently the hypothesis of adiabaticity of the proton component breaks down. We have therefore considered the equations of energy balance of both protons and electrons with a coupling term describing the Coulomb interaction between the two populations. A full account of this work is given in Colpi *et al.* (1982, 1983).

The main result is that the proton temperature is everywhere smaller than in the adiabatic case. This is due in part to the fact that at large radii, where the coupling between protons and electrons is efficient, the radiative losses of electrons are important

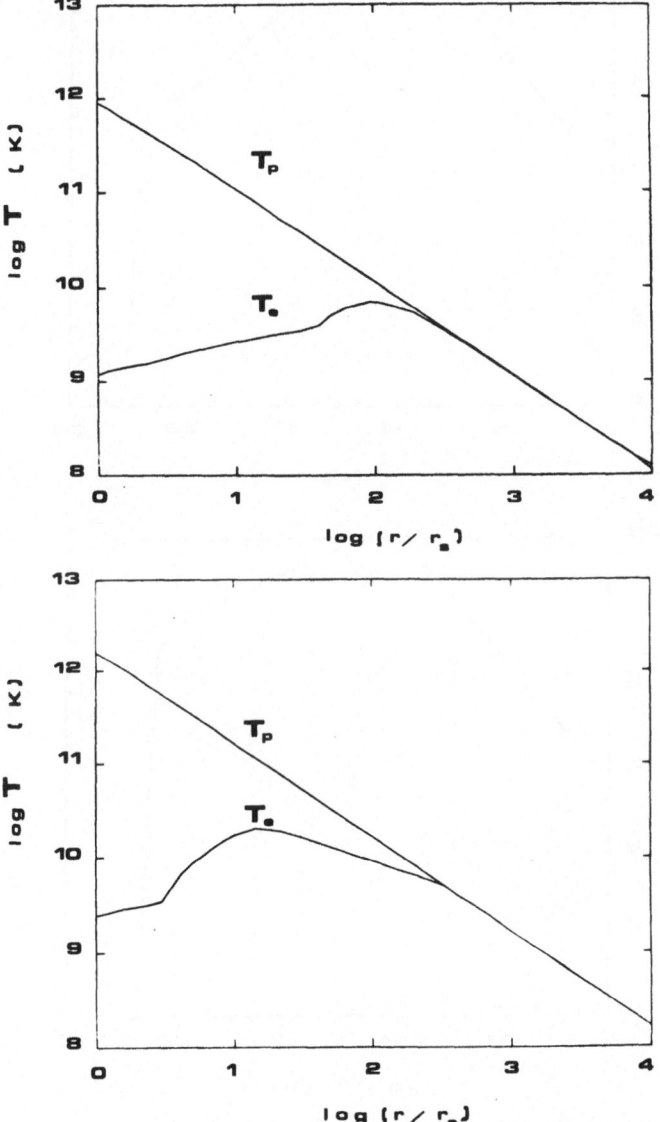

Fig. 1. Temperature profiles of electrons (T_e) and protons (T_p) accreting onto $10\,M_\odot$ black hole for different accretion rates. (a) $\dot{M} = 10^{16}\,\mathrm{g\,s^{-1}}$. (b) $\dot{M} = 10^{18}\,\mathrm{g\,s^{-1}}$.

and in part to the transfer of energy from protons to electrons in the inner region where the difference of the two temperatures is largest (see Figure 1). Since the dependence of the nuclear cross-section on the temperature is very strong in this regime, the computed γ-ray luminosity is much smaller than that of the previous authors (see Figure 2 and Table II). Because the deviation from adiabaticity increases with increasing

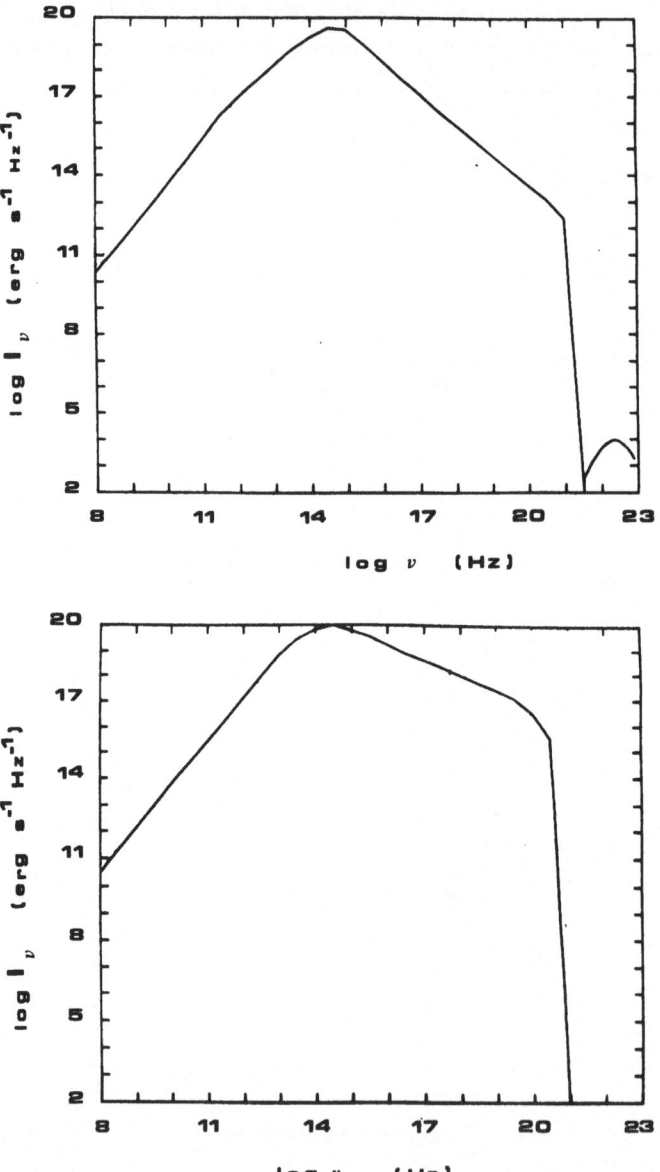

Fig. 2. Emission spectrum of a 10 $M_\odot$ black hole. (a) $\dot{M} = 10^{16}$ g s^{-1}. The high frequency component is due to π° decay. (b) $\dot{M} = 10^{18}$ g s^{-1}, here the γ-rays due to π° decay are degraded to lower energies because of photon photon pair production, and are not reported in the figure.

TABLE II

e_{max} and $T_{p_{max}}$ are the maximum temperatures reached by electrons and protons. ν^* is the self-absorption synchrotr(e)(n) equency at the gravitational radius. L_e is the emission from the electrons which extends from ν^* to $3kT/h$. L_γ is t(he) iminosity deriving from π° decay, $\tau_{e^+e^-}$ (50 MeV) is the optical depth for photon-photon interaction at 50 MeV (s)(ee) Equation (3)). L_γ coincides with the γ-ray luminosity L_γ^* effectively released by the system only if $\tau_{e^+e^-} \lesssim 1$.

		$T_{p_{max}}$ (K)	L_γ (erg s^{-1})	$T_{e_{max}}$ (K)	ν^+ (Hz)	L_e (erg s^{-1})	$\tau_{e^+e^-}$ (50 MeV)
$f = 10\,M_\odot$	$M = 10^{16}\,G^{-1}$	1.6×10^{11}	2.4×10^{30}	2×10^{10}	2.1×10^{15}	6.4×10^{34}	6.8×10^{-2}
	$M = 10^{18}\,G^{-1}$	9×10^{11}	1×10^{33}	7×10^{9}	10^{16}	5.3×10^{36}	26
$f = 10^4\,M_\odot$	$M = 10^{19}\,G^{-1}$	1.6×10^{12}	2.4×10^{33}	2.2×10^{10}	2.4×10^{14}	6.6×10^{37}	5.6×10^{-2}
	$M = 10^{21}\,G^{-1}$	9×10^{11}	1.4×10^{36}	7.4×10^{9}	5.5×10^{14}	5.5×10^{39}	25
$f = 10^8\,M_\odot$	$M = 10^{23}\,G^{-1}$	1.5×10^{12}	2.4×10^{37}	2.5×10^{10}	9×10^{12}	6.5×10^{41}	8×10^{-2}
	$M = 10^{25}\,G^{-1}$	9×10^{11}	1.4×10^{40}	8.3×10^{9}	10^{13}	5.7×10^{43}	32

accretion rate, the increase of L_γ with M is limited and in fact we find that, increasing M, L_γ reaches a maximum (see Figure 3). A further reduction of the γ-ray emission is due to the fact that the atmosphere may be opaque for photon-photon interaction. The optical depth for this interaction is approximately given by (Lightman *et al.* (1978))

$$\tau_{ee} = \frac{\sigma_T L(E_c)}{4\pi c r E_c} ,$$

(3)

where $E_c \simeq (m_c^2)^2/E$ is the threshold energy for pair production for an incident photon of energy E.

If $\tau_{ee} < 1$ the luminosity effectively released by the system L_γ^* coincides with that due to π° decay. Otherwise the higher photon energies will be cut-off and degraded.

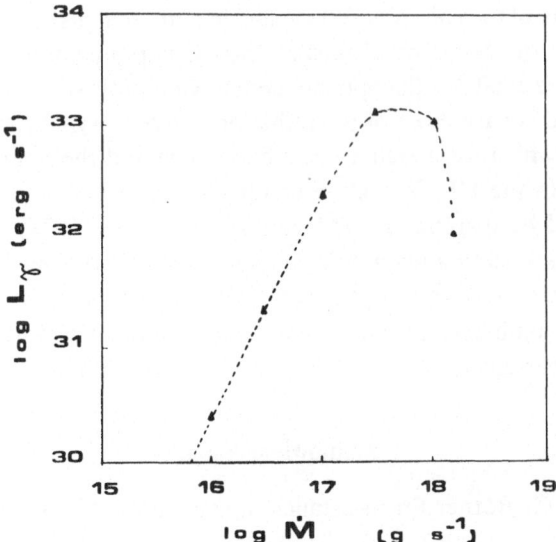

Fig. 3. Gamma-ray luminosity L_γ of a 10 $M_\odot$ black hole for various accretion rates.

Another aspect which is important for a comparison with the observations and was neglected before, is the computation of the spectrum of the radiation emitted by the electrons. If even a low (i.e. with energy density much smaller than the gravitational one) magnetic field is present in the plasma, the emission is dominated by cyclo-synchrotron radiation, which at low harmonics is self absorbed, and by Comptonization of the cyclo-synchrotron photons (Maraschi et al., 1982). In the absence of magnetic field the only emission mechanism is bremsstrahlung, which is less efficient in cooling the electrons. The high temperature reached in this case by the electrons requires to take into account pair production, which has not yet been included in the energy balance equation.

For the magnetized case the X-ray luminosity is much larger than the γ-ray luminosity. An example of X-ray spectrum is given in Figure 2 (see also Table II). Because of the low efficiency and high L_x/L_γ we feel that this model cannot be relevant for explaining the COS-B sources.

It appears that the only way of producing effectively γ-rays with a low L_x/L_γ ratio in an accretion model, is to consider a situation in which non thermal energy distribution functions for the particles are possible. This is very plausible in a scenario where the accreting plasma is magnetized and turbulent, as proposed by Mezsaros (1975). It is however difficult to estimate quantitatively what fraction of the particles will be suprathermal.

If particle acceleration is limited by radiation losses (Cavaliere and Morrison, 1980; Maraschi and Treves, 1977; Pacini and Salvati, 1978) it turns out that the maximum synchrotron emission frequency is ~ 20 MeV, independent of the model parameters. Compton scattering would extend the emission spectrum up to the maximum electron energy of about 10^{10} eV (which is weakly parameter dependent).

In an accretion model it seems likely that, even in the presence of a large runaway component, the bulk of the energy is still located in a thermal distribution. The condition of little emission in the X-ray band requires that Comptonization (multiple Compton scattering) should be small i.e. the optical depth to Compton scattering should be much smaller than 1. In this case most of the radiation is due to synchrotron emission from thermal electrons, which for a stellar mass black hole and the luminosities of interest should be peaked in the UV. Therefore a soft γ-ray and UV emission association is expected and could be searched for (Maraschi and Treves, 1977).

A scaling of the model to a black hole of $\sim 10^4 M_\odot$ would shift the frequency of the synchrotron emission peak ν^* to optical-infrared wavelengths (see Table II). Because of the possibly low luminosity in X-rays and large luminosity in soft γ-rays, a model of this type may be relevant for the source in the Galactic Center.

Acknowledgement

The editors thank G. Börner for assistance in evaluating this paper.

References

Apparao, K. M. V., Bignami, G. F., Maraschi, L., Helmken, H., Margon, B., Hjellming, R., Bradt, H. V., and Dower, R. G.: 1978, *Nature* **273**, 450.

Bennett, K., Bignami, G. F., Hermsen, W., Mayer-Hasselwander, H. A., Paul, J. A., and Scarsi, L.: 1977, *Astron. Astrophys.* **59**, 273.

Bignami, G. F., Maraschi, L., and Treves, A.: 1977, *Astron. Astrophys.* **55**, 155.

Boldt, E. and Kondo, Y. (eds.): 1976, *Proceedings of the GSFC Symposium on X-Ray Binaries*, Contributions to the Cyg X-3, Panel NASA SP-395.

Bradt, H. and McClintock, J. E.: 1983, *ARAA* **21**, (to appear).

Caraveo, P., Bennett, K., Bignami, G. F., Hermsen, W., Kanbach, G., Lebrun, F., Masnou, J. L., Mayer-Hasselwander, H. A., Paul, J. A., Sacco, B., Scarsi, L., Strong, A. W., Swanenburg, B. N., and Wills, R. D.: 1980, *Astron. Astrophys.* **91**, L3.

Cavaliere, A. and Morrison, P.: 1980, *Astrophys. J.* **238**, 63.

Colpi, M., Maraschi, L., and Treves, A.: 1982, Proc. Workshop *Accreting Neutron Stars*, Garching, July 1982, MPE, p. 177.

Colpi, M., Maraschi, L., and Treves, A.: 1983, submitted to *Astrophys. J.*

Dahlbacka, G. H., Chapline, G. F., and Weaver, T. A.: 1974, *Nature* **250**, 37.

D'Amico, N.: 1983, *Space Sci. Rev.* **36**, 195 (this volume).

Giovannelli, F., Karakula, S., and Tkaczyk, W.: 1982, *Astron. Astrophys.* **107**, 377.

Gregory, P. C. and Taylor, A. R.: 1978, *Nature* **272**, 704.

Hermsen, W.: 1983, *Space Sci. Rev.* **36**, 61 (this volume).

Illarionov, A. F. and Syunyaev, R. A.: 1975, *Astron. Astrophys.* **39**, 185.

Kolykhalov, P. I. and Syunyaev, R. A.: 1979, *Soviet Astron.* **23**, 189.

Lamb, R. C., Fichtel, C. E., Hartmann, R. C., Kniffen, D. A., and Thomson, D. J.: 1977, *Astrophys. J. Letters* **212**, L63.

Lightman, A. P., Giacconi, R., and Tananbaum, H.: 1978, *Astrophys. J.* **224**, 375.

Maraschi, L. and Treves, A.: 1977, *Astrophys. J.* **218**, L113.

Maraschi, L. and Treves, A.: 1979, *Nature* **279**, 401.

Maraschi, L. and Treves, A.: 1981a, *Monthly Notices Roy. Astron. Soc.* **194**, 1P.

Maraschi, L. and Treves, A.: 1981b, *Vistas Astron.* **137**, 225.

Maraschi, L. and Treves, A.: 1982, in P. W. Sanford, P. Laskarides, and J. Salton (eds.), *Galactic X-Ray Sources*, J. Wiley and Sons, Chichester.

Maraschi, L., Roasio, R., and Treves, A.: 1982, *Astrophys. J.* **253**, 312.

Maraschi, L., Traversini, R., and Treves, A.: 1983, *Monthly Notices Roy. Astron. Soc.* (in press).

Matteson, J. L.: 1982, in G. R. Riegler and R. D. Blanford (eds.), *The Galactic Center*, A.I.P. Conf. Proceedings, Vol. 83.

Meszaros, P.: 1975, *Astron. Astrophys.* **44**, 59.

Pacini, F. and Salvati, M.: 1978, *Astrophys. J.* **225**, L99.

Rees, M. J. and Gunn, J. E.: 1974, *Monthly Notices Roy. Astron. Soc.* **167**, 1.

Shapiro, S. L. and Salpeter, E. E.: 1975, *Astrophys. J.* **198**, 671.

Skinner, G. K., Bedford, D. K., Elsner, R. F., Leahy, D., Weisskopf, M. C., and Grindlay, J.: 1982, *Nature* **297**, 568.

Swanenburg, B. N., Bennett, K., Bignami, G. F., Buccheri, R., Caraveo, P., Hermsen, W., Konbach, G., Lichti, G. G., Masnou, J. L., Mayer-Hasselwander, H. A., Paul, J. A., Sacco, B., Scarsi, L., and Wills, R. D.: 1981, *Astrophys. J.* **243**, L69.

Taylor, A. R. and Gregory, P. C.: 1982, *Astrophys. J.* **255**, 210.

Van den Heuvel, E. P. J.: 1978, in R. Giacconi and R. Ruffini (eds.), *Physics and Astrophysics of Neutron Star and Black Holes*, North-Holland, Amsterdam.

Vladimirsky, B. M., Stepanian, A. A., and Fomin, V. P.: 1973, *Proc. 13th Int. Conf. Cosmic Rays* **1**, 456.

GAMMA RAYS FROM ACTIVE REGIONS IN THE GALAXY:
THE POSSIBLE CONTRIBUTION OF STELLAR WINDS*

CATHERINE J. CESARSKY and THIERRY MONTMERLE

Service d'Astrophysique, Centre d'Etudes Nucléaires de Saclay, 91191 Gif-sur-Yvette Cedex, France

Abstract. Massive stars ($> 20\,M_\odot$) release a considerable amount of mechanical energy in the form of strong stellar winds. A fraction of this energy may be transferred to relativistic cosmic rays by diffusive shock acceleration at the wind boundary, and/or in the expanding, turbulent wind itself. Massive stars are most frequently found in OB associations, surrounded by H II regions lying at the edge of dense molecular clouds. The interaction of the freshly accelerated particles with matter gives rise to γ-ray emission. In this paper, we first briefly review the current knowledge on the energetics of strong stellar winds from O and Wolf–Rayet stars, as well as from T Tauri stars. Taking into account the finite lifetime of these stars, we then proceed to show that stellar winds dominate the energetics of OB associations during the first 4 to 6 million years, after which supernovae take over. In the solar neighborhood, the star formation rate is constant, and a steady-state situation prevails, in which the supernova contribution is found to be dominant. A small, but meaningful fraction of the COS-B γ-ray sources may be fueled by WR and O stellar winds in OB associations, while the power released by T Tauri stars alone is perhaps insufficient to account for the γ-ray emission of nearby dark clouds. Finally, we discuss some controversial aspects of the physics of particle acceleration by stellar winds.

1. Introduction

Six years have elapsed since the publication of the first COS-B catalogue of γ-ray sources (Hermsen *et al.*, 1977). One of the most important informations brought by this catalogue, and subsequently confirmed by a more homogeneous and complete list of sources (Swanenburg *et al.*, 1981), is that these sources have a very narrow galactic latitude distribution, and therefore must be physically linked with the youngest objects in the Galaxy. This strongly suggests that some stellar associations, like OB associations or T associations, and/or their placental molecular clouds, may constitute a class of γ-ray sources. This is the type of γ-ray source discussed in this paper.

The following ingredients are required: a cosmic-ray 'factory', a confinement mechanism efficient enough to keep the cosmic rays within the vicinity of the factory for some time, and a concentration of interstellar matter, with which cosmic-ray protons and electrons can collide to produce high-energy γ-rays; alternatively, a strong radiation field leading to γ-ray emission via the inverse Compton mechanism.

Let us consider these ingredients in turn. Obvious concentrations of interstellar matter are molecular clouds, or cloud complexes; a typical molecular cloud has a mass $\sim 10^5\,M_\odot$, and a linear size ~ 10–100 pc. We attribute the confinement of cosmic rays close to the factory to resonant interactions of cosmic rays with Alfvén waves they have themselves generated, while streaming through the surrounding gas at a velocity greater that the local Alfvén velocity (see Wentzel, 1974, and references therein). These waves are damped strongly in dense, neutral clouds, but only weakly in ionized media, such

* Proceedings of the XVIII General Assembly of the IAU: *Galactic Astrophysics and Gamma-Ray Astronomy*, held at Patras, Greece, 19 August 1982.

Space Science Reviews **36** (1983) 173–193.   0038–6308/83/0362–0173$03.15.

as H II regions surrounding early-type stars, or the low-density, million-degree 'hot interstellar medium' (HIM). If a large flux of cosmic rays is released in a short time, the particles remain strongly confined within the vicinity of the acceleration region for a long time, even if neutral gas is present (see details in Kulsrud and Zweibel, 1975). If the cosmic rays are released over a long time, the net flux is lower, and so is the growth rate of the waves. When the cosmic-ray factory is surrounded by an H II region, particles are still efficiently scattered in the vicinity of the acceleration region; if it is embedded in a dense, neutral cloud, particles are nevertheless partially confined within the cloud, because of scattering in the surrounding HIM. Problems related to cosmic-ray confinement is dense clouds have been examined by Cesarsky and Völk (1978) and Zweibel and Shull (1982). Detailed self-consistent models of sources embedded in various media have been constructed by Montmerle and Cesarsky (1981) and Cesarsky and Montmerle (1983).

Supernovae (i.e., supernova shocks, or supernova remnants) are the most popular cosmic-ray factory. Indeed, it has been suggested (Montmerle, 1979; Montmerle and Cesarsky, 1980), that $\frac{1}{3}$ to $\frac{1}{2}$ of the COS-B sources can be identified with 'SNOBs' (Supernova remnants physically linked with OB associations, or with giant H II regions, containing early-type stars). It is evident that electrons are accelerated and trapped in supernova remnants, in view of their radio emission by the synchrotron mechanism. Quantitatively, it has been found that, for 8 SNOBs for which all the relevant information was available, the inverse Compton contribution to γ-ray emission is small, while the bremsstrahlung contribution can reasonably account for the bulk of the γ-rays observed, provided the electron spectrum extends down the energies as low as 10 or 20 MeV. Of course, this leaves room for a possible contribution of cosmic-ray protons, via direct $\pi^{\circ} \to 2\gamma$ decay, resulting from their collisions with H atoms. In the solar neighborhood, electrons and protons contribute about equally to the γ-ray emissivity above 100 MeV (Cesarsky et al. 1978; Lebrun and Paul, 1979), in spite of the fact that, at 1 GeV, electrons are 100 times less numerous than protons. More recent work on the relation between supernovae exploding in or close to molecular clouds, and γ-ray sources, is discussed by several people in this Symposium.

Our own task is to consider another potential cosmic-ray factory, the stellar winds (see Cassé and Paul, 1980), and to assess their role in a possible connection between stellar associations and γ-ray sources. In Section 2, we summarize some of the relevant information available on stellar winds from massive OB and Wolf–Rayet (WR) stars, and from low-mass, T Tauri stars. In Section 3, we compute the overall energetics of supernovae and stellar winds on various scales, with the conclusion that winds in general do not play a major role, but may be quite significant in some interesting cases. In Section 4, we address some of the physics underlying the results of Section 3: diffusive shock acceleration at the wind boundary, possibility of injection of particles from the thermal pool or by stellar flares, etc. ... We conclude in Section 5 by a brief outlook on some developments needed to firmly establish the links between stellar winds and γ-ray sources.

2. Properties of Mass-Losing Stars

A. MASSIVE STARS

Data gathered at various wavelengths (mainly in the UV and radio ranges) have shown that massive O and B stars ($M > 20\,M_\odot$) shed a considerable amount of mass in the form of stellar winds (e.g., de Loore, 1980). The mass-loss from O stars is on the order of $10^{-7}\,M_\odot\,\mathrm{yr}^{-1}$, Of stars reaching $10^{-6}\,M_\odot\,\mathrm{yr}^{-1}$. The winds blow at highly supersonic velocities (2000 to 3500 km s^{-1}). The corresponding kinetic 'luminosities' are large, $\gtrsim 10^{36}$ to 10^{37} erg s^{-1}. Integrated over the lifetime of these stars (typically a few million years), the energy released is therefore of the same order as that of a supernova explosion. This energy will be released in most cases *within* an OB association, since 70% to 80% of the known O stars belong to associations; the remainder are runaway stars (Cruz-González et al., 1974).

The maximum rates of mass-loss presumed to last on a significant time scale are observed in WR stars related to Pop. I stars, and are on the order of $\sim 3 \times 10^{-5}\,M_\odot\,\mathrm{yr}^{-1}$. About 140 such stars are known (Van der Hucht et al., 1981). They are thought to be a late, but comparatively brief (a few 10^5 yr, e.g., Maeder and Lequeux, 1982) evolutionary stage of O stars, perhaps immediately preceded by an Of phase (Conti et al., 1979). Several models exist for this transition (Maeder, 1982; de Loore, 1980), but it is thought that all O stars above $\sim 23\,M_\odot$ display the 'WR phenomenon' before becoming supernovae (Maeder and Lequeux, 1982). However, the WR stars appear to be linked somewhat less frequently than O stars with OB associations (or giant H II regions as their tracers), since at least $\sim 60\%$ are isolated objects (Vand der Hucht et al., 1981).

Furthermore, the OB associations in which there are WR stars are quite rare (see Humphreys, 1978): they have therefore specific energetic properties, in much the same way SNOBs do (see discussion in Section 3).

A feature worth mentioning is that the galactic distribution of WR stars displays a steep galactocentric gradient near the solar radius: compared with the blue supergiants, they are 3 times more frequent between 7 and 9 kpc than between 9 and 11 kpc, and 10 times more than between 11 and 13 kpc (Maeder et al., 1980). This will have consequences for the contribution of the winds to the overall energetics on a galactic scale (Section 3).

Another property of massive stars in OB associations, which is particularly helpful for cosmic-ray confinement, is their ability to produce extended H II regions around them. However, the number of ionizing Lyman continuum photons is a strong function of the spectral type: for instance, it is 10^{50} s^{-1} for O4 stars like in the Carina Nebula, and only 10^{49} s^{-1} for O6 stars, the earliest type found in the Orion Nebula (M42), and as low as 10^{45} s^{-1} for B2 stars (see Panagia, 1973). As a result, the sizes of the H II regions are strongly dependent on the stellar content of the OB associations: 50 pc in radius for Carina, down to ~ 2.5 for Orion, and less for associations having later-type stars. (The figure for Carina includes the contribution of the associated WR stars; this contribution is small because WR stars have optically thick envelopes.)

B. LOW-MASS STARS

Recent progress has also been made as regards another class of mass-losing stars: the T Tauri stars. These are low-mass stars ($M \lesssim 2$–$3\,M_\odot$), which usually cluster in associations (T associations), and are linked to small molecular clouds ($\sim 10^3$–$10^4\,M_\odot$), usually called 'dark clouds'. Many observational methods are used to derive the mass-loss characteristics of T Tauri stars: radio emission, Hα emission line width measurements, etc. ... As discussed by De Campli (1981), the mass-loss rates derived are affected by large uncertainties (up to 3 orders of magnitude in some cases), but rates on the order of up to a few $10^{-8}\,M_\odot\,\mathrm{yr}^{-1}$ are consistent with the observational data and not inconsistent with recent theoretical suggestions. This does not exclude the possibility of *eruptive* mass loss, leading to time-averaged values 10 times higher (Silk, 1983; see also Section 3.C). On the other hand, the terminal velocities are moderate, though still supersonic, $\sim 250\,\mathrm{km\,s}^{-1}$. As a result, the mechanical energy release per star is not enormous ($\sim 10^{32-33}\,\mathrm{erg\,s}^{-1}$), but the T Tauri stars are plentiful (up to several tens in the ρ Oph dark cloud for instance) and, at least in their earliest stages, appear to be concentrated and buried within the external layers of their parent cloud.

While T Tauri stars have spectral types (K5 to M5) corresponding to cool photospheres ($\sim 3500\,\mathrm{K}$), recent UV observations have shown that a significant part of their surface is covered by hot emission regions analogous to, but much more extended than, solar plages (Giampapa *et al.*, 1982). They are not likely, however, to drive extended H II regions − a significant difference, in our context, with respect to hot stars.

3. Stellar Winds: Contenders of Supernovae as Generators of Cosmic Rays and Gamma Rays?

A. MECHANICAL POWER RELEASED BY MASSIVE STARS

The lifetime of an OB association is typically 2×10^7 yr, after which the association is dispersed because of random star motions. Such an association may be made up of several well-separated sub-associations (e.g., Blaauw, 1964; Reeves, 1978). The stars more massive than several $M_\odot$ end their lives in the form of supernovae.

The formation time scale for massive stars in associations (i.e., on a small scale) is much smaller than the stellar lifetimes, hence we can approximate the birth of an association (or of a sub-association, as the case may be) by a localized *burst*. In the solar neighborhood (i.e., on a larger scale), births and deaths of massive stars average out, resulting in a *steady state*. On this scale, the star formation rate must have been constant for more than 3×10^8 yr, at least for A stars (Grosbøl, 1978).

Lequeux (1979) has derived the Initial Mass Function (IMF), using the stellar population of the solar neighborhood ($\lesssim 2.5$ kpc), for masses 2.5 to 100 $M_\odot$. Claudius and Grosbøl (1980) derived the IMF of individual young OB associations: their results, at least in the range 2.5–10 $M_\odot$, are compatible with those of Lequeux, and have been essentially confirmed by more recent work, covering a wider mass range (1.25–14 $M_\odot$; Tarrab, 1982). We emphasize that the situation for higher masses is much less clear. Consequently, we first use Lequeux's IMF (which is comparatively poor in stars with masses above $20\,M_\odot$), and then investigate the consequences of using another, more recent IMF, which is richer in high-mass stars (Garmany et al., 1982).

We adopt, for the rate of star formation $\dot{N}$, and for the number of stars formed simultaneously, N, per unit area (with masses in $M_\odot$), the following expressions:

Steady state:

$$\frac{d\dot{N}}{d\ln M} = \zeta' M^\alpha \begin{cases} \zeta' = 1.3 \times 10^{-3}\,\mathrm{yr}^{-1}\,\mathrm{kpc}^{-2}. \\ \alpha = -2.0 \end{cases} \tag{3.1}$$

Burst:

$$\frac{dN}{d\ln M} = \zeta M^\alpha, \qquad \alpha = -2.0. \tag{3.2}$$

ζ being a normalisation factor, which may vary from one association to the next; it is related to the 'strength' of the burst. In both cases, the IMF extends up to some $M_{\max}$.

The lower limit M_p to the mass of the progenitors of type II supernovae is still under debate, and we take $M_p = 4\,M_\odot$ or $M_p = 8\,M_\odot$. The upper limit is $M_{\mathrm{II}} \sim 23\,M_\odot$, since supernovae from WR stars ($\mathrm{SN}_{\mathrm{WR}}$) must form an observationnally distinct supernova class, owing to the absence of hydrogen from the core of WR stars. (This may be the case of Cas A; see Maeder, 1983a.) On the other hand, the progenitors of type I supernovae are not well understood; they are possibly low-mass stars in binary systems. It is therefore impossible to derive a SN I rate using IMF considerations, but observations of supernova explosions in external galaxies of type similar to the Milky Way suggest that the rates of SN I and SN II explosions are about equal (Tammann, 1981). We assume that the same is true in the solar neighborhood. The kinetic energy released by each supernova explosion is taken to be $\overline{E}_s = 10^{51}$ ergs for SN I, SN II, and $\mathrm{SN}_{\mathrm{WR}}$.

As for stellar winds, the rate of mass-loss $\dot{M}$ for OB stars is observed to depend mainly on the bolometric magnitude M_{bol} of the star. The influence of other parameters, such as gravity, temperature, etc. ... cannot be clearly disentangled, given the observational uncertainties (e.g., de Loore, 1980) although several attempts have been made (e.g., Lamers, 1981). Using a theoretical HR diagram, it is possible to derive an empirical relation between $\dot{M}$ and the luminosity L (Lamers, 1981), and, going one step further, and $\dot{M}-M$ relation (since for high masses, the luminosity remains approximately constant throughout the evolution), in the form:

$$\dot{M} = \lambda M^\mu \quad \left.\begin{array}{l} \lambda = 10^{-8}\,M_\odot\,\mathrm{yr}^{-1} \\ \mu = 1.6 \end{array}\right\} \quad M > M_{\min} = 20\,M_\odot$$

$$\lambda = 0 \qquad\qquad\qquad M < 20\,M_\odot, \tag{3.3}$$

The wind terminal velocity is, on average:

$$\langle v_\infty \rangle = 2500\,\mathrm{km\ s}^{-1}.$$

The theoretical lifetime $\tau(M)$ of stars is approximated by:

$$\tau(M) \simeq \theta_1 M^{\gamma_1}, \qquad M > 15 \, M_\odot \tag{3.4}$$

with $\theta_1 = 5.7 \times 10^7$ yr, $\gamma_1 = -0.7$ (from de Loore, 1980), and

$$\tau(M) \simeq \theta_2 M^{\gamma_2}, \qquad 4 \, M_\odot < M < 15 \, M_\odot \tag{3.5}$$

with $\theta_2 = 9.4 \times 10^8$ yr, $\gamma_2 = -1.73$ (from Miller and Scalo, 1979).

To compute the total power released by WR stars, it is best to use directly the statistical data, since their relation to other stars is not fully understood. The surface density σ of WR stars in the solar neighborhood is $\sigma \sim 1.8$ kpc^{-2} (Hidayat et al., 1981). Also, we take $\dot{M} = 3 \times 10^{-5} \, M_\odot$ yr^{-1} and $v_\infty = 2500$ km s^{-1} for all WR stars.

(i) Steady state

In the steady-state case, the average mechanical powers released by supernovae and stellar winds per unit area are:

$$\bar{P}_s(\mathrm{I}) = \bar{P}_s(\mathrm{II}),$$

$$\bar{P}_s(\mathrm{II}) = \bar{E}_s \int_{M_p}^{M_{\mathrm{II}}} \frac{d\dot{N}}{dM} \, dM,$$

$$\bar{P}_s(\mathrm{WR}) = \bar{E}_s \int_{M_{\mathrm{II}}}^{M_{\mathrm{max}}} \frac{d\dot{N}}{dM} \, dM, \tag{3.6}$$

$$\bar{P}_w(\mathrm{OB}) = \int_{M_{\mathrm{min}}}^{M_{\mathrm{max}}} \tfrac{1}{2}\dot{M} \langle v_\infty \rangle^2 \frac{dN}{dM} \tau(M) \, dM,$$

$$\bar{P}_w(\mathrm{WR}) = \tfrac{1}{2}(\sigma \dot{M} v_\infty^2)_{\mathrm{WR}}.$$

With the IMF of Lequeux and $M_{\mathrm{max}} = 120 \, M_\odot$, one has, altogether:

$$\begin{cases} \bar{P}_s = 2 \times 10^{52} \text{ erg kpc}^{-2} (10^6 \text{ yr})^{-1} & \text{if } M_p = 8 \, M_\odot, \\ \bar{P}_s = 8 \times 10^{52} \text{ erg kpc}^{-2} (10^6 \text{ yr})^{-1} & \text{if } M_p = 4 \, M_\odot; \end{cases}$$

$$\bar{P}_w(\mathrm{OB}) = 1.3 \times 10^{51} \text{ erg kpc}^{-2} (10^6 \text{ yr})^{-1}; \tag{3.7}$$

$$\bar{P}_w(\mathrm{WR}) = 3.5 \times 10^{51} \text{ erg kpc}^{-2} (10^6 \text{ yr})^{-1}.$$

Therefore, WR stars dominate the energetics of stellar winds, not only individually, but also collectively. Still, the total mechanical power released by supernovae exceeds that of stellar winds by a fairly large factor:

$$\left. \begin{array}{ll} \bar{P}_s/\bar{P}_w = 5 & \text{if } M_p = 8 \, M_\odot \\ \bar{P}_s/\bar{P}_w = 20 & \text{if } M_p = 4 \, M_\odot \end{array} \right\}. \tag{3.8}$$

These results depend only weakly on M_{max}, provided, of course, that $M_{max} \gg 20\,M_\odot$. For instance, if $M_{max} = 60\,M_\odot$ instead of $120\,M_\odot$, $\overline{P}_w$ is lower by 20%, $\overline{P}_w$ (WR) by 10%, $\overline{P}_s$ is practically unchanged. The results are not very sensitive either with respect to the slope of the IMF, if different at high masses from that at low masses. For instance, if $d\dot{N}/dM \propto M^{-2.6}$ above $20\,M_\odot$, as proposed by Garmany et al. (1982), the ratio $\overline{P}_s/\overline{P}_w$ of Equation (3.8) is decreased by 20%.

These results are in excellent agreement with those obtained recently by Abbott (1982), by summing up directly the individual contributions of the stars, type by type. This may be taken as a confirmation of the validity of our adopted formalism.

However, the results of Equations (3.8) are perhaps not valid beyond the solar neighborhood. Given the magnitude of the observed WR/OB gradient (Section 2) as a function of galactocentric distance (a factor of 10 increase from 13 to 7 kpc) we cannot rule out the intriguing possibility that winds from WR stars shed more mechanical energy than supernovae in the inner Galaxy.

(ii) Burst

Let us turn now to the energetics of a region where a *burst* of star formation has just taken place. During the liftime of an OB association, some field stars may explode as SN I; they are so rare that we disregard their contribution. The mechanical energies released are given by:

$$
E_s\,(\mathrm{II}) = p_s \overline{E}_s \int\limits_{M(t)}^{M_{max}} \frac{dN}{dM}\,dM \begin{cases} p_s = 0 & \text{if } t < \tau(M_{max}), \\ p_s = 1 & \text{if } t \geq \tau(M_{max}), \end{cases}
$$

$$
E_s\,(\mathrm{I}) \ll E_s\,(\mathrm{II}),
$$

$$ (3.9) $$

$$
E_w\,(\mathrm{OB}) = p_w \int\limits_{0}^{t} dt \int\limits_{M_{min}}^{M^*} \tfrac{1}{2} M \langle v_\infty \rangle^2 \frac{dN}{dM}\,dM,
$$

$$
\begin{cases} p_w = 1, & M^* = M_{max} & \text{if } t < \tau(M_{max}), \\ p_w = 1, & M^* = M(t) & \text{if } \tau(M_{max}) \leq t \leq \tau(M_{min}), \\ p_w = 0, & & \text{if } t > \tau(M_{min}). \end{cases}
$$

$$ (3.10) $$

For WR stars, it is not possible to make a similar evaluation without some additional assumptions about their genesis. We then assume that all stars having more than $M_{II} = 23\,M_\odot$ become WR stars near the end of their evolution, and that this stage lasts $\sim 4 \times 10^5$ yr (Maeder and Lequeux, 1982). The total energy supplied by a WR star is then $\overline{E}_{WR} \simeq 7 \times 10^{50}$ ergs, comparable to that of a SN explosion. Since this energy is released on a comparatively short timescale, the contribution of WR stars can be calculated like that of SN (Equation (3.9)), with $p_{WR} = 1$ for $\tau(M_{max}) \leq t \leq \tau(M_{II})$, and $p_{WR} = 0$ elsewhere.

To calculate the normalization factor ζ appearing in Equation (3.2), we will consider that OB associations having very early-type stars (O4, O3) i.e. $M_{max} \simeq 120\,M_\odot$, contain roughly 40 stars with masses above $15\,M_\odot$ (O stars and supergiants; see Humphreys, 1978). Then $\zeta = 2 \times 10^4$.

The corresponding (absolute) powers $\overline{P} = E$ are represented as a function of time on Figure 1 for OB and WR winds and for SN explosions. With $M_{max} = 120\,M_\odot$, the total power turns out to be approximately constant in the wind-dominated phase of the OB association, $1{-}2 \times 10^{38}$ erg s^{-1}. Winds dominate during the first ~ 5 million years, SN II thereafter. Such a configuration could be representative of the Carina Nebula, which contains 3 WR stars and several O3 years (and probably no supernova, see discussion in Montmerle *et al.*, 1982), or of the Cygnus X complex, featuring 7 WR stars

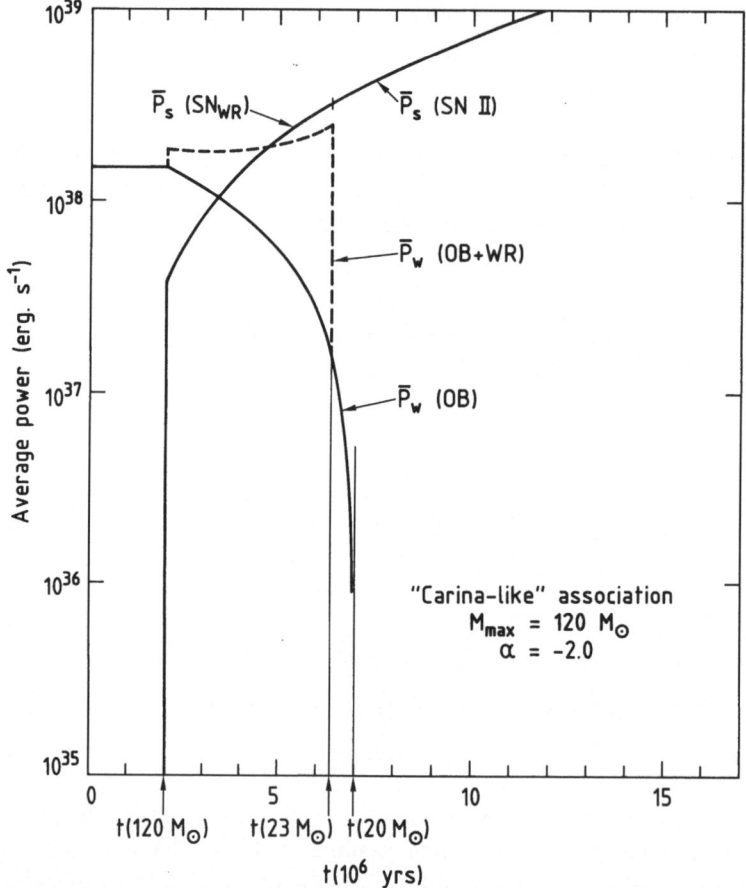

Fig. 1. The energetics of a 'Carina-like' OB association as a function of time. Such associations are characterized by a high mass cut-off in the IMF (slope $\alpha = -2.0$, after Lequeux, 1979) at $M_{max} = 120\,M_\odot$. Wolf–Rayet stars (WR) are here supposed to be a late evolutionary stage of all stars more massive than $23\,M_\odot$, immediately preceding their explosion in the form of supernovae. The minimum mass for a star to have a strong stellar wind is taken as $20\,M_\odot$. (The average power is normalized so that the association comprises about 40 stars between $15\,M_\odot$ and $120\,M_\odot$, and matches approximately the actual mechanical power observed to be released by the OB and WR stars in the Carina Nebula.)

and several O3–O4 stars in its two youngest associations. The mechanical power determined from the mass-loss actually measured by IUE from the early-type and WR stars present in the Carina Nebula is $\sim 3 \times 10^{38}$ erg s^{-1} (see discussion in Montmerle, 1981; and Abbott, 1982), which compares favourably with the results shown on Figure 1.

In reality, M_{max} is probably different from one association to the next. Take for instance, the Orion OBId association (Trapezium cluster, ionizing M42). Its age is estimated, from the bluest Main-Sequence stars, at $\lesssim 5 \times 10^5$ yr (see discussion in Warren and Hesser, 1978; and Reeves, 1978), whereas it contains no star above $\sim 30\,M_\odot$ (earliest type O6) which are able to live up to 4×10^6 yr. The stellar content is therefore very different from that of the 'Carina-like' associations mentioned earlier.

Taking now $M_{max} = 30\,M_\odot$ as representative of 'Orion-like' associations, we obtain Figure 2. Orion is in the pre-WR stage, and Figure 2 gives $P_w = 4.5 \times 10^{37}$ erg s^{-1}, as compared with 5×10^{37} erg s^{-1} obtained by summing the actual contribution of stars

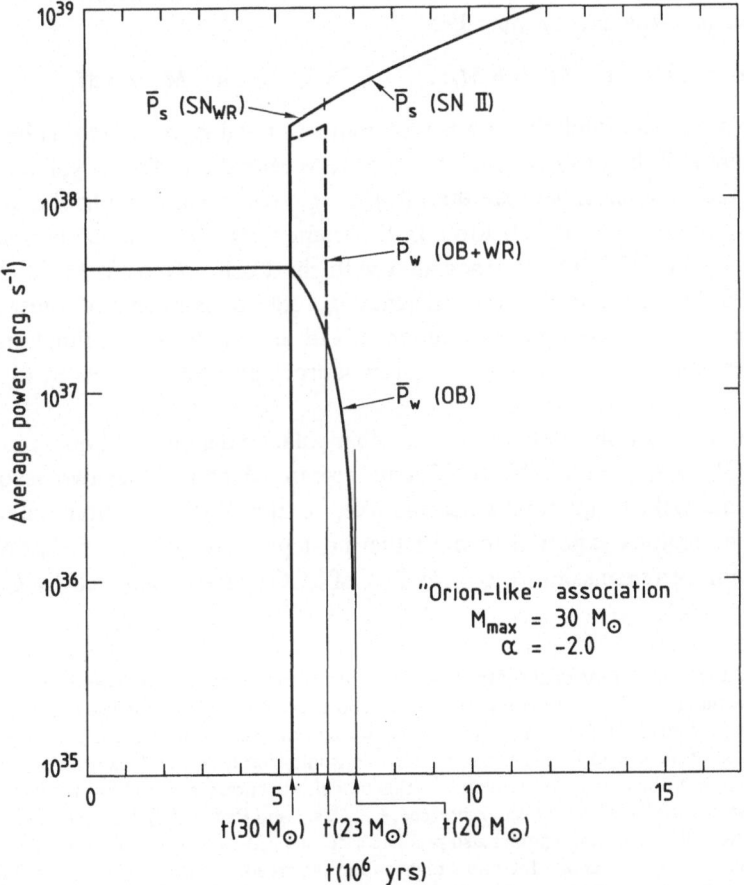

Fig. 2. Same as Figure 1, for 'Orion-like' associations, characterized by $M_{max} = 30\,M_\odot$.

with known mass-loss (Montmerle, 1981). Once the WR stage is reached, it lasts only ~ 1/10 of the total association lifetime. This explains – qualitatively – why there are relatively few associations with WR stars in them. Note also that SNRs dominate the energetics as soon as the first mass-losing stars die: this may also explain why there are more SNOBs than WR-dominated associations.

Another interesting case is the Gould Belt (Stothers and Frogel, 1974), an expanding ring of gas clouds and young stars, lying ~ 150–500 kpc around the Sun and thought to be about 30 million years old (see Olano, 1982). It probably originated in a huge burst of star formation, followed by other, smaller events. Its estimated age indicates that, as a whole, this part of the solar neighborhood has been in the SN-dominated phase for the last several million years*.

B. STELLAR WINDS, γ-RAYS, AND COSMIC RAYS

The power required to maintain the observed cosmic-ray pool in the solar neighborhood is ~ 2.3×10^{51} erg kpc^{-2} $(10^6$ yr$)^{-1}$, (e.g., Blandford and Ostriker, 1980). If supernovae are the factories of galactic cosmic rays, the efficiency of conversion of mechanical power into cosmic-ray energy must be:

$$\eta_s = 10\% \quad \text{if} \quad M_p = 8\,M_\odot\,, \qquad \eta_s = 2.5\% \quad \text{if} \quad M_p = 4\,M_\odot\,.$$

Local stellar winds can fulfill the power requirement only if $\eta_w \simeq 1$. On this basis, local supernovae are still the prime candidates for the acceleration of local galactic cosmic rays. For further reference, we note that, if $\eta_w \simeq \eta_s$, stellar winds contribute a fraction $f_w \simeq \frac{1}{5}$ to $\frac{1}{20}$ of the particles (see below). In the framework of the self-consistent model of the γ-ray source 2CG 288 – 0 associated with the Carina Nebula, Montmerle and Cesarsky (1981) have found that the efficiency η_w required is indeed of a few percent.

More generally, considering the evolution of OB associations as a function of time leads to other consequences as regards γ-ray sources and peculiarities in the cosmic rays.

In addition to the Carina Nebula, several other galactic regions are known to contain WR and/or Of stars, most notably the Cygnus region, which includes two associations (Cyg OB 1 and OB 2), apparently Carina-like, in the WR-wind dominated phase (Figure 1). The regions expected to have the most powerful winds are known to be associated with γ-ray emission: the γ-ray flux of 2CG 288 – 0 measured by COS-B is

---

* A very different scenario for the origin of the Gould Belt has been presented by Strauss *et al.* (1979). These authors consider that the Belt is a self-gravitating system, born as a result of the collapse of a massive clump of gas (~ $10^7\,M_\odot$), about 50 million years ago. The gas would have been almost entirely turned into stars, and the expansion of the Gould Belt in the galactic plane would simply be the dynamical consequence of the vertical collapse of the clump on the plane. In this model, the energy released by stellar winds in the solar neighborhood must have been considerable (see, e.g., Cassé and Paul, 1980). However, if an IMF such as that of Lequeux (1979) holds, the model also predicts a density of 1 to 5 $M_\odot$ stars – no higher-mass star is still alive after 5×10^7 yr – about 150 times higher than observed. Models of the Gould Belt such as recently proposed by Olano (1982), in which the expansion is driven by stellar winds and SN explosions, are much more satisfactory in this respect.

1.6×10^{-6} photons (> 100 MeV) cm^{-2} s^{-1} (Swanenburg *et al.*, 1981) whereas the Carina Nebula lies at ~ 2.5 kpc from the Sun. As for the Cygnus complex, ~ 1.8 kpc away, it is also a strong γ-ray emitter, but has a structure more complex than a single source.

Using the catalogue of Humphreys (1978), and assuming that the acceleration and confinement properties are as in the case of the Carina Nebula, it is possible to predict which of the known associations should be visible in γ-rays. The result is that only some of the associations featuring WR stars lie above the visibility threshold of COS-B (1.0×10^{-6} ph cm^{-2} s^{-1} for a localized source): Cyg OB 2, which is part of a bright γ-ray complex, Sco OB 1, ~ 2 kpc away, unfortunately near the galactic center direction, hence buried in a strong galactic γ-ray background, and the Carina associations. Six other associations featuring WR and/or Of stars, are either not powerful enough or too distant to be visible (Cassé *et al.*, 1981; see also Abbott, 1982).

What about Orion-like associations in the wind-dominated phase? If, again, efficiencies of acceleration and confinement are the same than in the Carina Nebula, we would clearly expect them to be below the visibility threshold of COS-B, if at a typical COS-B source distance, ~ 2 kpc. This is also confirmed by Cassé *et al.* (1981).

But Orion itself should emit a flux of $\sim 10^{-5}$ ph cm^{-2} s^{-1}, whereas no γ-ray source is found in this region. We attribute this lack of observed γ-ray emission to a much lower confinement efficiency by the ionized gas, considering that M42 is a much smaller H II region than the Carina Nebula (Cesarsky and Montmerle, 1982).

In the supernova-dominated phase, Orion-like associations release much more power, and thus are in a much better position to power a γ-ray source; hence the possible identification of SNOBs with a significant fraction of the γ-ray sources (Montmerle, 1979).

In short, OB associations may be 'typical' COS-B sources *only* if they are powered by WR winds or supernovae. (This is not inconsistent with the assertion of Wolfendale (1982), that most molecular clouds – hence, OB associations – are 'inert' γ-ray emitters, i.e., dominated by ambient galactic cosmic rays.)

If true, the very fact that an association is a γ-ray source indicates that the cosmic rays must have traversed a grammage X not small with respect to the proton interaction length, ~ 70 g cm^{-2}. This is most easily done while associations are still young and embedded in a dense gaseous medium. In the case of the Carina Nebula, taking confinement by ionized regions into account, we find $X \simeq 40$ g cm^{-2} (Montmerle and Cesarsky, 1981). Most nuclei will then be broken up by spallation reactions, while antiprotons (in addition to γ-rays) will be copiously produced as secondaries resulting from inelastic collisions of protons with the cloud particles. This type of 'thick source' may explain the high $\bar{p}/p$ ratio observed at a few GeV by Golden *et al.* (1979), and Bogomolov *et al.* (1979), but not the $\bar{p}$ flux detected around 300 MeV by Buffigton *et al.* (1981), see Cesarsky (1982). The number of such $\bar{p}$ sources required is consistent with the number of γ-ray sources and peaks of γ-ray emission in the galactic plane observed by COS-B (Cesarsky and Montmerle, 1981; Cowsik and Gaisser, 1981). Given the values of $f_w = \frac{1}{5}$ to $\frac{1}{20}$ found above, it may well be that a sizeable fraction of the

(γ-ray + $\bar{p}$) sources are related to stellar winds embedded in dense, ionized regions.

Still, mass-losing stars are not found exclusively in associations, or in large cloud complexes: for instance, we have seen that many WR stars or runaway O stars do not belong to associations. In this context, Meyer (1981) has shown that the observed overabundances of both ^{22}Ne and C in cosmic rays could be explained if a fraction f_c $\sim 1/50$ of the galactic cosmic ray material originated in quiescent He-burning stellar layers (in which all the CNO is turned into ^{22}Ne). Cassé and Paul (1982) have proposed that carbon-rich WR stars (WC type), which make up about half of all WR stars (see Van der Hucht et al., 1981), provide a plausible site for extraction of this material. Detailed models for the evolution of massive stars recently developed by Maeder (1983b) lead to a similar value of f_c. We note that this fraction of cosmic rays should not however traverse more than the usual 7 g cm^{-2}.

C. Gamma Rays Associated With Low-Mass Stars?

Compared with the energy output of winds from massive stars and SN, the mechanical power associated with T Tauri stars seem minute, about 3×10^{32} erg s^{-1} at most per star (perhaps 10 times larger if the possibility of eruptive mass loss is taken into account).

However, molecular clouds contain a large number of these stars (T Tauri or related pre-Main-Sequence objects), lying often within the boundaries of the cloud. A powerful tool to detect them is through their highly variable X-ray emission: in a recent *Einstein* survey of the ρOph cloud (Montmerle et al., 1983), ~ 160 pc distant and associated with the γ-ray source 2CG 353 + 16, about 60 such stars were found. This is more than twice the previously known number of such objects. The total mechanical power released $P_{w,\,tot}$ is therefore on the order of

$$P_{w,\,tot} \simeq 2 \times 10^{34} \text{ erg s}^{-1} .$$

Assume further that the rate of conversion of gas into stars is $\sim 10\%$ (which is reasonable for ρOph if the PMS stars have $\sim 3\,M_\odot$ on average, with a cloud mass of $\sim 2000\,M_\odot$, see discussion in Montmerle et al., 1983). For clouds having $M \sim 10^5\,M_\odot$, we get about 3000 stars, i.e., a total power $\sim 5 \times 10^{35}$ erg s^{-1}. This remains small with respect to the contribution of massive stars.

Now, if we assume that the confinement properties of the Carina Nebula and of the ρOph cloud are identical, scaling for the wind powers, 'ρOph-like' clouds should not be visible at the level of 10^{-6} ph cm^{-2} s^{-1} further away than 32 pc. This indicates that either the confinement is even more efficient, or that an energy source other than the PMS star winds is present. For the specific case of ρOph, there may be also up to 9 massive B2 stars present (see discussion in Montmerle et al., 1983), boosting the wind power to 4×10^{35} erg s^{-1}. The 'visibility range' then becomes ~ 140 pc, in satisfactory agreement with the distance of the cloud. This visibility range may be increased to ~ 170 pc if *all* PMS stars display eruptive mass loss.

An alternative proposal has been made by Morfill et al. (1980), in terms of a chance collision between the ρOph cloud and a fraction of an old supernova remnant believed to be associated with the North Polar Spur, and visible in soft X-rays. If true, the source 2CG 353 + 16 would then fall into the SNOB class, even though the original supernova is not genetically linked with the cloud.

4. Making a Gamma-Ray Souce Out of Stellar Winds

A. A HANDY MECHANISM: DIFFUSIVE SHOCK ACCELERATION

During the past years, the theory of particle acceleration by shock waves in a diffusive medium has developed rapidly (see recent reviews by Axford, 1981; Drury, 1983; also Webb, 1983). This mechanism relies on the fact that fast particles of velocity v increase their momentum by a relative amount $\sim w/v$ every time they cross a shock of velocity w. If particles are scattered efficiently on both sides of the shock, they remain trapped for some time in the shock vicinity, and, on average, cross the shock v/w times. But a few particles remain around for a longer time, as in any Fermi-type mechanism, so that a power-law spectrum of cosmic rays develops. The most attractive feature of such a mechanism is that, in the case of an infinite plane shock, and in the time-independent limit, the spectral index depends only on the compression ratio of the shock, ρ(downstream)/ρ(upstream) (Bobalsky, 1977a, b, 1978a, b), as long as the angle φ between the magnetic field direction and the shock normal is not too close to 90° (tan $\varphi < v/w$ in the shock rest frame). In the context of galactic cosmic-ray acceleration, it has been applied to supernova shocks, and to stellar wind terminal shocks, which separate the wind from the external medium. Stellar wind terminal shocks are like inverted supernova shocks, the shocked gas lying outside of the shock (Weaver et al., 1977).

Depending on the value of the diffusion coefficient K in the vicinity of the shock, different results are obtained. If $K(R) \gtrsim wR$ (when R is the shock radius), the adiabatic losses suffered by the particles while they are diffusing in the stellar wind (in the absence of local acceleration, see next section), hinder seriously the efficiency of the acceleration mechanism. In that case, even in the time-independent, linear limit (i.e., neglecting the back-reaction of cosmic rays on the shock), the problem is extremely cumbersome. Webb et al. (1981), Forman et al. (1981), and Webb (1983) were able to solve the problem analytically when the diffusion coefficient K is assumed to be independent of momentum, and proportional to the distance to the star, while Drury (1983) gives a solution valid when $K/(wR)$ is small, but not negligible. For a given rate of particle injection, the maximum yield in cosmic rays is obtained when $K/(wR)$ is very small, in which case the shock can be considered as planar, allowing to recover the simple power-law spectrum predicted by the elementary theory.

The limit $K \ll wR$, which is thus the most favourable for stellar wind acceleration of cosmic rays, has been adopted in recent discussions of this problem by Montmerle and Cesarsky (1981), Völk and Forman (1981), and Cesarsky and Montmerle (1982). In the remainder of this paper, we will discuss the most controversial issues regarding this type of model.

B. CURRENT OBJECTIONS TO THE PLAUSIBILITY OF STELLAR WIND ACCELERATION, AND POSSIBLE WAYS OUT

The initial framework for SW acceleration (Cassé and Paul, 1980) involved the possibility of injecting MeV particles by stellar-flare-like events, to fulfill a possible requirement of pre-existing non thermal particles injected into the diffusive shock mechanism.

Two major objections were put forward against this approach:

(1) Adiabatic losses suffered by flare particles, during their transport out to the border of the wind cavity, typically a few million times the stellar radius, must be enormous, thus bringing efficiently the initially non-thermal particles back into the thermal pool (Völk, 1981; Völk and Forman, 1982).

(2) The magnetic field lines, anchored to a mass-losing star, are in the form of Archimedes spirals if the star is rotating (Parker, 1958). As a result, far from the star, the magnetic field is azimuthal, and diffusive shock acceleration does not work any more.

Several ways out of these difficulties may be suggested. While flare particles should indeed suffer enormous adiabatic losses, they may also gain energy as they pass through the wind. A first possibility is that, as Montmerle and Cesarsky (1981) pointed out, stellar flares particles may be re-accelerated directly by encounters with shocks while they traverse the stellar wind cavity; indeed, recent observations have shown that interplanetary acceleration does occur in the solar wind, and, apparently the farther from the Sun, the more efficiently (McDonald, 1981). Another possibility is to consider that the strong winds of OB and WR stars are likely to be highly turbulent, as implied by recent models (see below). The flare particles would then be *continuously* accelerated by a second-order Fermi mechanism, the acceleration time being:

$$t_a = 3\Lambda v/v_A^2 \,, \tag{4.1}$$

where Λ is the mean free path of particles having a velocity v, and v_A is the Alfvén velocity in the wind. At a distance r from the star, the adiabatic loss time is

$$t_l = 3r/w \,. \tag{4.2}$$

Acceleration overcomes the adiabatic losses if $t_a \ll t_l$, hence when the ratio $\beta = v_A^2 r/(\Lambda v w)$ is $\gg 1$.

Typical values of the parameters for O stars are $v_\infty = 2500$ km s^{-1}, the stellar radius $R_* = 10^{12}$ cm. To compute the Alfvén velocity, we need to know the wind density $n(r)$ and the magnetic field $B(r)$. Assuming flux conservation for the wind gives $n(r) = n_*(R_*/r)^2$, n_* being the density in the upper atmosphere of the star, where the wind starts to get accelerated; it is reasonable to take $n_* = 10^{10}$ cm^{-3}. In a Parker-type wind, the radial component of B dominates up to $r = R_B$, the azimuthal component thereafter. For a stellar rotation velocity $v_* \sim 250$ km s^{-1}, $R_B \sim (v_\infty/v_*)R_* \sim 10\,R_*$. For $r < R_B$, $B(r) \simeq B_*(R_*/r)^2$; for $r > R_B$, $B(r) \simeq B(R_B)(R_B/r)$. Values of B of the order of 100 G are implied by the variability in the X-ray flux observed in some O stars

(Vaiana, 1981), if interpreted in terms of $\gtrsim 10^7$ K flares (see discussion in Montmerle *et al.*, 1983). Therefore $B(R_B) \simeq 1$ G and $B(R_1) \simeq 3 \times 10^{-6}$ G at the terminal shock, situated at a typical distance $R_1 \simeq 1$ pc from the star. (This is probably a lower limit, since the turbulence itself may add some dynamo effect within the wind.) As a result, far from the star $(r > R_B)$, v_A is constant:

$$v_A = \frac{B(R_B)}{(4\pi m_p n_*)^{1/2}} \frac{R_B}{R_*} . \tag{4.3}$$

With the parameters above, one finds $v_A \simeq 200$ km s^{-1}. The diffusion mean free path will be taken as $\Lambda \simeq r_g/a$, where r_g is the gyroradius of the particles, and $a < 1$ is a parameter which, in the framework of the quasi-linear theory, is roughly equal to the ratio of the energy density in waves resonating with the particles to the energy density of the magnetic field. For flare-originating particles of typical energy ~ 1 MeV $(v \sim 1.5 \times 10^4$ km s$^{-1})$:

$$\Lambda = \Lambda_1/aB \tag{4.4}$$

with $\Lambda_1 \simeq 1.5 \times 10^5$ cm, and B in gauss. Therefore, using $w \simeq v_\infty$ far from the star:

$$\beta = 8.8 \times 10^4 \, a . \tag{4.5}$$

Hence, acceleration overcomes adiabatic losses if $a \gg 10^{-5}$ throughout the cavity. This is not unreasonable in a turbulent wind: in the interplanetary medium, at 1 AU, $a \simeq 0.1$–0.3 (see, e.g. Fisk, 1979).

Now the question arises whether the necessary turbulence can be sustained throughout the wind cavity. Recent advances in the theory of mass loss from massive stars suggest that this may be the case.

One should first realize that the mechanisms driving the winds of these stars must be *very different* from those driving the solar wind, to account for the huge mass flux, higher by a factor of up to 10^7. The basic idea is that the mass loss must somehow be related to the very large radiative energy output $(L_{bol} \lesssim 10^{39}$ erg s$^{-1})$.

An interesting model, in our framework, has been put forward recently by Lucy and White (1980), and Lucy (1982a, b). In essence, the radiation-driven mass outflow is subject to Rayleigh–Taylor-type instabilities (Nelson and Hearn, 1978), which presumably lead to the formation of inhomogeneities ('blobs'). The calculations show that the radiation pressure from UV resonance lines is able to accelerate these blobs, which drag the surrounding gas along at supersonic velocities. The shock waves they create heat the wind to X-ray temperatures. Beyond a certain radius R_0, the line-of-sight to the star crosses so many blobs that a 'shadowing' effect occurs. As a result, the radiation pressure goes to zero, but the wind velocity has by then already reached almost its terminal value v_∞. For a wind velocity of the form $w = v_\infty(1 - R_*/r)$, one has $R_0 \simeq 10\, R_*$. This picture is supported by various observations (e.g., ζ Pup, see discussion in Lucy, 1982a; ζ^1 Sco, see Burki *et al.*, 1982; the unusual X-ray source 4U 1700–37, see White *et al.*, 1983).

Beyond the 'radiative piston region' (RPR, i.e., from R_* to R_0), it is possible to show that, under certain conditions, the blobs are able to maintain supersonic velocities throughout the wind cavity (i.e., out to $R_1 \simeq 10^6 R_*$ or more); the relative inefficiency of the braking is due mainly to the r^{-2} dependence of the wind density (Cesarsky and Montmerle, 1983). The shock waves continuously generated throughout the passage of the blobs in the wind cavity may maintain a sufficiently high level of turbulence everywhere. This is a complex problem, however, and one must await more detailed calculations on turbulence generation and dissipation in the wind to draw definitive conclusions as to whether indeed $\beta \gg 1$ everywhere.

An alternative possibility is that, instead of originating in flares, the particles participating the acceleration process are ions picked directly out of the tail of the thermal plasma, as suggested by Eichler (1979), Krymsky (1980), and Ellison et al. (1982). This approach has been studied recently by Völk and Forman (1982). Rather than elaborating a self-consistent scheme of shock-regulated injection, they assume, in analogy with the Earth's bow shock, that $\sim 1\%$ of the stellar wind ions are injected into the process. They argue that the acceleration can only be intermittent, occurring along small parts of the therminal shock where the magnetic field lines are at a finite angle to the shock, for a time on the order of a stellar rotation period at most, $\sim 10^5$–10^6 s. In such short times, the ions can reach only a few MeV. As a result, according to these authors, stellar winds could at best be associated with sources of nuclear γ-rays, but not of γ-rays in the COS-B range, which require protons above ~ 1 GeV.

However, in view of the fact that the blobs have radial supersonic velocities with respect to the mean flow, the field must have also a significant component perpendicular to the shock, so that the angle φ is close, but not quite equal, to $90°$. Acceleration at the wind boundary can then occur only for particles which can satisfy the injection threshold requirement $v > w \tan \varphi$. These are mainly particles from stellar flares having overcome adiabatic losses because of turbulence, as opposed to the wind particles themselves. The number of such particles available is probably not high enough to affect the shock structure so that the linear theory of shock acceleration is valid.

Provided the acceleration is not intermittent, and in the optimum case, the highest energies that cosmic rays of charge Z can attain at stellar wind terminal shock are:

$$E_{max} = 4 \times 10^6 \, Z (B/10^{-5} \, \text{G}) \, (w/2.5 \times 10^8 \, \text{cm s}^{-1})^2 \, \text{GeV}$$

whereas for supernova shocks, under similar conditions:

$$E_{max} < 10^5 \, Z (B/10^{-6} \, \text{G}) \, \text{GeV} \, ,$$

(Cesarsky and Lagage, 1981). Stellar winds are better than supernovae to boost particles to very high energies for two reasons: stellar wind shocks are bounded on both sides by a highly turbulent medium (and therefore short acceleration times are expected), and the shock velocity remains high significantly longer than for supernova shocks.

D. WINDS FROM T TAURI STARS

As compared with O stars, the wind driving mechanisms for T Tauri stars seem to be much less well understood. Of course, radiation is by far insufficient, in energy and in wavelength. What is sure, however, is that the winds from T Tauri stars are very different from the acoustic wave-driven solar wind. In particular, if it were not so, the X-ray luminosity that one would expect from T Tauri stars should be several orders of magnitude higher than observed (De Campli, 1981; Montmerle et al., 1983).

A class of current models is based on Alfvén waves, assumed to originate in the 'shaking' of the surface magnetic field by the convective zone which exists just below. The matter is driven by hydromagnetic pressure, and mass losses up to $\sim 10^{-8} M_\odot$ yr^{-1} may be explained in this way. (De Campli (1981) estimates that higher mass-loss rates cannot be explained and that the observational uncertainties are compatible with the conclusion that they do not exist although Silk (1983) argues that higher mass losses, $\sim 10^{-7} M_\odot$ yr^{-1}, may have an eruptive origin.) As such, this mechanisms does not generate shocks, and therefore does not appear to be able to lead to particle acceleration within the wind, as in the case of massive stars, although wave driven shocks may be a possibility. But the stability of such a flow has not been studied, and it is known observationnally that, at least in some cases (YY Ori stars), there is evidence for a competition between mass outflow and accretion (see Appenzeller, 1982), leading to Rayleigh–Taylor instabilities at the interface between the wind and the surrounding dense medium. Cases of non-spherical mass loss are also known in Herbig–Haro objects, related to T Tauri stars (see Mundt and Hartmann, 1982).

A good deal more work seems necessary at present to fully understand T Tauri star winds, and therefore the situation as to the adiabatic losses of particles injected by the giant flares known to be present (Montmerle et al., 1983), or as to the magnetic field configuration, must still be considered as open. Even binarity, as recently discovered in the case of T Tauri itself (Dyck et al., 1982), may play a role.

5. Conclusions

Stellar winds can play a significant role in various areas of galactic astrophysics. This role is in general short-lived, since massive O stars ($> 20 M_\odot$) live only a few million years, WR stars a few 10^5 yr, and T Tauri stars several 10^5 yr. Stellar winds probably dominate the energetics of the earliest stages of evolution of OB associations, giving rise to γ-ray sources observed by COS-B only if WR stars are present. They are also perhaps an original clue to some intriguing problems in galactic cosmic-ray astrophysics, such as the origin of the ^{22}Ne excess, the high $\bar{p}/p$ ratio, or the origin of some very high energy cosmic rays.

On the other hand, time is working for supernovae. Indeed, since all stars with masses above 4 or 8 $M_\odot$ end up as supernovae, they largely outnumber the stars able to have strong stellar winds. In the long run, then, the victory of supernovae over stellar winds seems unavoidable ... The situation, for various galactic environments, is summed up in Table I.

Of course, the possible relatively small, albeit physically meaningful, role of stellar winds relies entirely on their assumed ability to accelerate particles to relativistic

TABLE I

Contribution of stellar winds and supernovae to cosmic rays and gamma rays in the Galaxy

Scale	Medium (distance)	Stellar winds important for:	Supernovae important for:	Remarks
Very small ($\lesssim 1$ pc)	Dark clouds ($\lesssim 200$ pc)	T associations, if CR confinement strong enough: γ-ray sources?	if chance collision with field SNRs: γ-ray sources	ρOph cloud only known possible example
Small (~ 10–100 pc)	Molecular clouds ($\lesssim 3$ kpc)	OB associations, if WR present (Carina, Cygnus): γ-ray sources $\bar{p}$ in CR very high-energy CR?	OB associations, if SN present (SNOBs): γ-ray sources $\bar{p}$ in CR	Average OB associations ('Orion-like') invisible as γ-ray sources
Medium ($\lesssim 1$–2 kpc)	Solar neighborhood ($\lesssim 2.5$ kpc) Gould Belt ($\lesssim 500$ pc)	^{22}Ne excess in CR from isolated WC; diffuse γ-ray features	Local CR; diffuse γ-ray features	$\bar{P}_s/\bar{P}_w = 5$ or 20 (depending on SN progenitors)
Large	Galaxy	dominant contribution to GCR from WR in the inner galaxy? part of diffuse γ-ray emission	probable major contribution to GCR; part of diffuse γ-ray emission	gives SN acceleration efficiency: $\eta_s = 2.5$ to 10%

energies. On theoretical grounds, this has not been clearly demonstrated yet. But the prospects of such a demonstration look rather promising for O and WR stars. The situation, unfortunately, is much less clear for low-mass pre-Main Sequence stars, for which the very phenomenon of mass loss is still poorly understood.

Can we expect some advances on the observational side? Obviously, one of the best ways is to look at OB associations in γ-rays, but a great leap forward is required from the experiments, both in angular resolution and sensitivity. The angular resolution must first help in assessing the identification of OB associations with a class of γ-ray sources, and must be such that it becomes possible to separate out possible γ-ray 'hot spots' linked with OB associations in a molecular complex. For Orion as well as for Carina, at least a few arc minutes must be reached, a factor of 10 better than COS-B. The gain required in sensitivity should be large enough to allow observations of as many associations as possible in detail, in particular to check the links between confinement efficiency and extent of the ionized regions. A gain in sensitivity of a factor of 10 would allow to observe Orion-like associations out to ~ 1.5 kpc, while an improved angular resolution would help in increasing the contrast with a possible galactic diffuse γ-ray emission on the same line of sight.

Even then, it should be stressed that the observational answer to the problem of CR acceleration by stellar winds may not be clearcut, as it is very difficult to be sure that, in an association or in a molecular cloud, no SN lies hidden somewhere. For instance, at face value, the diffuse X-ray flux from the Carina Nebula region could be explained in terms of a SNR a few 10^5 yr old (Seward and Chlebowski, 1982); we have seen that a SNR, associated with the North Polar Spur, may be at work to explain the γ-ray source in the direction of the ρOph dark cloud.

Perhaps another distinct possibility is to look for non-thermal radio emission from T Tauri stars. At frequencies corresponding to a good angular resolution, one should choose preferably T Tauri stars with weak mass loss, otherwise the free-free emission associated with the stellar wind (e.g., Berthout and Thum, 1982) might bury the possible non-thermal emission. But only (comparatively) strong fluxes, on the order of a few tenths of mJy, could be detected with the best instrument to date, namely the Very Large Array. Such a research is now in progress for the ρOph dark cloud, in which there are several groups of PMS stars, these stars being separated typically by a few arc minutes. Observations at lower frequencies ($\lesssim 100$ MHz) are also a possibility, provided the stars are sufficiently apart.

It thus seems that, for some time, we will have to rely mostly on theory to decide whether or not stellar winds are able to accelerate particles. "The answer, my friend, is blowing in the wind", but we do not know yet how to listen to it ...

Acknowledgements

We thank Michel Cassé, James Lequeux, Leon Lucy, André Maeder, Jean-Paul Meyer, and Joe Silk for helpful discussions. We also thank Gary Webb for useful remarks on the manuscript, and Greg Morfill for his patience during the preparation of this paper.

The editors thank G. Webb for assistance in evaluating this paper.

References

Abbott, D. C.: 1982, *Astrophys. J.* **263**, 723.
Appenzeller, I.: 1982, *Fund. Cosmic Phys.* 7, 313.
Axford, W. I.: 1981, *Proc. 17th Int. Cosmic Ray Conf., Paris* **12**, 155.
Axford, W. I., Leer, E., and Skadron, G.: 1977, *Proc. 15th Int. Cosmic Ray Conf., Plovdiv* **11**, 132.
Bell, A. R.: 1978, *Monthly Notices Roy. Astron. Soc.* **182**, 147.
Bertout, C. and Thum, C.: 1982, *Astron. Astrophys.* **107**, 368.
Blaauw, A.: 1964, *Ann. Rev. Astron. Astrophys.* **2**, 213.
Blandford, R. D. and Ostriker, J. P.: 1978, *Astrophys. J. Letters* **221**, L29.
Blandford, R. D. and Ostriker, J. P.: 1980, *Astrophys. J.* **237**, 793.
Bogomolov, E. A., Lubyanaya, N. D., Romanov, V. A., Stepanov, S. V., and Shulakova, M. S.: 1979, *Proc. 16th Int. Cosmic Ray Conf., Kyoto* **1**, 330.
Buffington, A., Schindler, S. M., and Pennypacker, R.: 1981, *Astrophys. J.* **248**, 1179.
Burki, G., Heck, A., Bianchi, L., and Cassatella, A.: 1982, *Astron. Astrophys.* **107**, 205.
Cassé, M. and Paul, J. A.: 1980, *Astrophys. J.* **237**, 236.
Cassé, M. and Paul, J. A.: 1982, *Astrophys. J.* **258**, 860.
Cassé, M., Montmerle, T., and Paul, J. A.: 1981, in G. Setti, G. Spada, and A. W. Wolfendale (eds.), 'Origin of Cosmic Rays', *IAU Symp.* **94**, 323.
Cesarsky, C. J.: 1982, *International School of Cosmic-Ray Astrophysics*, Third Course, Erice (in press).
Cesarsky, C. J. and Lagage, P. O.: 1981, *Proc. 17th Int. Cosmic Ray Conf., Paris* **2**, 335.

Cesarsky, C. J. and Montmerle, T.: 1981, *Proc. 17th Int. Cosmic Ray Conf., Paris* **1**, 173.
Cesarsky, C. J. and Montmerle, T.: 1983, in preparation.
Cesarsky, C. J. and Völk, M. J.: 1978, *Astron. Astrophys.* **70**, 367.
Cesarsky, C. J., Paul, J. A., and Shukla, P.: 1978, *Astrophys. Space Sci.* **59**, 73.
Claudius, M. and Grosbøl, P. J.: 1980, *Astron. Astrophys.* **87**, 339.
Cowsik, R. and Gaisser, T. K.: 1981, *Proc. 17th Int. Cosmic Ray Conf., Paris* **2**, 218.
Conti, P. S., Niemela, V. S., and Walborn, N. R.: 1979, *Astrophys. J.* **228**, 206.
Cruz-González, C., Recillas-Cruz, E., Costero, R., Peimbert, M., and Torres-Peimbert, S.: 1974, *Rev. Mex. de Astron. y Astrophys.* **1**, 211.
De Campli, W. M.: 1981, *Astrophys. J.* **244**, 124.
De Loore, C.: 1980, *Space Sci. Rev.* **26**, 113.
Drury, L. O'C.: 1983, *Rep. Progr. Phys.*, in press.
Dyck, H. M., Simon, T., and Zuckerman, B.: 1982, *Astrophys. J. Letters* **255**, L103.
Eichler, D.: 1979, *Astrophys. J.* **229**, 419.
Ellison, D. C., Jones, F. C., and Eichler, D.: 1981, *J. Geophys.* **50**, 110.
Fisk, L. A.: 1979, in C. F. Kennel, L. J. Lanzerotti, and E. N. Parker (eds.), *Solar System Plasma Physics*, Vol. 1, North-Holland, Amsterdam, p. 177.
Forman, M. A., Webb, G. M., and Axford, W. I.: 1981, *Proc. 17th Int. Cosmic Ray Conf., Paris* **9**, 238.
Garmany, C. D., Conti, P. S., and Chiosi, C.: 1982, *Astrophys. J.* **263**, 777.
Giampapa, M. S., Calvet, N., Imhoff, C. L., and Kuhi, L. V.: 1982, *Astrophys. J.* **251**, 113.
Golden, R. L., Horan, S., Mauger, B. G., Badhwar, G. D., Lacy, J. L., Stephens, S. A., Daniel, R. R., and Zipse, J. E.: 1979, *Phys. Rev. Letters* **43**, 1196.
Grosbøl, P. J.: 1978, *Astron. Astrophys. Suppl.* **32**, 409.
Hermsen, W., Swanenburg, B. N., Bignami, G. F., Boella, G., Buccheri, R., Scarsi, L., Kanbach, G., Mayer-Hasselwander, H. A., Masnou, J. L., Paul, J. A., Bennett, K., Higdon, J. C., Lichti, G. G., Taylor, B. G., and Wills, R. D.: 1977, *Nature* **269**, 494.
Hidayat, B., Supelli, K., and Van der Hucht, K.: 1981, *Contr. Bosscha Obs.*, No. 68.
Humphreys, R. M.: 1978, *Astrophys. J. Suppl.* **38**, 309.
Krymsky, G. F.: 1977, *Dokl. Akad. Nauk SSSR* **234**, 1306.
Krymsky, G. F.: 1980, *Proc. 7th European Cosmic Ray Symposium*, Leningrad.
Kulsrud, R. M. and Zweibel, E.: 1975, *Proc. 14th Int. Cosmic Ray Conf. Munich* **2**, 465.
Lamers, H. J. G. L. M.: 1981, *Astrophys. J.* **245**, 593.
Lebrun, F. and Paul, J. A.: 1979, *Proc. 16th Int. Cosmic Ray Conf., Kyoto* **12**, 13.
Lequeux, J.: 1979, *Astron. Astrophys.* **80**, 35.
Lucy, L. B.: 1982a, *Astrophys. J.* **255**, 286.
Lucy, L. B.: 1982b, *Astrophys. J.* **255**, 278.
Lucy, L. B. and White, R. L.: 1980, *Astrophys. J.* **241**, 300.
Maeder, A.: 1982, *Astron. Astrophys.* **105**, 149.
Maeder, A.: 1983a, *Astron. Astrophys.*, **120**, 113.
Maeder, A.: 1983b, *Astron. Astrophys.*, **120**, 130.
Maeder, A. and Lequeux, J.: 1982, *Astron. Astrophys.* **114**, 409.
Maeder, A., Lequeux, J., and Azzopardi, M.: 1980, *Astron. Astrophys.* **90**, L17.
Meyer, J. P.: 1981, *Proc. 17th Int. Cosmic Ray Conf., Paris* **2**, 265.
McDonald, F. D.: 1981, *Proc. 17th Int. Cosmic Ray Conf., Paris* **13**, 199.
Miller, G. E. and Scalo, J. M.: 1979, *Astrophys. J. Suppl.* **41**, 513.
Montmerle, T.: 1979, *Astrophys. J.* **231**, 95.
Montmerle, T.: 1981, *Phil. Trans. Roy. Soc. London* **A301**, 505.
Montmerle, T. and Cesarsky, C. J.: 1980, *Proc. COSPAR Symp. on Non-Solar Gamma Rays*, Bangalore, India, *Adv. Sp. Expl.* **7**, 61.
Montmerle, T. and Cesarsky, C. J.: 1981, *Proc. Int. School and Workshop on Plasma Astrophysics*, Varenna, ESA SP-161, p. 319.
Montmerle, T., Cassé, M., and Paul, J. A.: 1982, *Astrophys. J.*, submitted.
Montmerle, T., Koch-Miramond, L., Falgarone, E., and Grindlay, J. E.: 1983, *Astrophys. J.* **269**, 182.
Morfill, G. E., Völk, H. J., Drury, L. O'C., Forman, M., Bignami, G. F., and Caraveo, P.A.: 1981, *Astrophys. J.* **246**, 810.
Mundt, R. and Hartmann, L.: 1982, *Astrophys. J.*, in press.

Nelson, G. D. and Hearn, A. G.: 1978, *Astron. Astrophys.* **65**, 223.

Olano, C. A.: 1982, *Astron. Astrophys.* **112**, 195.

Panagia, N.: 1973, *Astron. J.* **78**, 929.

Parker, E. N.: 1958, *Astrophys. J.* **128**, 664.

Reeves, H.: 1978, in T. Gehrels (ed.), *Conf. on Protostars and Planets*, Univ. of Arizona Press, Tucson, p. 399.

Seward, F. D. and Chlebowski, T.: *Astrophys. J.* **256**, 530.

Silk, J.: 1983, *Proc. N.Y. Acad. Sci.* **395**, 257.

Strauss, F. M., Poeppel, W. G. L., and Vieira, E. R.: 1979, *Astron. Astrophys.* **71**, 319.

Stothers, R. and Frogel, J. A.: 1974, *Astron. J.* **79**, 456.

Swanenburg, B. *et al.*: 1981, *Astrophys. J. Letters* **243**, L49.

Tammann, G.: 1981, in M. J. Rees and R. J. Stoneham (eds.), *Supernovae*, D. Reidel Publ. Co., Dordrecht, Holland, p. 371.

Tarrab, I.: 1982, *Astron. Astrophys.* **109**, 285.

Vaiana, G. S.: 1981, *Space Sci. Rev.* **30**, 151.

Van der Hucht, K. A., Conti, P. S., Lundström, I., and Stenholm, B.: 1981, *Space Sci. Rev.* **28**, 227.

Völk, H. J.: 1981, *Izv. AN SSSR, ser. fiz.* **45**, (7), in press.

Völk, H. J. and Forman, M.: 1982, *Astrophys. J.* **253**, 188.

Warren, W. H., Jr. and Hesser, J. E.: 1978, *Astrophys. J. Suppl.* **36**, 497.

Webb, G. M.: 1983, *Astrophys. J.*, in press.

Webb, G. M., Axford, W. I., and Forman, M. A.: 1981, *Proc. 17th Int. Cosmic Ray Conf., Paris* **2**, 309.

Wentzel, D. G.: 1974, *Ann. Rev. Astron. Astrophys.* **12**, 71.

White, N. E., Kallman, T. R., and Swank, J. H.: 1983, *Astrophys. J.* **269**, 264.

Wolfendale, A. W.: 1982, *Proc. XXIV COSPAR Meeting*, Ottawa, in press.

Zweibel, E. G. and Shull, J. M.: 1982, *Astrophys. J.* **259**, 859.

SEARCH FOR PULSARS AS COUNTERPARTS
OF THE GALACTIC GAMMA-RAY SOURCES*

N. D'AMICO

Istituto di Fisica dell'Università, via Archirafi 36, 90123 Palermo, Italy

Abstract. A sensitive search for pulsars inside a sample of gamma-ray source error boxes has been carried out using the Arecibo and Parkes radiotelescopes. The paper describes the motivation of this search and the characteristics of the experiments used. As a preliminary result, new pulsars have been discovered and some of them are possibly candidates to be the counterparts of the gamma-ray sources.

1. Introduction

The point spread function of present gamma-ray telescopes is orders of magnitude broader than that of the telescopes at lower energies, and the concept of a point source in gamma-ray astronomy is quite rough. In fact, the intrinsic angular extent of unresolved gamma-ray sources could be upt to 1–2 degrees (Swanenburg *et al.*, 1981). It is difficult to derive informations about the compactness of a gamma-ray source from a non parametric analysis of its time variability: the observed flux is too low and short scale variations are confused by low counting statistics. The association of a gamma-ray source with a known compact object on a pure positional basis is of course also difficult to be explored since the typical error box for the source position is of the order of several square degrees.

In fact, there are no observational arguments for the *unidentified* gamma-ray sources to be compact. However, compact gamma-ray emitters are believed to exist (Meszaros, 1975; Massaro and Salvati, 1979; Cavallo and Pacini, 1980); in this sense, the gamma-ray error boxes have simply the role of general indicators of those regions of the sky where these objects could be found.

The observed gamma-ray emission from the Crab and Vela Pulsars (the only two identified gamma-ray sources) is the only observational example of compact gamma-ray emitters. In principle, the fact that the only two identified gamma-ray sources are pulsars, could be simply related to observational reasons, because their periodical signature is the unique tool for identification at gamma-ray energies. From this point of view, the gamma-ray emission from pulsars could be just a sporadic effect, leaving open the problem of the nature of all the other gamma-ray sources. On the other hand, if the gamma-ray emission is in some way a feature of pulsars, or better, it is related to a particular stage of their evolution, pulsars may have a major role as counterparts of gamma-ray sources (Massaro and Salvati, 1979; D'Amico and Scarsi, 1979; Buccheri *et al.*, 1978).

* Proceedings of the XVIII General Assembly of the IAU: *Galactic Astrophysics and Gamma-Ray Astronomy*, held at Patras, Greece, 19 August 1982.

Space Science Reviews **36** (1983) 195–205.   0038–6308/83/0362–0195$01.65.

The Crab and Vela pulsars are two peculiar objects if compared with the general characteristics of all the other known pulsars:

they are the two fastest free pulsars;

they are two strong gamma-ray sources;

they are the only two firm associations between a pulsar and a SNR.

This coincidence suggests a physical link between these three characteristics. The limited evidence for this link to be systematic is not surprising: a deficit of short period pulsars is observed and a lack of pulsar-SNR associations exists, indicating that the observational panorama about the young stage of a pulsar is poor.

On the other hand, it is possible that, because of selection effects on previous pulsar surveys, some other fast pulsars have been missed. Coupling these facts with the observed properties of Crab and Vela, the gamma-ray error boxes could be suggested, at the same level of SNRs, as indicators for the presence of missed Crab-like objects.

It is not obvious whether or not there is a unique link between fast pulsars, SNRs and the *observed* population of galactic gamma-ray sources. In this case in fact, an association between SNRs and gamma-ray sources should be already observable. In spite of several suggestions (Lamb, 1978; Van den Bergh, 1979), this association is not self-explanatory: besides Crab and Vela, there are five other cases of possible associations, and in some other cases a SNR is just outside a gamma-ray error box (see Table I). The fact that not all SNRs are associated with a significant gamma-ray flux, is not a negative argument if we take into account the beaming effects, as it is invoked

TABLE I

Associations SNR/gamma-ray source

Possible associations	
Gamma-ray source	SNR
2CG 184 – 05	CRAB (FIRM)
2CG 263 – 02	VELA (FIRM)
2CG 006 + 00	W28
2CG 078 + 01	γ-cigny
2CG 311 – 01	G311.5 – 0.3
321.0 – 1.2	G320.4 – 1.2
327.5 – 0.5	KES27

SNRs just outside a gamma-ray error box		
Gamma-ray source	SNR	Separation (degrees)
2CG 065 + 00	DA495	1.2
2CG 075 + 00	CTB87	1.2
2CG 356 + 00	MSH17 – 39	1.2
2CG 333 + 01	MSH16 – 51	1.4

for most of the lack of pulsar-SNR association. On the other hand, because of the number of observed SNRs and the large size of the gamma-ray error boxes, the suggested associations could be simply a coincidence by chance. Going back to the established facts, the association between pulsars and gamma-rays, and between pulsars and SNRs is supported by few results.

Figure 1 sketches the observational situation: the expected connection between pulsars and SNRs is obviously based on the fact that pulsars are believed to be formed in an event of SN explosion. The possibility of the gamma-ray emission as a further tracer of the young stage of a pulsar is attractive, but the only piece of this puzzle is the existence of Crab and Vela.

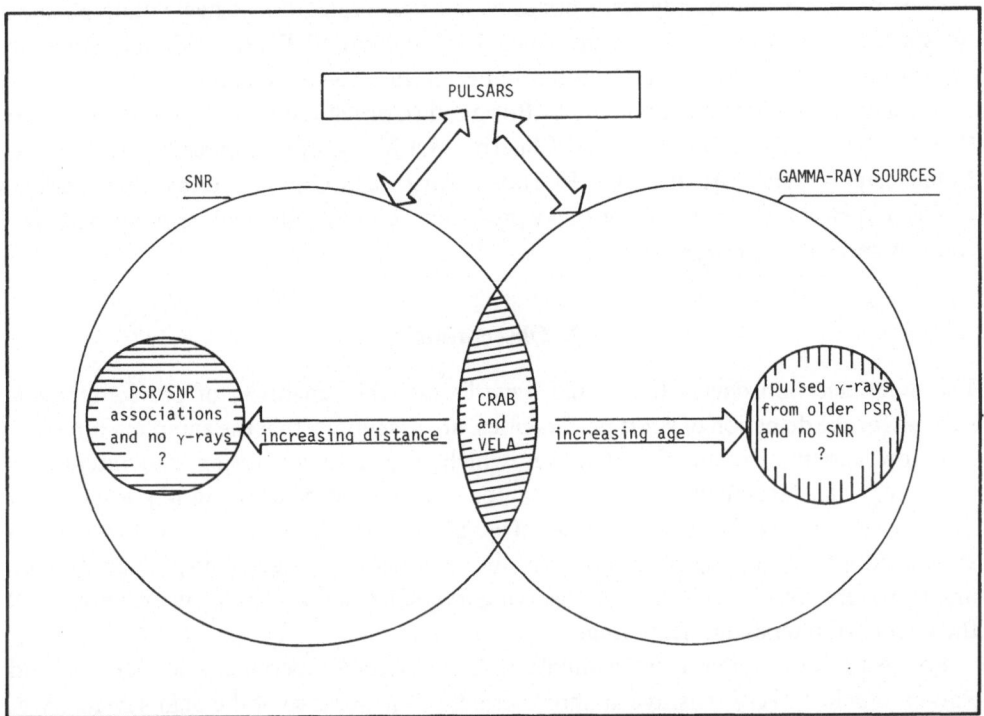

Fig. 1.   Observational situation about pulsars, SNRs and gamma-ray sources. We do expect a connection between pulsars and SNRs. A possible connection between pulsars and gamma-ray sources is suggested. Crab and Vela represent the only cases of pulsars firmly associated with both a SNR and a gamma-ray source.

Some other possible associations pulsar-SNR have been suggested (see for instance Manchester *et al.*, 1982a) in cases in which no obvious gamma-ray emission is present, and on the other side, a possible detection of pulsed gamma-ray emission from old pulsars has been reported (see for instance: Mandrou *et al.*, 1980), and in these cases no SNR is associated. Even if confirmed, these facts are not in conflict with the suggested connections: Crab-like objects would be undetectable at gamma-ray energies if more distant than ~ 4 kpc, while older pulsars where the associated SNR has been dispersed, could still exhibit a sporadic gamma-ray emission, detectable if they are sufficiently nearby. On the track of the properties of Crab and Vela, it is more realistic to believe that the gamma-ray emission is essentially related to the young stage of a pulsar, but the 'gamma-ray season' could have a different life time than the associated SNR. All these aspects have to be learned from future observations.

The general problem, as sketched in Figure 1, could be approached from different sides. On the other hand, in spite of several ideas and speculations, the lack of counterpart of the galactic gamma-ray sources persists. Following these considerations we believe that a pulsar search inside an *unbiased* sample of gamma-ray error boxes would be an useful investigation about the 'gamma-ray side' of the general sketch.

Looking at the gamma-ray error boxes as possible indicators of new young pulsars, we started a collaboration program using the Arecibo and Parkes radiotelescopes in order to make a sensitive search in a number of gamma-ray sources.

It is a massive program, involving V. Boriakoff (Cornell Univ., U.S.A.), R. Buccheri (IFCI/CNR, Italy), and F. Fauci (Palermo Univ., Italy) at Arecibo and R. N. Manchester (CSIRO, Australia), I. R. Tuohy (Australian Univ.), and myself at Parkes.

This paper is a status report of these programs. Final results will be published very soon by individual groups.

2. Observations

The main selection effects that could have limited the sensitivity of previous pulsar surveys for the detection of short periodicities, are essentially that the sampling time was too slow and the broadening of pulses which is due to dispersion and interstellar scattering. This last effect is rather crucial for short period pulsars and in particular for lines of sight along the galactic plane. Therefore, a sensitive search for fast pulsars should adopt a sampling time down to few milliseconds, a good dispersion system, and a proper compromise between observing frequency and sensitivity in order to avoid the effects of interstellar scattering.

Following these criteria, we estimate that the actual experiments at Arecibo and Parkes should be very sensitive at short periodicities. Because of the high sensitivity of these experiments and the large size of the gamma-ray error boxes, we do expect to discover also new pulsars possibly not connected with the gamma-ray sources. In order to estimate the association, similarities with the properties of Crab and Vela (age, energy loss) should be observed, and final conclusions can be reached *phasing* the gamma-ray data with the pulsar parameters.

This is a strong argument for this search to be considered more realistic if compared with other searches for counterparts of the gamma-ray sources based on a simple positional coincidence.

A. THE ARECIBO EXPERIMENT

Table II summarizes the main characteristics of the Arecibo experiment (Boriakoff

TABLE II

The ARECIBO experiment

ARECIBO antenna 300 m
Receiver 2 pol. chan.
Obs. freq. 318 and 430 MHz
31 × 2 dispersion chan.
Sampling time 4 ms
Integration time 10 min
Beam size 16' (318 MHz) and 9' (430 MHz)
System noise 120 K
Sensitivity ∼ 1 mJy

et al., 1983). The sensitivity is estimated to be about one mJy for periods down to 40–50 ms. Off-line analysis procedures are used: the individual dispersion channels are searched for periodicities after an FFT transformation; the time domain data are then folded and cross-correlated to check for a consistent dispersion shift.

Table III shows the gamma-ray error boxes observed. As it can be seen, the coverage is not complete and so the data analysis. The work is in progress but more observing sessions and routine analysis are required to complete the project.

TABLE III

ARECIBO observations

Gamma-ray source	Coverage (%)	Analysis (%)	Obs. freq. (MHz)
2CG 036 + 01	70	84	318
	10	15	430
2CG 054 + 01	70	65	318
	–	–	430
2CG 065 + 00	100	92	318
	51	68	430
2CG 195 + 04	50	100	318
	50	100	430

B. THE PARKES EXPERIMENT

Table IV shows the characteristics of the experiment used at Parkes (Manchester et al., 1982b, 1983). An observing frequency of 1.4 GHz was chosen in this case to better overcome the effects of interstellar scattering. The sensitivity of this experiment is 1 mJy down to period of 15–20 ms and DM values up to 1800 cm^{-3} pc.

Off-line analysis procedures were used. In this case the data were first dedispersed and then transformed using an FFT algorithm. A harmonic summing procedure was also introduced to search for periodicities. The time domain data were then folded at the preriodicities found and the significance of the effect was estimated.

TABLE IV

The Parkes experiment

Parkes antenna 64 m
FET cooled receiver 2 pol. chan.
Obs. freq. 1.4 GHz
4 × 2 dispersion chan.
Sampling time 2 ms
Integration time 18 min
Beam size 15'
System noise 40 K
Sensitivity 1 mJy

A complete sample of galactic and Magellanic Cloud SNRs, 3 gamma-ray error boxes and other objects have been observed (see again: Manchester *et al.*, 1983) during two observing sessions in November 1981 and March 1982. Table V shows the gamma-ray error boxes observed. In this case the coverage is quite complete and all the available data have been already analysed.

TABLE V

Parkes observations

Gamma-ray source	Coverage (%)	Analysis (%)
2CG 006 + 00	85	100
2CG 333 + 01	85	100
2CG 195 + 04	100	100

3. Results

Results from Arecibo are quite preliminary. There are some candidate pulsars and some of these at short period. A peculiar object has been discovered and confirmed within the 2CG 0036 + 00 error box. It is a pulsar relatively fast (period ~ 284 ms), with a structured light curve. We expect to measure very soon the period derivative in order to estimate the pulsar age and the rotational energy loss. If these parameters will be considered encouraging for this object to be the counterpart of the gamma-ray source, a timing program will be undertaken to obtain pulsar parameters with sufficient accuracy to fold the gamma-ray data. From Parkes, no significant signals have been found within the error box of 2CG 195 + 04. This source (the 'Geminga' gamma-ray source), has been extensively searched for periodicities also by other groups (Mandolesi *et al.*, 1978; Ozel *et al.*, 1980; Seiradakis, 1981). The present result is the most sensitive upper limit available up to now. A lack of obvious candidate counterparts for this strong gamma-ray source still exists. An X-ray source has been discovered within its error box using the

IPC on board the Einstein Observatory (Caraveo, 1983). However, there are no immediate arguments to claim an association between the two sources, and the nature of the X-ray source itself is unknown. No significant signals have been found within the 2CG 333 + 00 error box. The situation is more interesting for 2CG 006 + 00. This source is overlapping with the SNR W28. A possible connection between the gamma-ray source and the SNR has been studied recently (Andrews *et al.*, 1982) with the VLA and the HRI of the Einstein Observatory. Their results support a possible association: on W28 "the radio data have shown a flat spectrum central component reminiscent of Crab and Vela" and "the X-ray observations have revealed a compact source within W28, again reminiscent of both Crab and Vela".

Our coverage of 2CG 006 + 00 was quite uniform. We didn't detected any pulsation at the X-ray source position, which is already scheduled to be observed again at higher sensitivity. On the other hand, we discovered a new pulsar (PSR 1758 − 23) at the edge

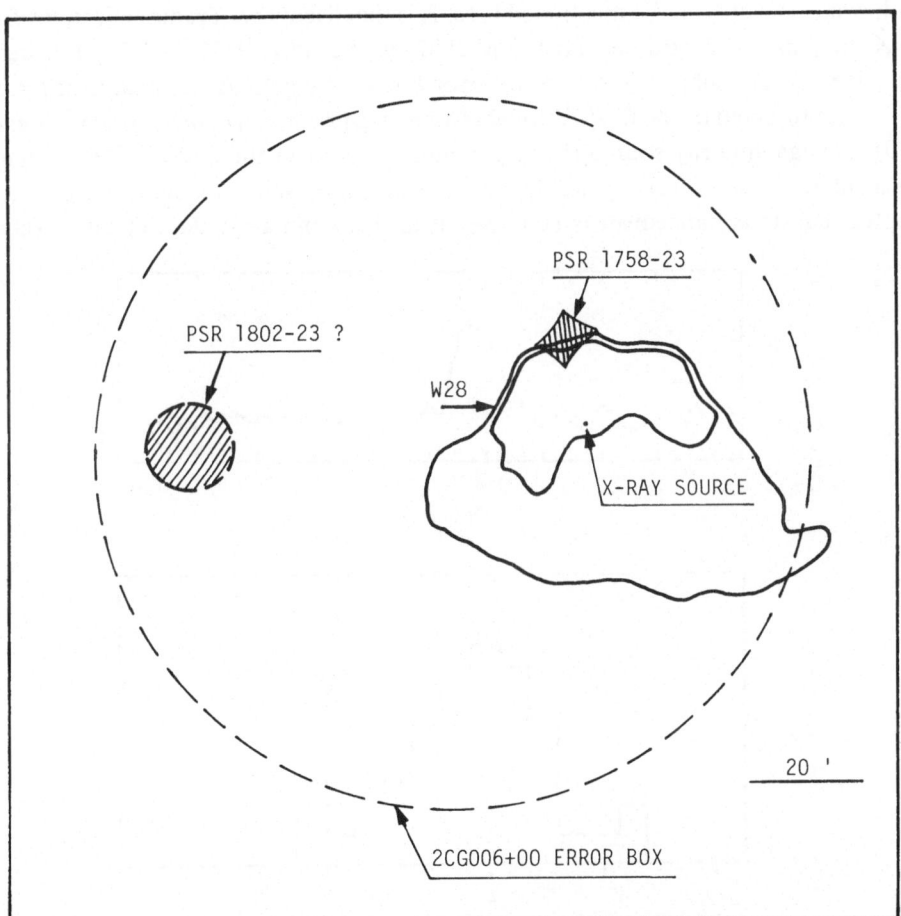

Fig. 2.   2CG 006 + 00 error box together with W28. The relative positions of the X-ray source and that of the new pulsars are also indicated. Contour lines of W28 are derived from Shaver and Goss (1970).

of W28. It is a medium period pulsar (period ~ 415 ms); its peculiarity is the dispersion measure of ~ 1050 cm^{-3} pc, the highest value observed up to now. It is not clear whether or not it is associated with W28, and the connection with the gamma-ray source is indeed not promising unless an extraordinary high period derivative will be discovered.

Within this error box we detected also a fast periodicity (PSR 1802−23?) but this result has to be confirmed. If real, it would be a good candidate counterpart for the gamma-ray source as the period is quite short (period ~ 112 ms). Figure 2 shows the gamma-ray error box together with W28; the position of these new pulsars and the X-ray source are also indicated.

The observational panorama of this gamma-ray source is quite interesting but still not clear. Deeper pulsar searches at the X-ray source position could be fruitful. A timing program on PSR 1758−23 and on the probable pulsar 1802−23 is also important to be undertaken very soon.

In the framework of the same experiment at Parkes, we obtained another interesting result which is worthy to be mentioned here. It is the discovery of radio pulsations from the X-ray pulsar (Seward and Harnden, 1982) in the SNR G320.4−1.2 (Manchester *et al.*, 1982b). This object is within the error box of the gamma-ray source at $l = 321$ $b = -1.2$, presented by the COS-B collaboration in a preliminary source list (Wills *et al.*, 1980). This gamma-ray source is not any more present in the 2nd COS-B catalogue (Swanenburg *et al.*, 1981), possibly because of more selective acceptance criteria adopted, but an enhancement is still present in the gamma-ray data at this position.

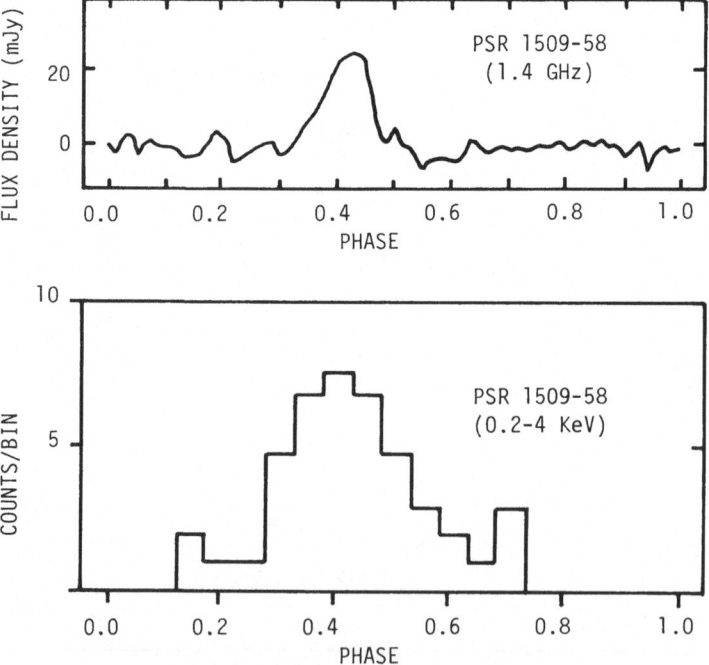

Fig. 3.   Light curves of PSR 1509−58 at radio and X-ray frequencies (Manchester *et al.*, 1982b; Seward and Harnden, 1982). The relative phase-shift is arbitrary.

Figure 3 shows the pulsar light curves at radio and X-ray frequencies. The pulsar period is ~ 150 ms and a peculiar characteristic of this object is the period derivative observed $(P \sim 1.5 \times 10^{-12}$ s s$^{-1})$, the highest value discovered up to now (see Figure 4). The estimated pulsar age is ~ 1500 yr.

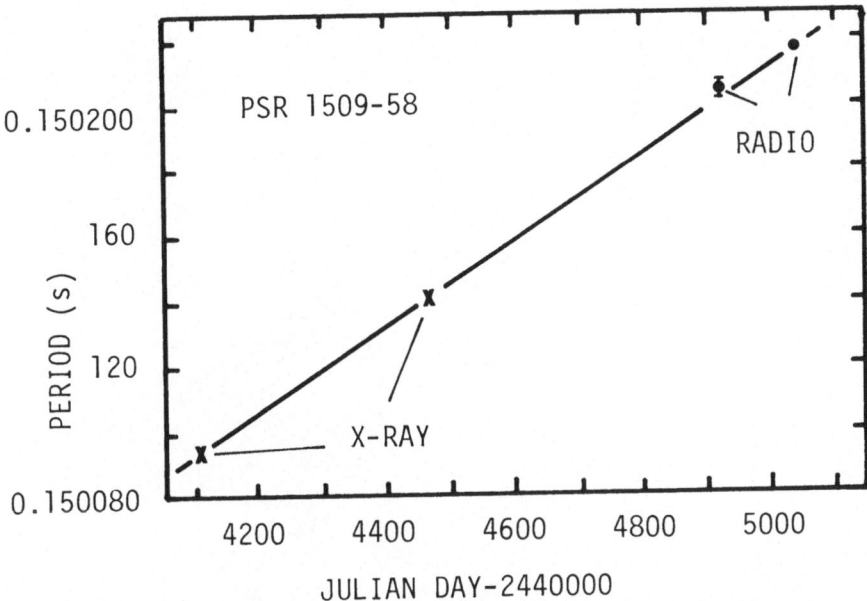

Fig. 4.   Period derivative of PSR 1509 – 58 (from Manchester *et al.* 1982b)

From all these characteristics and the association with the SNR, it is concluded that this object is Crab-like. It is the third pulsar observed in the young stage of its evolution: the presence of gamma-ray emission from this region stress the idea about the energy release in this frequency range to be a systematic effect of young pulsars. In fact, taking into account the observed gamma-ray flux and the estimated pulsar distance, this object would have a gamma-ray luminosity of the same order of that of the Crab pulsar.

Figure 5 shows the gamma-ray error box together with the SNR and the pulsar position. We suggest that this pulsar (PSR 1509 – 58) is the counterpart of the gamma-ray source at $l = 321$, $b = -1.2$. A deeper analysis of the COS-B data is important to be carried out, for both the gamma-ray source reliability and the detailed time analysis. Unfortunately, the long time elapsed from the last COS-B observation of this region (~ 5 yr) could affect the correct *phasing* of the gamma-ray data. As other gamma-ray experiments are planned for the near future, it is recommended to perform contemporary radio observations of PSR 1509 – 58 in order to reach final conclusions about its connection with the observed gamma rays.

Further X-ray observations, for instance with the EXOSAT observatory, would be also very important, especially for the determination of the phase relationship between the radio and X-ray pulses (and the gamma-ray pulses, if available).

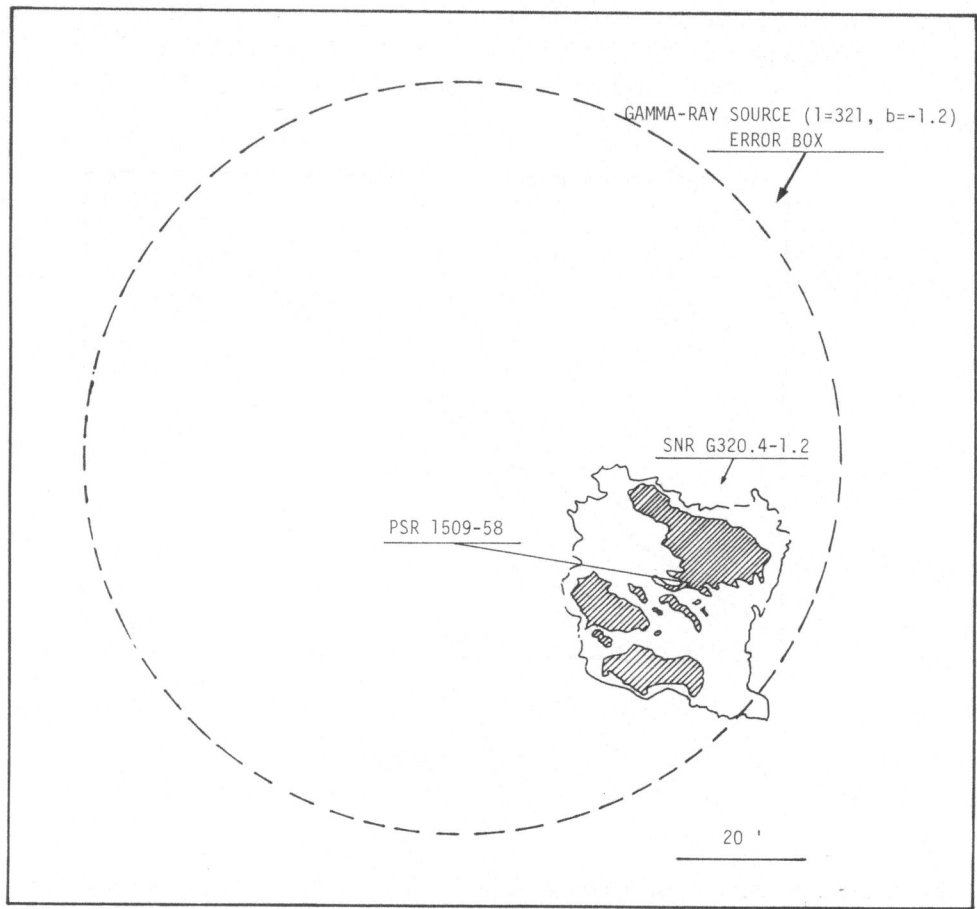

Fig. 5.   Relative positions of the gamma-ray source ($l = 321$, $b = -1.2$), the SNR G320.4 $-$ 1.2 and PSR 1509 $-$ 58. Contour lines for the SNR are derived from Caswell *et al.* (1981).

4. Conclusions

Even if preliminary, all these results encourage the idea about the high energy gamma-ray emission as a peculiar aspect of young pulsars.

As the general panorama about the young stage of pulsars is quite poor, this aspect gives to gamma rays an attractive role in order to define those regions of the sky where young pulsars could be found. This is an important point because sensitive searches for fast periodicities are hard to carried out, and all-sky surveys would be prohibitive. Of course, pulsars in the earlier stage of their evolution are already expected to be found within SNRs. In this sense, those cases of SNRs where a significant gamma-ray flux is observed, are the best candidates for a young pulsar to be there.

Depending on the number of short period pulsars which are expected to be observable within a reasonable distance ($\lesssim 7$ kpc), a fraction of the galactic gamma-ray sources could be pulsars, and this gives a natural input to the general problem about their nature.

Acknowledgement

The editors thank G. Kanbach for assistance in evaluating this paper.

References

Andrews, M. D., Basart, J. P., Lamb, R. C., and Becker, R. H.: 1982, *Astrophys. J.* **266**, 684.

Boriakoff, V., Buccheri, R., and Fauci, F.: 1983, submitted to *Astron. Astrophys.*

Buccheri, R., D'Amico, N., Massaro, E., and Scarsi, L.: 1978, *Nature* **274**, 572.

Caraveo, P.: 1983, *Space Sci. Rev.* **36**, 207 (this volume).

Cavallo, G. and Pacini F.: 1980, *Astron. Astrophys.* **88**, 367.

Caswell, J. L., Milne, D. C., and Wellington K. J.: 1981, *Monthly Notices Roy. Astron. Soc.* **195**, 89.

D'Amico, N. and Scarsi, L.: 1979, Proc. of Einstein Centenary Summer School 'Gravitational Radiation and Collapsed Objects', Perth, Australia.

Lamb, R. C.: 1978, *Nature* **272**, 429.

Manchester, R. N., Tuohy, I. R., and D'Amico, N.: 1982a, in 'SNR and Their X-Ray Emission', *IAU Symp.* **101**.

Manchester, R. N., Tuohy, I. R., and D'Amico, N.: 1982b, *Astrophys. J.* **262**, L31.

Manchester, R. N., D'Amico, N. and Tuohy I. R.: 1983, in preparation.

Mandolesi, N., Morigi, G., and Sironi, G.: 1978, *Astron. Astrophys.* **67**, L5.

Mandrou, P., Vedrenne, G., and Masnou, J. L.: 1980, *Nature* **287**, 124.

Massaro, E. and Salvati, M.: 1979, *Astron. Astrophys.* **71**, 51.

Meszaros, P.: 1975, *Astron. Astrophys.* **44**, 59.

Ozel, M. E., Dickel, J. R., and Webbere, J. C.: 1980, *Nature* **285**, 645.

Seiradakis, J. H.: 1981, *Astron. Astrophys.* **101**, 158.

Seward, F. D. and Hardnen, Jr. F. R.: 1982, *Astrophys. J.* **256**, L45.

Shaver, P. A., and Goss, W. M.: 1970, *Australian J. Phys.* **14**.

Swanemburg, B. N. *et al.*: 1981, *Astrophys J.* **243**, L69.

Van den Berg, S.: 1979, *Astron. J.* **84**, 71.

Wills *et al.*: 1980, Proc. of COSPAR Symp. on 'Non-Solar Gamma-Rays', Bangalore, 1980.

Cosmochemistry and the Origin of Life

Proceedings of the NATO Advanced Study Institute, held at Maratea, Italy, June 1–12, 1981

Edited by
CYRIL PONNAMPERUMA

1983, viii + 386 pp.
Cloth Dfl. 145,– / US $ 63.00
NATO ADVANCED STUDY INSTITUTES SERIES
C. Mathematical and Physical Sciences 101

ISBN 90-277-1544-0

Using an interdisciplinary approach, this volume brings together current knowledge, united under the single theme, *Cosmochemistry and the Origin of Life*, and includes some of the most recent discoveries related to interstellar molecules and the earth's earliest sediments. Beginning with the Big Bang and the synthesis of the chemical elements, the book explores the discovery of organic molecules in space, and comets and planetary environment before focusing on earth. In this section, attention is given to chemical composition and climatology of the earth's early atmosphere, the earliest known sediments, and molecular fossil record, all with an eye to the chemistry necessary for the origin of life. Students and researchers interested in this subject will find a wealth of valuable information in the volume.

Contents: Preface. Cyril Ponnamperuma: Cosmochemistry and the Origin of Life. Vic Viola: Synthesis of the Chemical Elements. J. Mayo Greenberg: Organic Molecules in Space. William M. Irvine and Åke Hjalmarson: Comets, Interstellar Molecules and the Origin of Life. Donald De Vincenzi: Planetary Environments and the Origin of Life. A. Henderson-Sellers: The Chemical Composition and Climatology of the Earth's Early Atmosphere. Stephen Moorbath and P. N. Taylor: The Dating of the Earliest Sediments of the Earth. Ei-Ichiro Ochiai: Inorganic Chemistry of the Earliest Sediments of the Earth. Manfred Schidlowski: Isotope Fractionation in the Earliest Sediments of the Earth. Geoffrey Eglinton: The Molecular Fossil Record.

D. Reidel Publishing Company

P.O. Box 17, 3300 AA Dordrecht, the Netherlands
190 Old Derby St., Hingham, MA 02043, U.S.A.

Kinematics, Dynamics and Structure of the Milky Way

Proceedings of a Workshop on "The Milky Way" held in Vancouver, Canada, May 17–19, 1982

Edited by
W. L. H. SHUTER, *Department of Physics, University of British Columbia, Vancouver, B.C., Canada*

1983, xii + 392 pp.
Cloth Dfl. 125,– / US $ 54.50   ISBN 90-277-1540-8
ASTROPHYSICS AND SPACE SCIENCE LIBRARY 100

Galactic Astronomy is at present in a state of turmoil. The reason for this is that observational astronomers are accumulating a wealth of new data, often from previously inaccessible regions of the electro-magnetic spectrum, such as the sub-millimetre range and the gamma radiation range, and at the same time, interpreters and theoreticians are developing new techniques and perspectives which are being applied to understand the new information and re-evaluate previously collected data. The papers in this volume present new data and ideas that have emerged since the last major symposium on the subject – published as 'The Large-Scale Characteristics of the Galaxy', edited by W. B. Burton (D. Reidel, 1979) – and should be of value to astronomers who are attempting to compile a coherent picture of the Galaxy.

Contents
Contributors include: Bart J. Bok, Carl Heiles and Thomas H. Troland, A. R. Upgren and E. W. Weis, Edward B. Jenkins, J. G. B. M. Bloemen, Sidney van den Bergh, C. Yuan, Ivan R. King, Per Olof Lindblad, M. V. Ovenden and J. Byl, M. W. Ovenden *et al*, M. H. L. Pryce, W. L. H. Shuter, Roland Wielen and Burkhard Fuchs, F. J. Kerr, Shrivinas R. Kulkarni *et al*, Carl Heiles, J. M. Dickey *et al*, David B. Sanders, Antony A. Stark, H. S. Liszt and W. B. Burton, Leo Blitz and Michel Fich, Jan Brand *et al*, W. H. McCutcheon *et al*, B. E. Turner, Frank Bash and Gabriella Turek, M. G. Jauser *et al*, Robert L. Brown, John N. Bahcall and Raymond M. Soneira, Douglas Forbes, H. A. Mayer-Hasselwander, G. R. Knapp, Jeremiah P. Ostriker and John A. R. Caldwell, Philip E. Seiden, William W. Roberts, Jr., C. C. Lin, John Haass, James W-K. Mark, Felix J. Lockman, Allan Sandage, K. C. Freeman, K. A. Innanen, Carlos S. Frenk and Simon D. M. White, D. Lynden-Bell, Harvey B. Richer *et al*, Judith S. Young, and Vera C. Rubin.

D. Reidel Publishing Company

P.O. Box 17, 3300 AA Dordrecht, the Netherlands
190 Old Derby St., Hingham, MA 02043, U.S.A.

Internal Kinematics and Dynamics of Galaxies

Proceedings of IAU Symposium No. 100, held in Besançon, France, August 9—13, 1982

Edited by

E. ATHANASSOULA, *Observatoire de Besançon, France*

1983, xvi + 432 pp.

Cloth Dfl. 115,— / US $ 49.50   ISBN 90-277-1546-7
Paper Dfl. 60,— / US $ 26.00   ISBN 90-277-1547-5
INTERNATIONAL ASTRONOMICAL UNION SYMPOSIA 100

This volume contains the proceedings of the IAU Symposium, 'Internal Kinematics and Dynamics of Galaxies', held in Besançon, France, in August 1982. The Symposium was attended by 166 scientists from 21 countries and, besides invited reviews and contributed papers, a large number of poster papers were presented. The scientific programme is divided into the following sections: Kinematics of Gas and the Underlying Mass Distribution; Spiral Structure; Warps-Barred Galaxies; Spheroidal Systems; Mergers; and Galaxy Formation. Attention is given to both observational data and theoretical developments. The question of the existence of a massive halo around spiral galaxies is widely debated and a large number of new results on spiral amplification, modes, and barred galaxies are presented. In addition, calculations from N-body cones on a variety of topics are compared with observations and interesting results on triaxiality are reported. Finally, the influence of mergers on the evolution of galaxies is discussed.

Contents

Foreword. List of Participants. **I. Kinematics of Gas and the Underlying Mass Distribution.** Systematics of HII Rotation Curves. HI Velocity Fields and Rotation Curves. The Distribution of Molecular Clouds in Spiral Galaxies. Gas at Large Radii. Vertical Motion and the Thickness of HI Disks: Implications for Galactic Mass Models. Mass Distribution and Dark Halos. **II. Spiral Structure.** Conflicts and Directions in Spiral Structure. **III. Warps.** Theories of Warps. **IV. Barred Galaxies.** Morphology. Stellar Kinematics and Dynamics of Barred Galaxies. Disk Stability. Theoretical Studies of Gas Flow in Barred Spiral Galaxies. Formation of Rings and Lenses. **V. Spheroidal Systems.** Dynamics of Early-Type Galaxies. Models of Ellipticals and Bulges. Interstellar Matter in Elliptical Galaxies. **VI. Mergers.** Observational Evidence for Mergers. Simulations of Galaxy Mergers. **VII. Galaxy Formation.** Dynamics of Globular Cluster Systems. The Formation of Galaxies. Galaxy Formation. Some Comparisons Between Theory and Observation. **VIII. Summaries.** Summary: Observational Viewpoint. Summary: Theoretical Viewpoint.

D. Reidel Publishing Company

P.O. Box 17, 3300 AA Dordrecht, the Netherlands
190 Old Derby St., Hingham, MA 02043, U.S.A.

SPACE SCIENCE REVIEWS

Volume 36 No. 3 1983

Published monthly.
Subscription prices, per volume: Institutions $92.00, Individuals $30.00.
Second-class postage paid at New York, N.Y. USPS No. 509–100.
U.S. Mailing Agent: Expediters of the Printed Word Ltd., 527 Madison Avenue (Suite 1217), New York, NY 10022.
Space Science Reviews is published by D. Reidel Publishing Company, Voorstraat 479–483, P.O. Box 17, 3300 AA Dordrecht, Holland, and 190 Old Derby Street, Hingham, MA 02043, U.S.A.
Postmaster: please send all address corrections to: c/o Expediters of the Printed Word Ltd., 527 Madison Avenue (Suite 1217), New York, NY 10022, U.S.A.

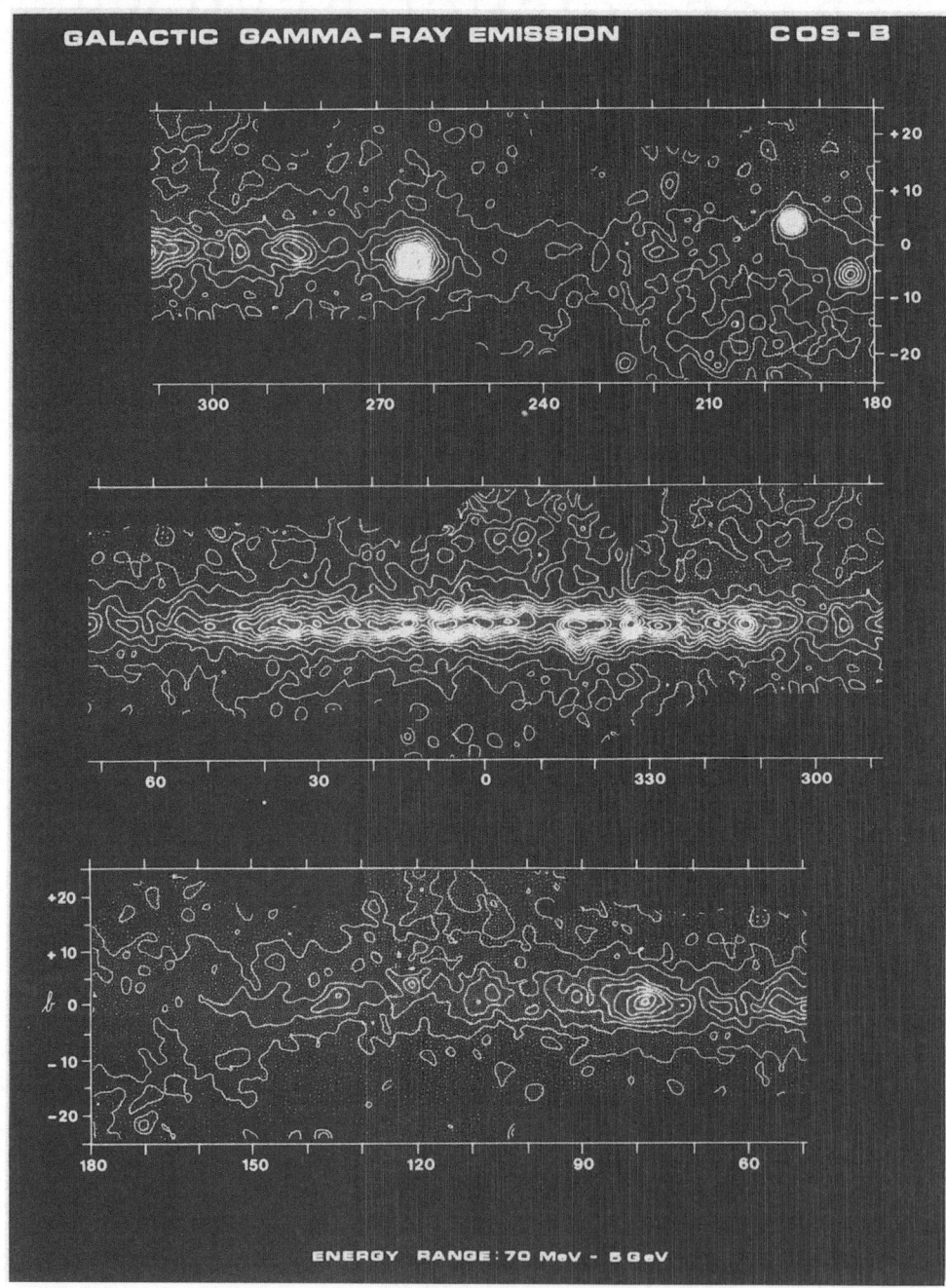

Presentation in galactic coordinates of the structure of the galactic gamma-ray emission as measured by the gamma-ray experiment of the 'Caravane' collaboration aboard the ESA satellite COS-B. In the map the surface fitted to the data is indicated by contour lines and a grey scale. Regions outside the accepted field of view are left blank. The contour levels are indicated at multiples of 3×10^{-3} *on-axis* counts s^{-1} sr^{-1}. A detailed discussion can be found in H. A. Mayer-Hasselwander *et al.*: 1982, *Astron. Astrophys.* **105**, 164–175.

A SEARCH FOR COUNTERPARTS OF
SELECTED COS-B GALACTIC GAMMA-RAY SOURCES:
THE X-RAY SURVEY AND OPTICAL IDENTIFICATIONS* **

PATRIZIA A. CARAVEO

Istituto di Fisica Cosmica del C.N.R., Via Bassini 15, 20133 Milano, Italy

Abstract. The Einstein X-ray Imaging Instruments have been used to explore, down to an unprecedented sensitivity, the X-ray behavior of 7 high-energy γ-ray sources discovered by the COS-B satellite.

32 low latitude ($|b| < 5°$) IPC fields, mosaic-arranged to cover the few-square-degrees COS-B error circles, yielded 30 soft X-ray sources, the fluxes of which range from $\sim 1/100$ to few UFU, and no diffuse features.

While the density of ~ 1 source/IPC field is consistent with the value found at higher latitudes, the percentage of 'stellar' identifications among these low-latitude sources is significantly higher than in non-galactic-biased samples. Unfortunately, the positional accuracy achieved with the IPC does not allow astronomical identification in the absence of obvious counterpart(s). However, after the exploratory coverage, the IPC data were used, when possible, to point out potentially interesting targets for the HRI instrument capable of an accuracy of ~ 3 arc sec.

Due to the misfortunes which occurred to the Einstein satellite, this time-consuming process was feasible only in two cases: within the error circle of 2CG 135 + 01, the radio variable star LSI 61.303 was pinpointed by the HRI, while the HRI exposure of the brightest X-ray source discovered in 2CG 195 + 04 (Geminga) positioned a source in an empty POSS field. The latter case will be presented and the nature of the X- and γ(?)-ray source briefly discussed.

1. Introduction

The publication of the first COS-B catalogue of γ-ray sources (CG catalogue, Hermsen *et al.*, 1977) unveiled the presence of unresolved sources (i.e. excesses of photon counts compatible with the Point Spread Function of γ-ray telescope) buried deep in the galactic ridge radiation. It was soon realised that the majority of such sources were not easily identifiable with obvious candidates such as strong X-ray sources molecular clouds, known pulsars and radio continuum features.

In spite of much effort devoted by many groups to this problem, the five years elapsed since then did not witness any major breakthrough on the subject, the secret nature of the γ-ray sources being effectively guarded by the very enormity of their error circles.

In the mean time, the longevity of the COS-B satellite allowed the family of γ-ray sources to grow up, yielding eventually the second catalogue of γ-ray sources (2CG Catalogue), including and superceeding the previous one, (Swanenburg *et al.*, 1981) of 25 objects, 21 of which share the property of being unidentified.

In our ignorance about these sources we can at least be sure that their very name of γ-ray sources is well deserved, since from their error circles (0.6 to 7 square degrees

---

* Proceedings of the XVIII General Assembly of the IAU: *Galactic Astrophysics and Gamma-Ray Astronomy*, held at Patras, Greece, 19 August 1982.

** The X-ray data used in this paper were collected as a part of the Guest Observer Program of the Einstein Observatory.

Space Science Reviews **36** (1983) 207–221. 0038–6308/83/0363–0207$02.25.

© 1983 *by D. Reidel Publishing Co., Dordrecht and Boston*

wide) in the sky we recorded a γ-ray flux energy-wise far more important than at other wavelengths. With no hint on the nature of γ-ray sources, the most straightforward approach seemed an unbiased in-depth coverage of the error circles looking for objects of potential interest (see also Bignami, 1981; and Caraveo, 1981; for a discussion of this philosophy). Having to cover few square degrees in the sky, an unbiased search for counterpart(s) implies the use of an instrument with a field of view of comparable size, coupled with good sensitivity and positional accuracy.

Such a fortunate combination appears to be possible in radioastronomy, both pulsar-oriented (D'Amico, 1983) and continuum-oriented (e.g. Sieber and Schlickeiser, 1982), as well as in X-ray astronomy where, since the beginning, it seemed natural to look. This window, already exploited in the 'pre-Einstein era' (see e.g. Bignami and Hermsen, 1982, for a review of the results), saw its potentialities dramatically improved by the advent of the Einstein Observatory (Giacconi et al., 1979). The Imaging Proportional Counter (IPC), on board the Einstein Observatory, appeared to be particularly well suited for this task having both a wide (1 square degrees) field of view and a good (1 arc min) positional accuracy (improvable to few arc sec through the High Resolution Imager). Moreover, few IPC pointings of medium length could cover, down to an unprecedent sensitivity, a typical COS-B error box.

An unbiased exploration of some of the unidentified COS-B sources was, then, undertaken in the framework of the Einstein Guest Observer Program. The task was pursued through a number of Guest Observer Proposals involving several groups and in some cases the whole of the Caravane Collaboration for the COS-B satellite. A total of 100 ksec of observing time has been granted for this exercise and in what follows we shall report on the general outcome of this search both X-ray Astronomy and γ-ray Astronomy wise.

This will be the first general presentation of our results, only partially given elsewhere (Caraveo et al., 1981; Bignami et al., 1981). For sake of clarity we shall limit our view to the γ-ray sources listed in the 2CG catalogue, i.e. 2CG 54 + 01, 2CG 95 + 04, 2CG 121 + 04, 2CG 135 + 01, 2CG 195 + 04, 2CG 218 − 00, 2CG 284 − 00, (Swanenburg et al., 1981) while the Einstein observations encompassing positions of previously reported γ-ray sources, now no longer in the 2CG catalogue, will be neglected in this context. However, in one case, the mosaic centered at $l = 295°$, $b = 0°$ yielded remarkable results as given in Lamb et al. (1980), Markert et al. (1981).

Furthermore, the present work is not based on the totality of the time dedicated by the Einstein satellite to γ-ray sources. Other γ-ray sources have been covered, at least partially in soft X-ray with somehow different phylosophy.

The reader is referred to Montmerle et al. (1983), for the investigation of 2CG 353 + 16, yielding the discovery of 50 variable stellar X-ray sources buried in the ρOph cloud, to Lamb and Markert (1981), for the study of SNR's near 2CG 311 − 01, CG 327 − 0, and CG 333 + 0, to Singh et al. (1982), for the investigation of weak Uhuru sources in CG 189 + 1 and 2CG 075 + 00, and to Seward and Chelebowski (1982) for a detailed study of the Carina nebula, inside the 2CG 288 − 00 error radius. Although the latter investigation was not prompted by the presence of the COS-B source, it is of particular interest since Montmerle (1981) and Cesarsky and Montmerle (1983) consider Carina as a prototype of a class of γ-ray emitters, where a cloud is energized by the stellar winds of the newly born stars inside it.

However, all of the investigations listed above were done either to check a model of γ-ray emission or to investigate previously known sources and, with the exception of one case (Montmerle *et al.*, 1983), they cover but a tiny fraction of the four 2CG sources mentioned.

2. The Einstein Pictures

The data cover 34 IPC fields, 3 of which have been followed by HRI pointings to improve the knowledge on the position of interesting new sources. Two IPC fields have been observed twice, for the remaining 30 only one pointing was done so that little information on the long term variability of the sources can be inferred.

The exposure time of the IPC fields varied between 1500 to 6000 s with an isolated maximum of 8000 s. The range of values spanned is, therefore, too big to allow a unique lowest detectable flux to be assumed for all the observations. Moreover, the impact of a varying galactic absorption on intrinsically different spectral shapes prevents the calculation of precise figures. Roughly speaking, a limit of few 10^{-13} erg cm^{-2} s^{-1}, in the energy range 0.15–4.5 keV, is achieved during pointings lasting between 2000 and 4000 s, while the limit of $\sim 10^{-13}$ erg cm^{-2} s^{-1} is reached during pointings lasting more than 4000 s.

Luckily enough, the exposure times remained reasonably constant within the coverage of each γ-ray source, so that a general view of the soft-X-ray appearance of a γ-ray source error circle can be gained by doing a collage of the fields. Thanks to the computer facility provided to Guest Observers at the Center for Astrophysics, the collages were easily made at the interactive screen, and, as a final touch, the COS-B error circles were superimposed to the resulting mosaics.

The final pictures, shown in Figures 1 to 7, eloquently describe our effort to keep the X-ray pointings up-to-date following the rapid evolution of the knowledge of the parameters of the γ-ray sources. The patient collaboration of F. Seward, in charge of baby-sitting Guest Observers, in many cases contributed to the success of our effort and Figures 1, 3, 4, 5, 6 present a satisfactory coverage of the γ-ray circles. In few cases, however, the observations took place before the updating of the γ-ray position, as happened for Figure 7, or their number was reduced due to the Einstein Observatory period of malfunctioning, as was the case for Figure 2, so that the resulting pictures look somewhat incomplete. The mosaics have been done in celestial coordinates using, unless otherwise stated in the caption, the parameters given in Swanenburg *et al.* (1981). The figures are not drawn to the same scale, and the reader is referred to the caption for the actual size of the error circle (ovalized in the reproduction) as well as for the average exposure of the fields.

The data have been searched for sources both with the 'standard' and the 'deconvolution' algorithm. Since the latter is believed to be more sensitive, we accept an excess

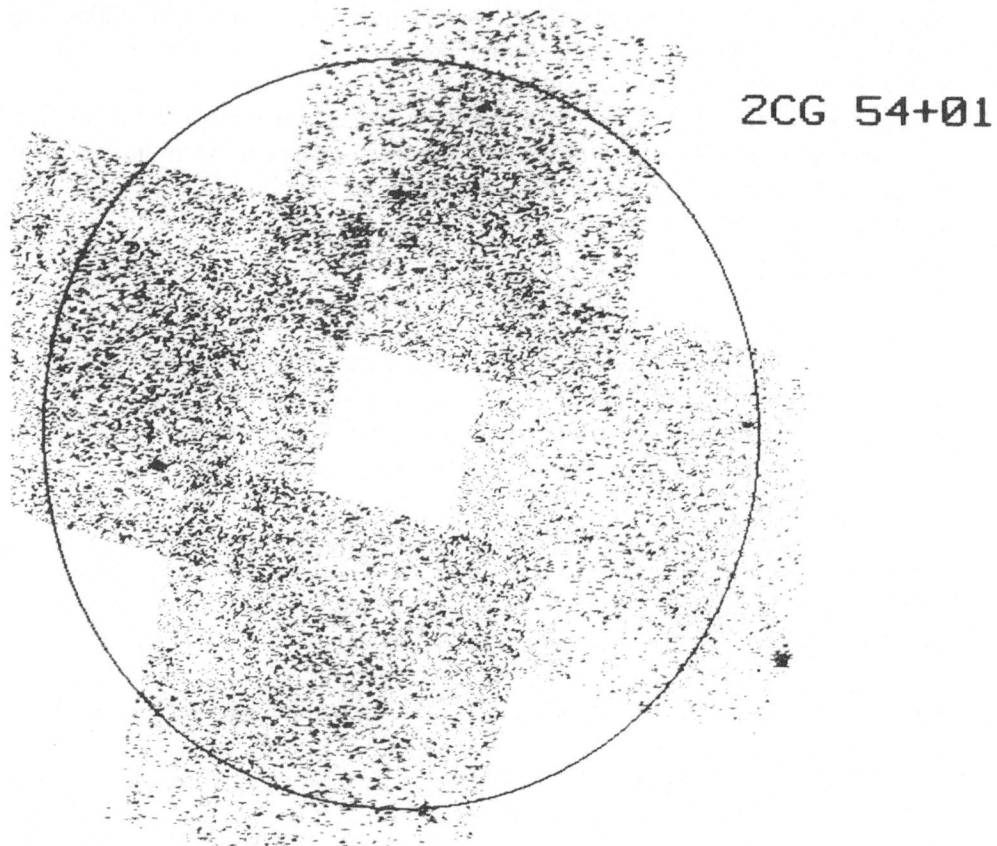

Fig. 1. Soft X-ray coverage of 2CG 54 + 01. The γ-ray error circle is centered at $l = 54°2$, $b = 1°7$ with a radius of $1°$. With an average exposure of 5000 s, 2CG 54 + 01 is the γ-ray source covered with greater sensitivity. Therefore, it is not surprising to find in this mosaic a density of X-ray sources greater than the average for the whole survey. Eight sources meet the acceptance criteria: 5 of them are identified with normal stars, while the brightest is unidentified. A search for radio continuum emission from the 8 serendipitous sources has been done using the Effelsberg radio telescope (Schlickeiser, private communication) and will be reported elsewhere.

to be a source if it reaches the 5σ level in the deconvolved picture. The flux of such excesses was computed from the raw image after substraction of the local background.

Thirty sources were detected within 32 independent fields. Since our pointings were 'non target' ones, none of the sources lies in the central few arcmin of the IPC fields. This makes it impossible to fully exploit the (limited) spectral measurement capability of the IPC instrument as well as to use the data from the Monitor Proportional Counter (MPC), which could give good spectral data on the brightest sources. The hardness ratio of a source can only be estimated studying the distribution of counts per channel of energy of the IPC instrument. A poor knowledge of the spectral shape, together with the lack of reliable determination of the amount of matter on the line of sight, cause a

2CG 95+04

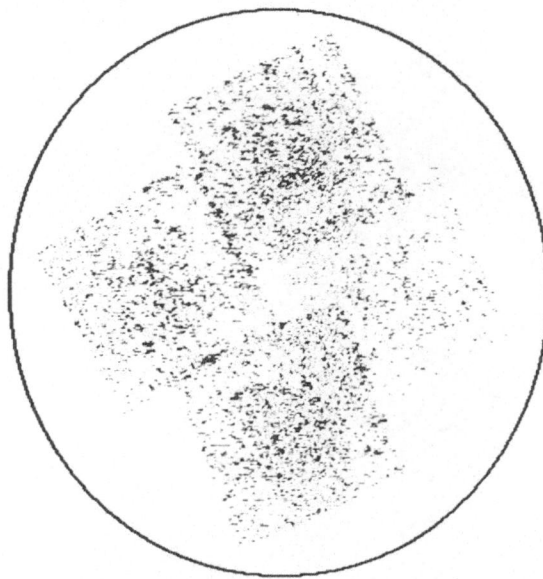

Fig. 2.   Partial soft X-ray coverage of 2CG 95 + 04. The γ-ray circle is centered at $l = 95°2$, $b = 4°2$ with a radius of $1°5$. At an average exposure of 2000 s, only one weak source is found in the upper field.

2CG 121+04

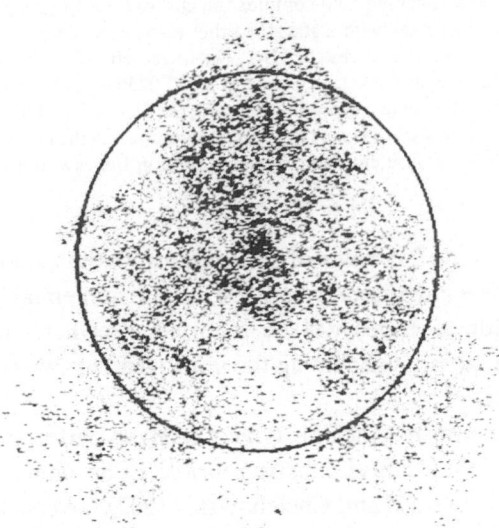

Fig. 3.   Soft X-ray coverage of 2CG 121 + 04. The γ-ray error circle is centered at $l = 121°0$, $b = 4°0$ with an error radius of $1°$. At an average exposure of 1800 s, 2 weak sources are found: one in the middle left field and one in the lower middle field. A search for radio continuum emission from the 2 sources has been done using the Effelsberg radio telescope (Schlickeiser, private communication) and will be reported elsewhere.

PATRIZIA A. CARAVEO

2CG 135+01

Fig. 4.   Soft X-ray coverage of 2CG 135 + 01. The γ-ray error circle is centered at l = 135°0, b = 1°5 with an error radius of 1°. The exposure is not uniform throughout the mosaic. While the uniformly grey fields have been exposed for 1800 s, the lower left field reaches 2500 s and the upper left 5000 s (being the sum of two observations). Seven sources are found in the mosaic. Caraveo *et al.* (1981) report the X-ray parameters as well as optical study of 6 of the 7 sources, the missing one being the source in the lower left field just outside the COS-B circle not visible in the 700 s exposure available to them (see Figure 2, field N. 5 of Caraveo *et al.*, 1981). The upper left field contains the QSO 0241 + 61 (the brightest source) together with 2 more sources, one positively identified with a star, the other also likely to be of stellar origin. The middle right field contains a normal star pinpointed by the HRI. The lower left field contains a bright X-ray source identified, through the HRI data, with the variable radio star GT 0236 + 61 shining via *NON* coronal X-ray emission, and a normal star at the south, also pinpointed by the HRI. The H II region IC 1805, proposed by Montmerle (1979) as the counterpart of the γ-ray source, encompasses the fields on the lower left, middle left and upper right, where no diffuse emission is visible both in the raw and in the smoothed data.

significant uncertainty on the flux value of the source and, thus, on its luminosity in the IPC energy window (see e.g. Bignami *et al.*, 1981). The numerical data on the sources (i.e. position, flux, hardness ratio) will be reported elsewhere in a more extended compilation of the data, possibly including the improvements awaited from the 'reprocessing' of the IPC data.

The search for sources was not the only type of analysis performed on the IPC fields. In the absence of structured diffuse emission visible to an unaided eye, like the one reported e.g. in Carina by Seward and Chelebowski (1982), we checked for the presence of weak diffuse emission in the COS-B circles by merging together the fields of each mosaic in order to equalize the exposure time and to suppress the unreal magnification of the counts in the overlapping regions.

No diffuse X-ray features were detectable in the merged pictures, neither in emission nor in absorption even when H II region are covered by the mosaic as discussed e.g. in

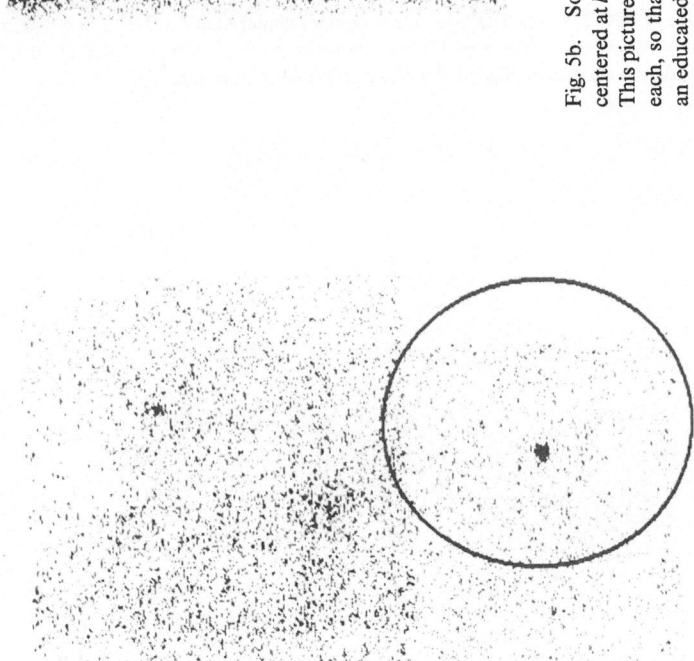

2CG 195+04 'GEMINGA'

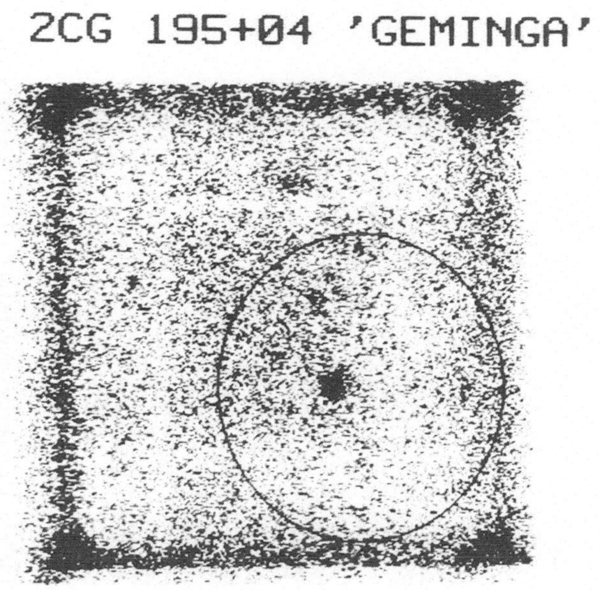

Fig. 5b. Soft X-ray coverage of 2CG 195 + 04. The γ-ray error circle is now centered at $l = 195°.1$, $b = 4°.2$ with an error radius of $0°.4$ (Masnou et al., 1981). This picture is the sum of two 'extended' IPC fields of 3000 and 8000 s exposure each, so that it is the deepest field in our survey. Four sources are visible to an educated eye, only two of them inside the COS-B error circle. The brightest source does not show either long or short term variability and its HRI position falls into a POSS empty field. Deeper optical exposures have already been taken and a full account of the work can be found in Bignami et al. (1983). The 2CG 195 + 04 error circle has been covered in radio continuum by Sieber and Schlickseir (1982) yielding a number of new radio sources only one of which is consistent with an X-ray source position. It is the source at the north of the field just above the shade of the rib. The long-term variability of this source (inferred from the comparison of the two observations), together with its radio double structure points towards and extragalactic identification. The source at the east of the field outside the COS-B error circle is probably an uncatalogued star ($m_v \sim 10$), while the other source inside the circle is unidentified.

2CG 195+04

Fig. 5a. Original mosaic (average exposure 3000 s) aimed at the coverage of 2CG 195 + 04 tuned to the coordinates given by Hermsen et al. (1977). Since then, the γ-ray position has been refined twice and the latest error circle (Masnou et al., 1981) encopasses only the lower IPC field, where a bright X-ray source is seen. This prompted an HRI observation of the bright X-ray source and a repetition of the IPC field.

2CG 218-00

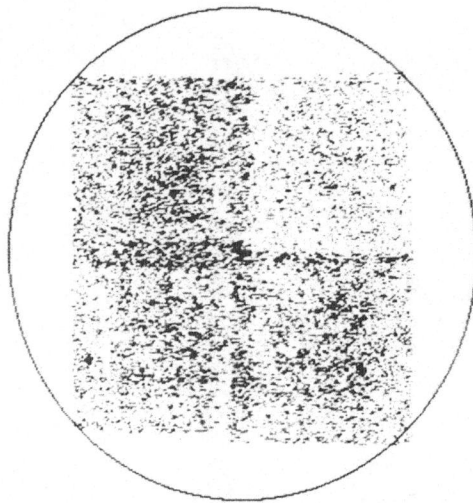

Fig. 6.   Soft X-ray coverage of 2CG 218 – 00. The γ-ray error circle is centered at $l = 218°.5$, $b = -0°.5$ with an error radius of $1°.3$. At an average exposure of 1700 s, 5 sources have been found, one of which is a cataloged star belonging to the class of VV Mon variables.

2CG 284-00

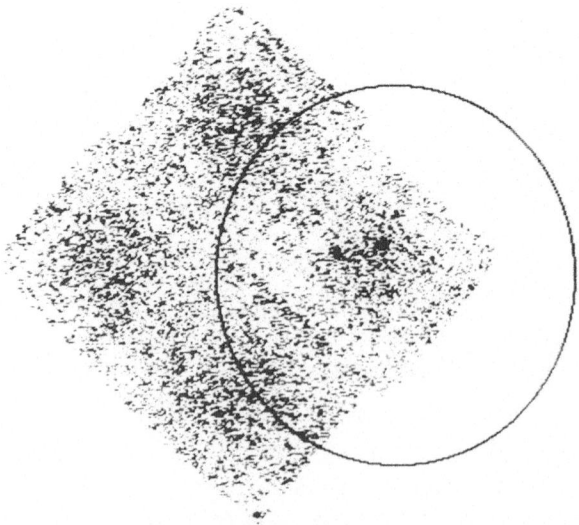

Fig. 7.   Partial soft X-ray coverage of 2CG 284 – 00. The γ-ray error circle is centered at $l = 284°.3$, $b = -0°.5$ with an error radius of $1°$. At the average exposure of 2000 s, 3 sources are found and they all lie inside the COS-B error circle. The southern one is identified with a star.

the caption to Figure 4. This is mildly surprising in view of the low galactic latitude of all the fields and will be discussed elsewhere.

3. On Identified Sources

3.1. STELLAR IDENTIFICATIONS

Normal field stars have been shown to represent a substantial fraction of the Einstein serendipitous sources (Helfand and Caillault, 1982, hereafter referred to as H & C; Vaiana *et al.*, 1981). Therefore, it was not surprising to collect 12 findings charts (out of 30) where a star brighter than $m_v \sim 11$ dominated the 1' radius error circle. In order to compute the probability for change coincidence, it is worth dividing such stellar identifications in two subsets, one containing three stars fainter than $m_v \sim 10$ pinpointed by the HRI and identification of which is beyond doubt, and the remaining nine brighter than $m_v \sim 10$ for which only IPC positions are available. For the latter subset the chance coincidence probability can be calculated as follows. On the galactic plane the probability of finding a star brighter than $m_v \sim 10$ in a 1 arcmin radius error circle is $\sim 10^{-2}$ (Allen, 1973), so that on the basis of the number of candidates we expect less than 0.1 spurious coincidence for the whole survey.

Twelve stars over a sample of 30 serendipitous X-ray sources represents $\sim 40\%$ star identifications: such a percentage appears significantly greater than the 15–20% found by H & C over a sample ten times bigger than ours. However, before claiming a significant difference, it is necessary to ensure that the flux limits (i.e. exposure times) and the acceptance criteria are in reasonable agreement. This is not the case, since H & C used the standard search algorithm taking a threshold of 3σ for source acceptance. In order to make our results comparable to theirs we selected sources meeting this requirement from our list and compared their density (i.e. number of sources per IPC field) with values calculated from H & C for non target sources in fields with exposures less than 6000 s. This compromise seemed good enough to allow for a straightforward comparison of the density values. The numbers are given in Table I where, for sake of completeness, the surface density of serendipitous sources in high latitude fields ($> 15°$) has been added (Chanan *et al.*, 1981). Although the last number is not comparable with the other, since it resulted from 3000 s fields with an acceptance level of 5σ in the search algorithm, it can be considered as a lower limit to nonstellar identifications in high-latitude fields.

The comparison of consistent figures (i.e. those above 3σ) shows that, indeed, we found more stars that H & C. Moreover it is interesting to note that of the 11 sources that do not meet the 3σ criterium, 6 are identified with stars. This suggests that the percentage of stellar identifications is increasing when going to lower values of the limiting flux. The present high percentage can be understood in terms of a bias introduced by the galactic distribution of the fields used. In our case the fields were *all* at low galactic latitude while H & C tried to avoid the galactic plane where the density of stars with $m_v \sim 10$ (i.e. potential X-ray emitters) is twice as big as the same value averaged over

TABLE I

	Serendipitous non target sources surface density (source/field)		% stellar identifications
	All	Nonstellar	
This work (galactic plane)	1	0.6	40%
This work $> 3\sigma$ (galactic plane)	0.6	0.4	30%
H & C $> 3\sigma$ $\tau < 6000$ s (all sky)	0.6	0.5	15%
Chanan et al. $> 5\sigma$, $\tau \sim 3000$ s (high latitudes)		0.6	

all latitudes (Allen, 1973). Such bias, however, does not apparently affect the total number of sources per field found by us and H & C. This probably means that galactic absorption is at work, suppressing sources otherwise visible at higher galactic latitudes. The same absorption effect could explain the increase of nonstellar identifications going from 0.4 per IPC field on the plane, to 0.5 for the average sky, to 0.6 for high galactic latitudes.

TABLE II

Spectral type	Number of detections
O	–
B	2
A	1
F	4
G	3
K	–
M	–

Having checked that our data do not show dramatic signals of inconsistency with more extended surveys, apart from a certain galaxy-induced overabundance of stellar identifications, one can look in more detail into the X-ray behavior of such stars. For 10 of the 12 stellar identifications, we know either by catalogues or by direct observations their spectral types and visual magnitudes. The spectral type distribution, given in Table II, shows a modest disagreement with the results of Topka et al. (1982) who found 18 F stars and only 3 G stars out of a sample of 33. Given the X-ray fluxes and visual magnitudes, the individual ratio f_x/f_v can be computed and compared with the value expected from each spectral class on the basis of the stellar survey of Vaiana et al.

(1981). The results are shown pictorially in Figure 8 where our data points have been superimposed to the original figure of Vaiana *et al.* (1981). Although we deal with star fainter than the ones in the sample of Vaiana *et al.* (1981), the numbers we found are all but one compatible, lying in the allowed range of values, slightly on the high side, as it is often the case for serendipitous stellar sources (see Vaiana *et al.*, 1981, for a complete discussion). The only one inconsistent with the f_x/f_v expected from coronal emission is that of the radio star GT 0236 + 61 ≡ LSI + 61°303 strongly radio variable with a 26.52 days period (Gregory *et al.*, 1979; Taylor and Gregory, 1982). IPC and

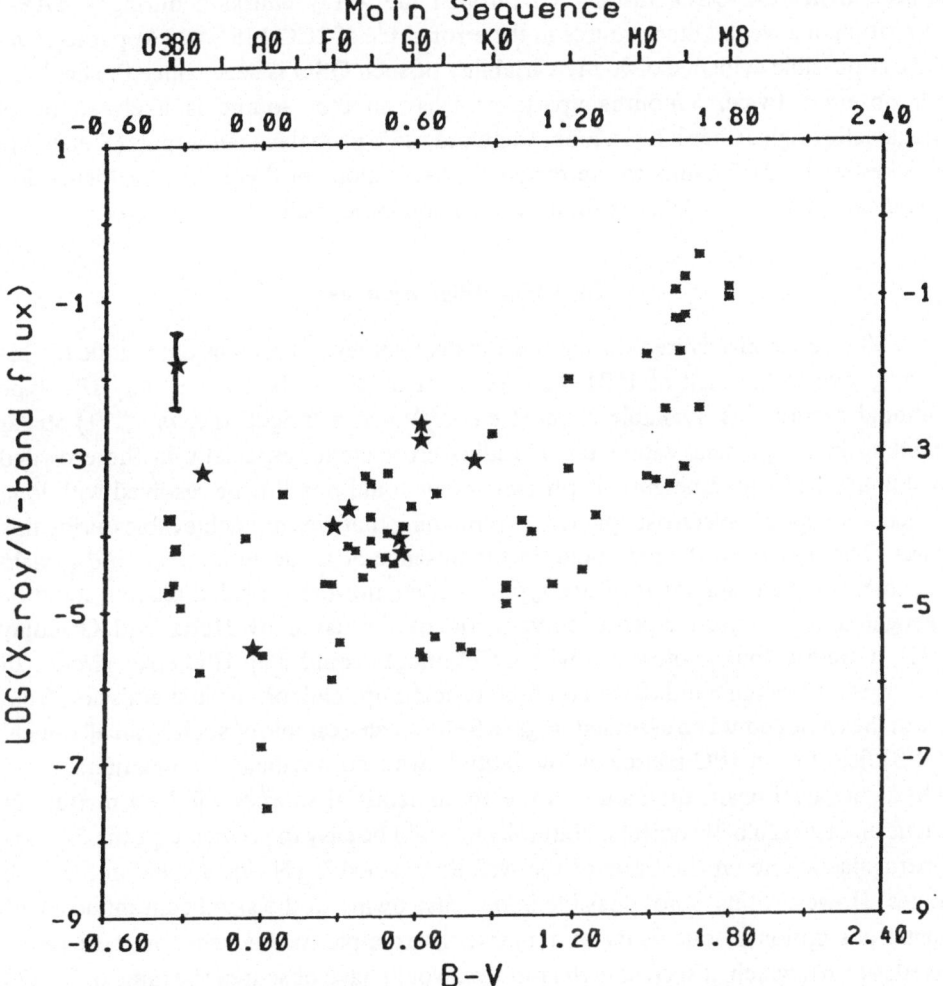

Fig. 8.   Comparison between the distribution of the values f_x/f_v derived from the CFA stellar survey by Vaiana *et al.* (1981) and the ones derived by us for 10 stars of known spectral class and magnitude (represented by★). Only a B type star, identified through the HRI with a variable radio star GT 0236 + 61; lies out of the allowed range for coronal emission. Since the region suffers from a non negligible amount of absorption ($A_v \sim 3.3$, estimated by Maraschi *et al.*, 1981, using UV data), the range of dereddened f_x/f_v is also plotted (see Bignami *et al.*, 1981, for the dereddened X-ray flux) showing that the absorption cannot account for the anomalously high value of the X-ray flux.

HRI data on this source have been presented by Bignami *et al.* (1981) with a discussion on the power requirements for the X-ray emission. More detailed theoretical considerations on GT 0236 + 61, often considered a possible counterpart of the γ-ray source 2CG 135 + 01, are discussed, e.g., by Maraschi and Treves (1983, and references therein).

3.2. EXTRAGALACTIC SOURCES

Amongst the identified sources one must mention the presence of the QSO 0241 + 61 the second nearest QSO, discovered through the X-ray emission during a SAS-3 observation of a weak Uhuru source in the error circle of 2CG 135 + 01 (Apparao *et al.*, 1978). A possible evidence of X-ray variability of such QSO is seen, since the field has been observed twice, 6 months apart. At least another source is likely to be of extragalactic origin, since it has a lobe structured radio probable counterpart (Sieber and Schlickeiser, 1982). Thanks to the repeated observations of this field (see Figure 5b), a significant X-ray variability is found in this source as well.

4. On Unidentified Sources

Of the 30 sources discovered during the Einstein survey, 16 remain unidentified. This is largely due to the lack of HRI data for most of the fields, so that only IPC-type positional accuracy is available in most cases. When a 'bright' (i.e. $m_v \lesssim 11$) stellar candidate is not present within the 1′ radius error circle, especially in the crowded low-latitude fields, an ambiguity is present which could possibly be resolved with long and accurate optical spectroscopic work, something that was not achievable during this project. The optical work that could, in the absence of an accurate positioning, yield some interesting candidates is of two types: limited, not-too-crowded regions could be investigated with objective-prism surveys, or, as discussed by Hertz and Grindlay (1981), a multi-colour photometry with CCD images could map IPC error circles. In both cases interesting candidates could be brought up, and/or coronal emission from stars in the circles could be assessed. Figure 9 shows an example of such 'typical' optical (POSS) field for an IPC source at low latitude with no obvious identification.

On a statistical basis, the nature of the 16 unidentified sources will be a mixture of galactic and extragalactic objects. Naturally it would be easy to predict, e.g., the fraction of extragalactic one on the basis of the well-established log N–log S relation for such sources. However, this is not possible in our case owing to the significant influence of galactic absorption in our fields. A remarkable example in this sense is the case of QSO 0241 − 61, which, if seen at high latitudes, would have obscured the fame of 3C 273 as an X-ray and optical QSO.

For the fraction of unidentified X-ray sources of galactic origin, one notes that while their X-ray fluxes are not, on average, smaller than those of the identified stellar sources, their optical magnitude certainly is. Thus, one is dealing with sources with a higher f_x/f_v, on average, which points to a possible difference in the type of the parent population. For instance, 'hot' stars (O to F) can be excluded beyond $f_x/f_v = 10^{-2}$ in favour of

N

↑

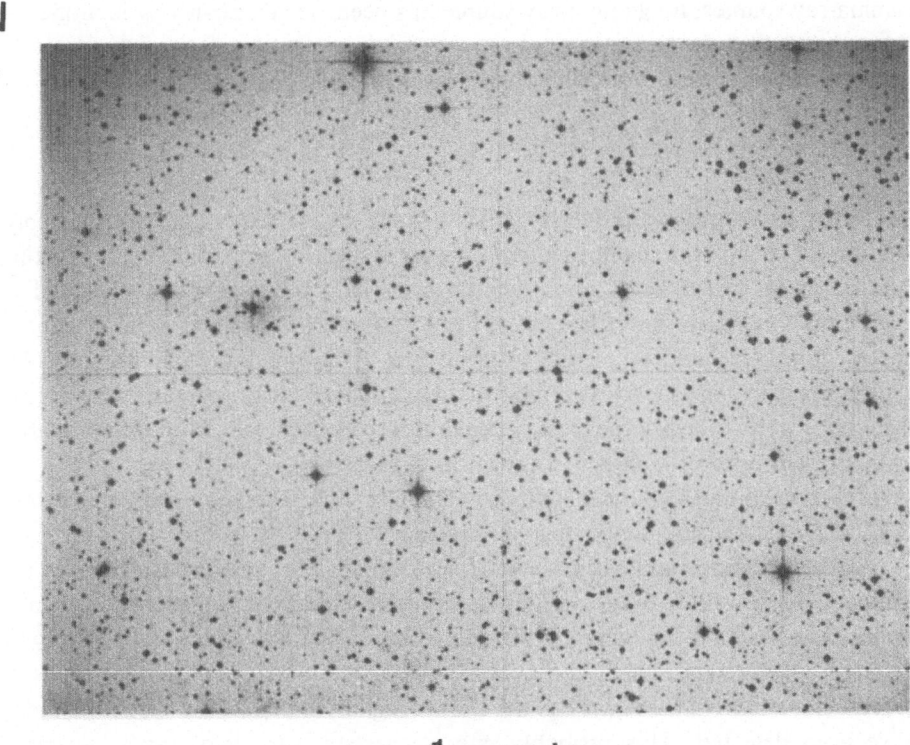

⊢——⊣ **1 arcmin**

Fig. 9.   'Typical' finding chart of an unidentified IPC source. Being in the galactic plane, the 1' radius error circle includes a non negligible amount of candidates, generally fainter than $m_v \sim 12$–13.

cooler, Main Sequence stars, where to our X-ray fluxes of ~ 1–5×10^{-13} erg cm^{-2} s^{-1} could correspond magnitudes as faint as $m_v \sim 15$–16. On the other hand, non-coronal emission is likely to be a more frequent phenomenon in these serendipitous sources with relatively faint optical counterparts.

Amongst the unidentified sources only one has been positioned with the HRI. It is the brightest source in the IPC field, shown in Figure 5b, covering the error circle of 2CG 195 + 04 (often called 'Geminga'), the brightest of the unidentified γ-ray sources. As mentioned in the caption, the final HRI position falls in an empty field on the POSS prints, both red and blue. Assuming $m_v \sim 20$ as a conservative value of the limiting magnitude of the PSS, the f_x/f_v ratio of this source ~ 100. Such a high value, although not usual, is not a novelty in X-ray astronomy. Low-mass binary systems (see e.g. Bradt and McClintock, 1982, for a review) as well as isolated radio pulsars emitting unpulsed (to few % limit) X-ray radiation (Helfand, 1981) are known to have an average L_x/L_v of several hundreds. The choice between the two possibilities is somewhat dependent on the optical limit achievable with deep CCD exposures and will be discussed in the light of the optical results (Bignami *et al.*, 1983).

It is interesting to note that, while young pulsars have no difficulties in being gamma-ray sources, no gamma-ray source has been, as yet, positively identified with a binary system.

It would be the first unpulsing pulsar discovered so far.

5. Discussion

According to Chanan *et al.* (1981) AGN's make up for about half, and possibly $\frac{2}{3}$, of the non-identified sources in medium exposure, high-latitude, IPC fields. Taking this number as an upper limit, owing to the non-negligible absorption expected in our low-latitude fields, we find that, out of the 17 non-stellar and non-catalogued sources, at least 6 are not likely to be due either to nearby normal stars or to distant AGN's. This lower limit happens to be similar to the number of γ-ray sources investigated, and, if anything, it may give some confidence that few 'non-trivial' X-ray emitters are likely to be present inside the mosaics shown in Figures 1 to 7.

A fraction of these sources could be somehow linked to the γ-ray ones if the ratio L_γ/L_x is to be kept below about one thousand.

The numerous mechanisms proposed for γ-ray emission, and discussed in this volume, can be divided in three main groups: emission from compact objects (free or in bound systems), emission from 'inert' clouds and emission from 'energized' clouds. The X-ray data presented here are easily compatible with the first two groups, while the systematic presence of an active cloud (as e.g. the Carina region) in the 2CG boxes explored would have been detected. This probably only suggests a variety of origin scenarios for gamma-ray sources. To systematically see 'inert' clouds in our X-ray data would have implied peculiar distance-opacity combinations, especially if such clouds are requested to have the mass range necessary ($\gtrsim 10^4$–$10^5 \, M_\odot$) for interesting γ-ray production.

The compact object remains thus the most viable option. However it is certainly not possible to distinguish between bound and free systems, owing to the short data coverage available for orbital period searches and to the complex behaviour observed for free radio pulsars as unpulsing X-ray sources.

Acknowledgement

The editors thank T. Montmerle for assistance in evaluating this paper.

References

Allen, C. W.: 1973, *Astrophysical Quantities*, 3rd ed., Athlone Press, London.

Apparao, K. M. V., Bignami, G. F., Maraschi, L., Helmken, H., Margon, B., Hjellming, R., Bradt, H. V., and Dower, R. G.: 1978, *Nature* **275**, 298.

Bignami, G. F. and Hermsen, W.: 1982, to appear in *Ann. Rev. Astron. Astrophys.* **21**.

Bignami, G. F., Caraveo, P. A., Lamb, R. C., Markert, T. H., and Paul, J. A.: 1981, *Astrophys. J.* **247**, L85.

Bignami, G. F., Caraveo, P. A., and Lamb, R. C.: 1983, *Astrophys. J. Letters*, in press.

Bradt, H. V. D. and McClintock, J. E.: 1982, to appear in *Ann. Rev. Astron. Astrophys.* **21**.

Caraveo, P. A.: 1981, *Phil. Trans. Roy. Soc. London* **A301**, 523.

Caraveo, P. A., Bignami, G. F., Paul, J. A., Marano, B., and Vettolani, J. P.: 1981, *Space Sci. Rev.* **30**, 301.

Cesarky, C. J. and Montmerle, T.: 1983, *Space Sci. Rev.* **36**, 173 (this volume).

Chanan, G. A., Margon, B., and Downes, R. A.: 1981, *Astrophys. J.* **243**, L5.

D'Amico, N.: 1983, *Space Sci. Rev.* **36**, 195 (this volume).

Giacconi, R., Branduardi, G., Briel, U., Fabricant, D., Feigelson, E., Forman, W., Gorenstein, P., Grindlay, J., Gursky, J., Harnden, F. R., Henry, J. P., Jones, C., Kellogg, E., Koch, D., Murray, S., Schraier, E., Seward, F., Tananbaum, H., Topka, K., van Speybroeck, L., Holt, S. S., Becker, R. H., Boldt, E. A., Serlemitsos, P. J., Clark, G., Canizares, C., Markert, T., Novick, R., Helfand, D., and Long, K.: 1979, *Astrophys. J.* **230**, 540.

Gregory, P. C., Taylor, A. R., Crampton, D., Hutchings, J. B., Hjellming, R. M., Hogg, D., Hvatum, H., Gottlieb, E. W., Feldman, P. A., and Kwok, S.: 1979, *Astron. J.* **84**, 1030.

Helfand, D. J.: 1981, in W. Sieber and R. Wielebinsky (eds.), 'Pulsars', *IAU Symp.* **95**, 343.

Helfand, D. J. and Caillault, J. P.: 1982, *Astrophys. J.* **253**, 760.

Hermsen, W., Bennett, K., Bignami, G. F., Boella, G., Buccheri, R., Higdon, J. C., Kanbach, G., Lichti, G. G., Masnou, J. L., Mayer-Hasselwander, H. A., Paul, J. A., Scarsi, L., Swanenburg, B. N., Taylor, B. G., and Wills, R. D.: 1977, *Nature* **262**, 494.

Hertz, P. and Grindlay, J. E.: 1981, *Bull. Am. Astron. Soc.* **13**, 787.

Lamb, R. C., Markert, T. H., Hartman, R. C., Thompson, D. J., and Bignami, G. F.: 1980, *Astrophys. J.* **239**, 651.

Maraschi, L. and Treves, A.: 1983, *Space Sci. Rev.* **36**, 161 (this volume).

Maraschi, L., Tanzi, E. G., and Treves, A.: 1981, *Astrophys. J.* **248**, 1010.

Markert, T. H., Lamb, R. C., Hartman, R. C., Thompson, D. J., and Bignami, G. F.: 1981, *Astrophys. J.* **248**, L17.

Masnou, J. L., Bennett, K., Bignami, G. F., Bloemen. J. B. G. M., Buccheri, R., Caraveo, P. A., Hermsen, W., Kanbach, G., Mayer-Hasselwander, H. A., Paul, J. A., and Wills, R. D.: 1981, *17th International Cosmic Ray Conference* **1**, 177.

Montmerle, T.: 1979, *Astrophys. J.* **231**, 95.

Montmerle, T.: 1981, *Phil. Trans. Roy. Soc. London* **A301**, 505.

Montmerle, T., Koch-Miramond, L., Falgarone, E., and Grindlay, J. E.: 1983, *Astrophys. J.* **269**, 182.

Seward, F. D. and Chelebowski, T.: 1982, *Astrophys. J.* **256**, 530.

Sieber, W., and Schlickeiser, R.: 1982, *Astron. Astrophys.* **113**, 314.

Singh, K. P., Apparao, K. M. V., and Manchanda, R. K.: 1982, *Astrophys. Space Sci.* **82**, 477.

Swanenburg, B. N., Bennett, K., Bignami, G. F., Buccheri, R., Caraveo, P., Hermsen, W., Kanbach, G., Lichti, G. G., Masnou, J. L., Mayer-Hasselwander, H. A., Paul, J. A., Sacco, B., Scarsi, L., and Wills, R. D.: 1981, *Astrophys. J. Letters* **243**, L69.

Taylor, A. R. and Gregory, P. C.: 1982, *Astrophys. J.* **255**, 210.

Topka, K., Avni, Y., Golub, P., Harnden, F. R., Rosner, R., and Vaiana, G. S.: 1982, *Astrophys. J.* **259**, 677.

Vaiana, G. S., Cassinelli, J. P., Fabbiano, G., Giacconi, R., Golub, L., Gorenstein, P., Haisch, B. M., Harnden, F. R., Johnson, H. M., Linsky, J. L., Maxson, C. W., Mewe, R., Rosner, R., Seward, F., Topka, K., and Zwaan, C.: 1981, *Astrophys. J.* **245**, 163.

LARGE-SCALE STRUCTURE OF THE GALAXY
AND HIGH-ENERGY GAMMA-RAY OBSERVATIONS*

H. A. MAYER-HASSELWANDER

*Max-Planck-Institut für Physik und Astrophysik, Institut für Extraterrestrische Physik,
D-8046 Garching, F.R.G.*

Abstract. Detailed information on the high-energy gamma-ray emission from our Galaxy has become available through the two dedicated satellite missions SAS-2 and COS-B. The consistency of the two datasets is discussed; while a satisfying general agreement is observed, a few distinct discrepancies point to possible time variations within the compact source component of the total galactic emission. The bulk of emission appears very well correlated to the column density of the total interstellar gas, as traced by radio observations of H I and CO. The gamma-ray observations exclude the possibility that H_2 dominates in the inner Galaxy, its mass should not exceed the mass existing in the form of H I. Neither a significant galactocentric gradient of the (high-energy) cosmic-ray flux density is suggested inside the solar circle (outside a decrease is needed), nor a linear coupling between the cosmic rays and the gas is indicated by the gamma-ray data. The systematic variation with longitude of the spectrum of the gamma-ray emission points to an increased flux of cosmic-ray electrons in the 100 MeV to 1 GeV energy range in regions where dense clouds are concentrated. The variation could as well be due to the largely unresolved population of compact gamma-ray objects.

1. Introduction

The gamma emission of our Galaxy in the energy range 70 MeV to 5 GeV has been surveyed by two successfull satellite missions during the last decade: by the experiment on the SAS-2 satellite (Fichtel *et al.*, 1975), in operation for 7 months in 1972, and by the experiment on the COS-B satellite (Bignami *et al.*, 1975; Scarsi *et al.*, 1977), operating for an exceptionally long time, from August 1975 to April 1982. Both experiments used the wire-sparkchamber technique with magnetic-core storage to record the photons, to determine their angle of incidence and to discriminate them against charged-particle background events. While in the SAS-2 experiment only a rough energy estimate could be derived from the sparkchamber picture data, in COS-B an additional large Cs-I energy calorimeter provided detailed spectral information.

Already the pioneering SAS-2 experiment could observe a large part of the galactic disc and made possible first studies of the galactic structure in the gamma-ray spectral range; yet the accuracy of the data was severely limited by photon statistics, only about 8000 photons were recorded. COS-B has completed the picture of the Galaxy and, by recording about 100 000 gamma rays from the disc, has removed or drastically narrowed down the statistical uncertainties. It has added also a new quality by providing spectral information which at present is made available through surveys in the three energy bands: 70–150 MeV, 150–300 MeV, and 300–5000 MeV.

---

* Proceedings of the XVIII General Assembly of the IAU: *Galactic Astrophysics and Gamma-Ray Astronomy*, held at Patras, Greece, 19 August 1982.

Space Science Reviews **36** (1983) 223–247.   0038–6308/83/0363–0223$03.75.

For COS-B due to strong technical arguments a highly excentric orbit had been selected. As a consequence the experiment operated outside the radiation belts and was exposed to the full cosmic-ray (CR) flux density which created a considerable instrumental gamma-ray background; above 70 MeV this background is of the order 8×10^{-5} ph cm^{-2} s^{-1} sr^{-1}. As the SAS-2 satellite had a low equatorial orbit, in which the Earth's magnetic field effectively shielded the experiment from CR's, this experiment had a drastically lower instrumental background than COS-B. So, even if the COS-B data base outdates the SAS-2 dataset with respect to statistical accuracy and spectral information, for investigations, where a low instrumental background is essential, the SAS-2 dataset still has its merits. As this dataset has been accumulated at an earlier epoch than the COS-B data base, it is also very valuable when secular time variations are investigated.

The SAS-2 dataset has been fully exploited (Hartmann *et al.*, 1979, and references therein) while for COS-B the final dataset, containing 66 observations (pointings), is not available yet. A major part of the COS-B data, including the first 44 observations, recently has been presented with an emphasis on the *large-scale* galactic structure by Mayer-Hasselwander *et al.* (1982). Concerning the relation of the observed emission to the *local* galactic environment, on the basis of the COS-B data the proportionality between the gamma-ray intensity and the column density of the total interstellar gas (as derived from galaxy counts) has been demonstrated: by the work of Lebrun *et al.* (1982) and Strong *et al.* (1982), within about 500 pc from the Sun and averaged over the line of sight, the total gas appears to be completely penetrated by a constant CR flux, making gamma rays a direct tracer of the total interstellar gas. This result is a necessary prerequisite, if the galactic emission is to be analyzed on a larger scale, where various relevant parameters, as for example the CR flux density, must be expected to vary.

The observed emission after subtraction of the instrumental background consists of contributions from:

– Diffuse processes; here these processes are always interactions of two components (e.g. CR's and gas) on scales of parsecs and larger (molecular clouds irradiated by the ambient CR flux are obviously included).

– A possible population of 'point-like' gamma-ray sources with angular sizes smaller than 1 to 2 degrees which cannot be resolved by the COS-B instrument. This type of hypothetical sources include star-like objects as well as e.g. gas clouds which might be immersed into an enhanced CR flux from an associated CR-source and therefore do not fall into the category quoted first.

– The gamma-ray pulsar population; the pulsar contribution has been estimated by Harding and Stecker (1981) to be 15%, but their model strongly depends on the luminosity of the only two observed gamma-ray pulsars (Crab and Vela) and does not account for the fact that 10 pulsars are predicted by their model to be visible in gamma rays but are not detected. Taking into account this lack of detection, Buccheri *et al.* (1983) have estimated a contribution of only 1%. These estimates assume a relatively conventional behaviour of pulsars and also do not consider selection effects in the pulsar catalogue, for example very fast pulsars are not unlikely to be missed in pulsar searches or as an other possibility there might exist gamma-ray pulsars whose radio beams are not visible at earth (fast pulsed emission in gamma rays as a consequence of the low count rate is impossible to be detected if the period is not known from e.g. radio measurements). So a contribution between 1 and 30% is not unlikely and is not contradicted by the data. As the distribution especially of young pulsars is not significantly different from that of the interstellar molecular hydrogen, their gamma-ray emission on a large scale also would appear very similar to that of H_2.

Neglecting the presence of the quantitatively unknown contributions from gamma-ray point source populations therefore has the consequence, that values for the diffuse emissivity have a tendency to result too high and better are considered as upper limits.

The diffuse production processes in our Galaxy which are expected to contribute to the observed emission are well known (e.g. Stecker, 1971; Fichtel and Trombka, 1981); the present question is: "What is the gamma-ray emissivity (photons s^{-1} per volume element or per surface element) at a given site within the Galaxy and what is the relative contribution from the various possible production processes to this emissivity?" The processes which, on the basis of our present knowledge of the physical conditions in our Galaxy, can contribute a significant amount to the emissivity are interactions of:

– CR-nucleons with the nucleons of the interstellar medium (ISM) (spallation, π°-decay);
– CR-electrons with the nucleons of the ISM (Bremsstrahlung);
– CR-electrons with radiation fields (Inverse Compton process).

These processes produce different gamma-ray spectra, so in principle their relative contributions can be deduced if the gamma-ray spectrum at a given site is known with sufficient accuracy.

When dealing with the large-scale galactic distribution of the emissivity, we have to consider a few observational constraints relevant to gamma-ray astronomy:

– Absorption can be neglected, even for radiation from the other side of the Galaxy.
– No direct distance indicator is available, line-of-sight integrals over the spatially distributed emissivity are observed, a projected view is obtained.
– The angular resolution (point-spread function) is strongly energy dependent (1.2 degrees HWHM at energies above 300 MeV, several degrees at lower energies). Therefore beam dilution effects, strongly depending on energy, have to be accounted for in the presently available data.
– Spectral information is available in the COS-B data, however, its use for the analysis of the large-scale diffuse emissivity from the experimental side is restricted by the (energy-dependent) beam-dilution problem, by count statistics and by variations of the instrumental background.

So, unfortunately, the gamma-emissivity at a given site usually cannot be measured directly and one has to resort to more indirect approaches involving other already known information. There are basically three ways of analysis, which were followed in the past:

(a) Comparison of the observed projected gamma-ray distribution with that predicted from Galaxy models which are constructed on theoretical concepts and make use of observed parameters of our Galaxy ('Simonson' model, 'Georgelin' model).

(b) Analysis of the direct correlations between the gamma-ray survey and projected surveys of other galactic structure tracers (21 cm, CO-line, dark clouds, 408 MHz, infrared).

(c) Comparison of 'derived' characteristic parameters describing large-scale structure (galactocentric variation, z-scaleheight).

All these approaches depend on the availability of complete surveys of the various galactic structure tracers (incomplete CO-line surveys are the main handicap here) and suffer from the relatively bad angular resolution of the gamma-ray data.

2. The Gamma-Ray Data

Figures 1 and 2 show the COS-B gamma-ray survey as recently presented by Mayer-Hasselwander *et al.* (1982), where a detailed discussion can be found. Since much work in the past was based on the SAS-2 dataset, it is interesting to investigate if and to which extent the two datasets are in agreement.

Here longitude profiles integrated over $|b| < 10°$ are compared. As the SAS-2 data are presented for energies above 100 MeV, the COS-B intensities derived for energies above 150 MeV were, on the basis of the observed average spectrum, extrapolated to the range > 100 MeV to make them directly comparable with the SAS-2 results. The instrumental background of COS-B was subtracted, but the isotropic celestial gamma-ray background measured by SAS-2 was not subtracted from the SAS-2 nor from the COS-B data. The COS-B profiles, which in Mayer-Hasselwander *et al.* (1982) were presented without correction for measured-energy dispersion, were suitably corrected. The remaining line intensities were integrated over all longitudes to define a normalisation factor accounting for systematic differences due to calibration errors or different definitions of the intensity maps. The integrated intensity measured by SAS-2 is 1.17 the value obtained by COS-B. In view of the various possible sources for systematic errors in the determination of the absolute sensitivity such a difference is expected; the difference could in part also be real if significant time variability of compact sources were involved. For the comparison the hypothesis that systematic errors are responsible is preferred and the profiles normalised to the COS-B level are shown in Figure 3.

Figure 3 demonstrates very good agreement within statistical errors over wide areas, but also shows a few regions where significant discrepancies exist. More detailed investigations using the two-dimensional datasets are desirable. But already from this comparison, given the experiment's field-of-view and angular resolution, it appears that the differences in the regions around 60°, 90°, and 20°–30° are likely to be due to instrumental effects while the differences at 330°, 75°, and 185° are as well attributable to variability in the observed gamma-ray intensities. The knowledge of the degree of consistency and of the discrepancies of the two independent datasets is a valuable help, when the correlation of the data with model predictions or other tracers is discussed.

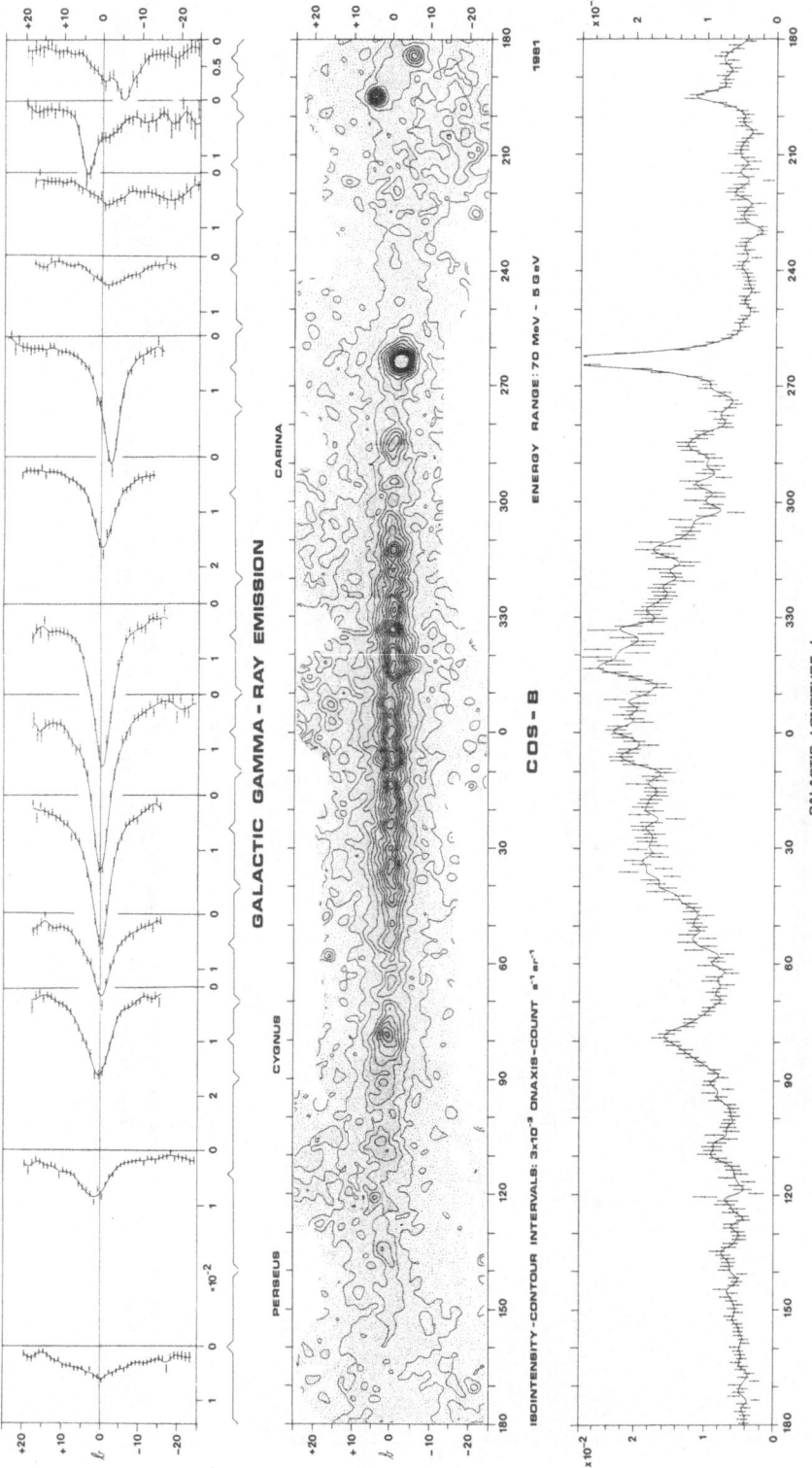

Fig. 1. Presentation in galactic coordinates of the structure of the galactic gamma-ray emission as measured by COS-B. In the map the surface fitted to the data matrix is indicated by contour lines and a grey scale. Regions outside the accepted field of view are left blank. The profiles along longitude and latitude show the data points with statistical errors and the fitted surface (solid line). In the longitude profile the data are averaged over $\pm 5°$ in latitude. For the latitude profiles the ranges for averaging over longitude are indicated by brackets. The map and the profiles show the parameter *on-axis* count s^{-1} sr^{-1}. The contourlevels are indicated at multiples of 3×10^{-3} *on-axis* count s^{-1} sr^{-1}.

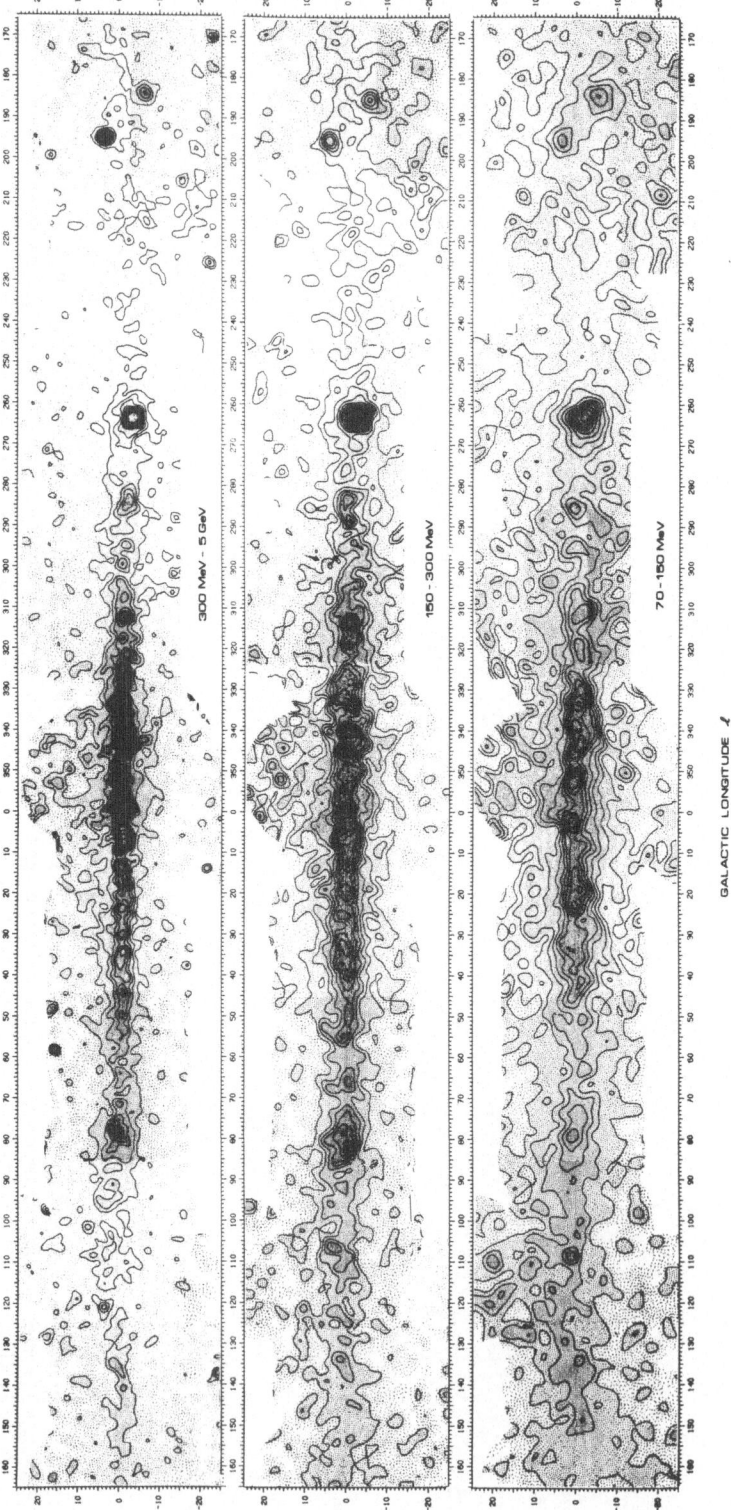

GALACTIC LONGITUDE *l*

Fig. 2.   Presentation of the galactic gamma-ray intensity in three energy intervals. The smoothed maps show the parameter photon intensity (photon cm^{-2} s^{-1} sr^{-1}). The contour levels are indicated at multiples of 5×10^{-5} for 70 MeV–150 MeV, 3×10^{-5} for 150 MeV–3000 MeV and 4×10^{-5} for 300 MeV–5 GeV. Regions outside the accepted field of view are left blank.

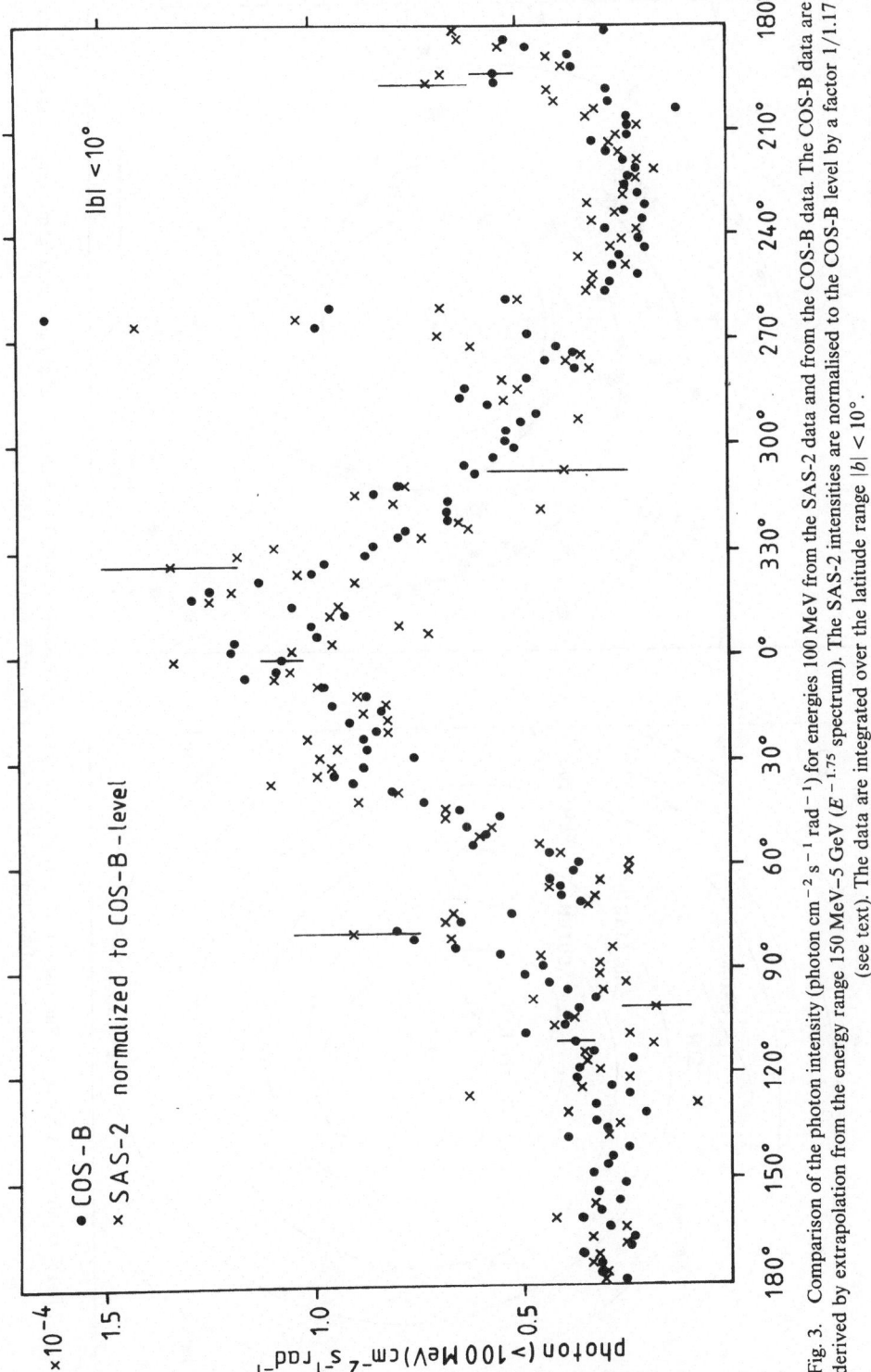

Fig. 3. Comparison of the photon intensity (photon cm^{-2} s^{-1} rad^{-1}) for energies 100 MeV from the SAS-2 data and from the COS-B data. The COS-B data are derived by extrapolation from the energy range 150 MeV–5 GeV ($E^{-1.75}$ spectrum). The SAS-2 intensities are normalised to the COS-B level by a factor 1/1.17 (see text). The data are integrated over the latitude range $|b| < 10°$.

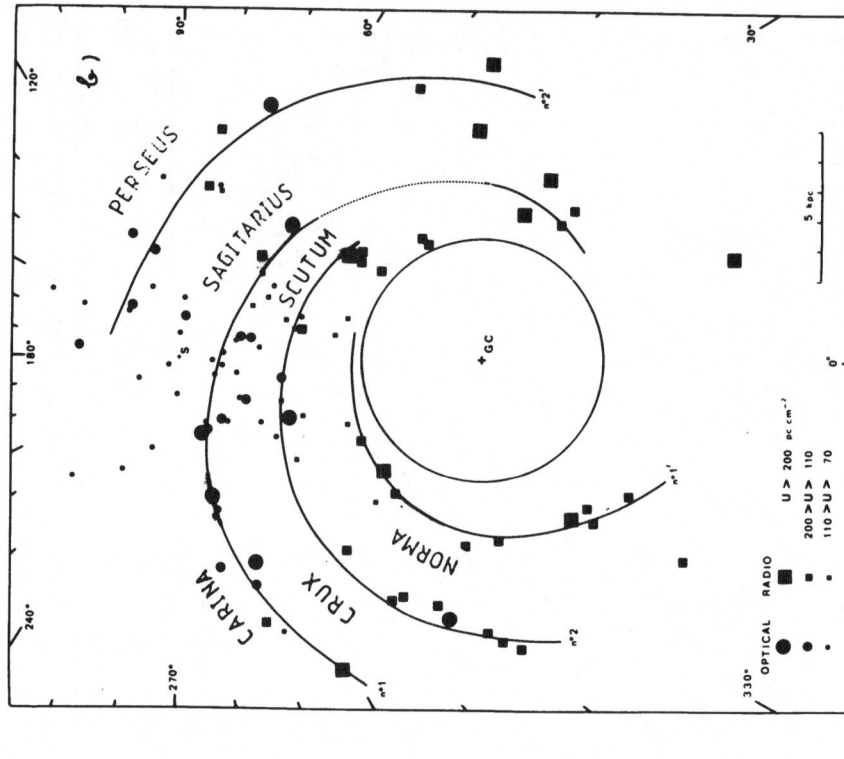

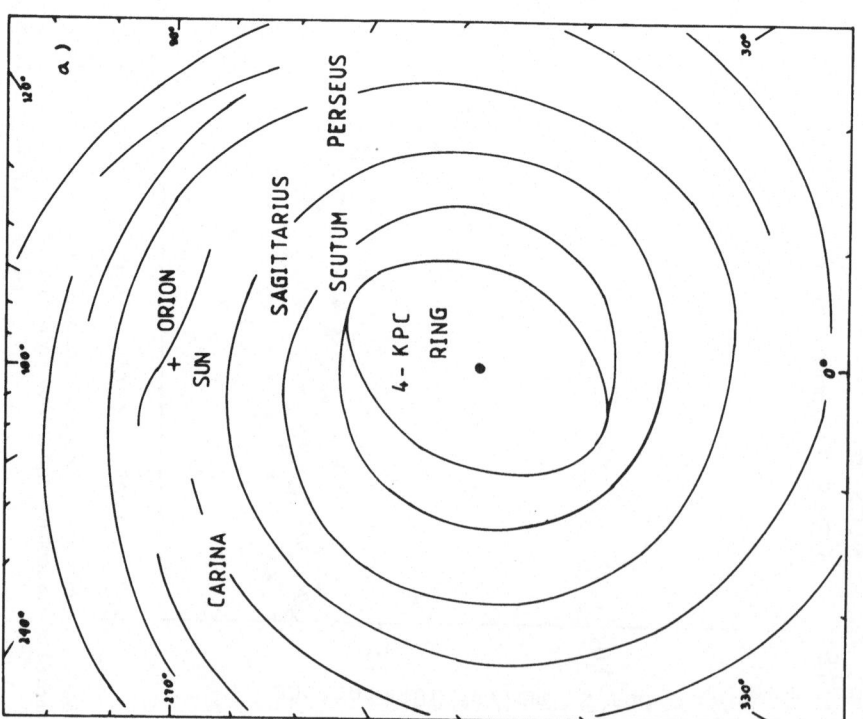

Fig. 4.   The spiral structures derived by Simonson (1976) (a) and by Georgelin and Georgelin (1976) (b), which are used as basis for the gamma-ray emission models of Kniffen and Fichtel (1981) and by Fichtel and Kniffen (1982).

3. Comparison with Galaxy Models

Two gamma-ray emission models have been developed, one by Kniffen and Fichtel (1981) based on the 'Simonson' Galaxy model (Simonson, 1976) and the other by the same authors (Fichtel and Kniffen, 1982; Fichtel, 1982) on the basis of the 'Georgelin' Galaxy model (Georgelin and Georgelin, 1976); both structures are shown in Figure 4. The total predicted emission based on these models is compared with the COS-B data (because of their beter statistics) in Figure 5. In both models a linear coupling between gas density and CR-flux density on the scale of galactic arms is assumed, but the z-scaleheight of the CR's is taken large compared to that of the gas. The contributions from inverse Compton interactions (blackbody-, infrared-, and optical radiation fields) are taken into account.

Both models, as a consequence of appropriately chosen emissivity parameters, agree with each other and with the gamma-ray data as far as the very-large-scale geometry of the Galaxy is concerned. The agreement becomes only a partial one when smaller structures like galactic arms are considered. The models, involving large-scale symmetry assumptions, cannot be expected to describe the emission from very nearby or 'local' features, which due to their small distance (no beam dilution) may be prominent in the gamma-ray picture.

In the following the correlation is discussed in some detail:

(a) The appropriate contrast between emission from the central region ($300° < l < 40°$) and from regions outside the solar circle is essentially obtained by selection of appropriate values for the molecular hydrogen content and under the assumption of a CR gradient outside the solar circle. Clearly a lower amount of H_2 (the $[H_2]/[CO]$ ratio is very uncertain) together with a positive gradient of the CR-flux density inside the solar circle would give a very similar prediction. Also the assumed coupling between gas density and CR-flux density could be replaced by a galactocentric gradient in the CR-flux density without significantly changing the prediction.

(b) The '4 kpc ring' of the models has a counterpart in the data only in the fourth quadrant, but not in the first. A central peak, extended about 10 degrees in longitude around the galactic center, is indicated by the models in agreement with the data. In the tangential direction of the 'Crux' spiral-arm at $315°$ (Simonson model), a well correlated peak is observed. A correlation can also be observed in the case of the Carina arm ($285°$) of the 'Georgelin' model, and a very marginal one for the counterparts of these arms in the first quadrant: Sagittarius at $45°–50°$ and Scutum at $35°$. For the Norma arm no correlated emission is observed; it is interesting to note that this tangential point is the farthest from the Sun.

(c) The local ('Orion') arm, which appears as a prominent emission feature in the direction of Cygnus ($70°–90°$), is not predicted by the models.

So a detailed satisfying correlation with the models is not found, there are several reasons which can account for that:

– The models make too much use of assumptions on large-scale symmetries, therefore relatively nearby irregularities of the Galaxy, not properly described by the models,

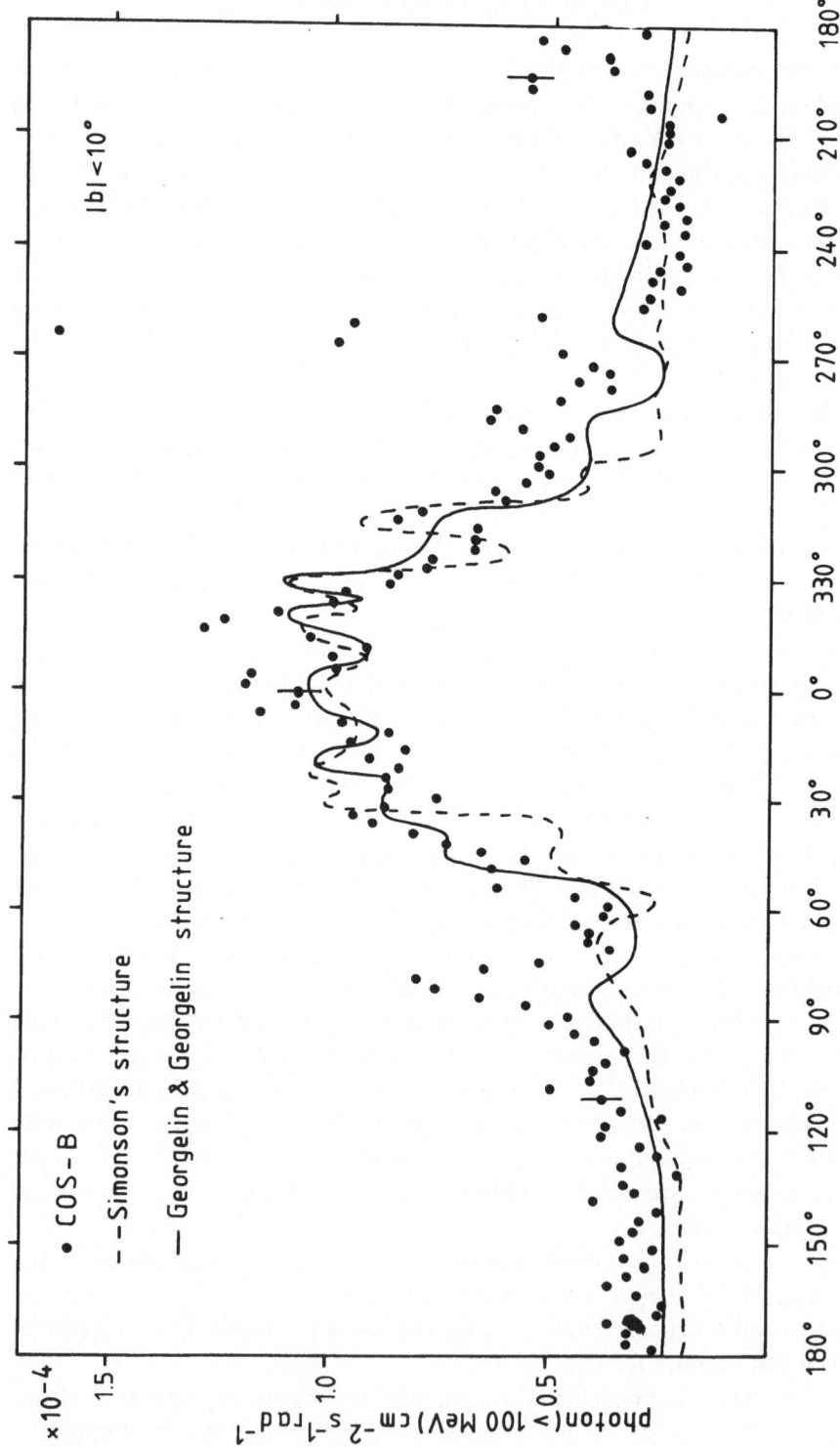

Fig. 5.   Comparison of the gamma-ray photon intensity (photon cm^{-2} s^{-1} rad^{-1}) longitude distribution as observed by COS-B with the emission predicted by the models of Kniffen and Fichtel (1981) (Simonson structure) and of Fichtel and Kniffen (1982) (Georgelin and Georgelin structure). The data are given for the energy range > 100 MeV and are integrated over the latitude range $|b| < 10°$. The model predictions are normalised to the COS-B intensity level.

are possibly seen as dominating features in gamma rays as they don't suffer much beam dilution due to their small distance.

– The two models indicate spiral features which especially in the fourth quadrant do not concur.

– Emission from nearby and possibly time-variable compact gamma-ray objects may hide in some regions the correlation with the diffuse emission.

– A limited number of compact gamma-ray objects, correlated with the galactic arm structure, could, if their number is small enough to yield the appropriate statistical fluctuations, cause one arm to appear prominent, the other one to be weak.

As a consequence especially the assumption of a linear coupling between gas and CR's, which is used in the models, cannot be proven, as such a proof would have to rest on a close correlation of the data with the predicted arm features, which instead is not observed (see Figure 5).

4. Correlation with Other Tracers

The diffuse constituents of the Galaxy which are involved in gamma-ray production can also be investigated by several other tracers: atomic hydrogen (H I) is traced by its 21 cm-line emission, molecular hydrogen by observation of the 2.6 mm-line emission of the associated CO molecules. The CR-electron flux density is involved in the emission of the radio continuum (e.g. 150 MHz, 408 MHz). All those tracers, like gamma rays, can be used on a galactic scale: in case of the usually optically thick CO-line individual clouds, for which the probability of coincidence is small, are observed; in the other cases absorption is negligible or relatively small (for H I less than 30%) and can be reliably corrected for. These tracers are observed with much finer angular resolution, so for the direct comparison the velocity integrated surveys are convoluted with the pointspread function (angular resolution) of the gamma-ray survey.

A comparison between the radio continuum emission at 408 MHz and the COS-B gamma-ray survey was made by Haslam *et al.* (1981). The analysis basically confirms that the z-scaleheight of the synchrotron emission and consequently that of the CR electrons and galactic magnetic fields is much larger than that of the gamma-ray emissivity. This supports the expectation that the dominant gamma-ray contribution stems from interactions between the gas, which has a comparatively much smaller z-scale height, and the CR's. Clearly at least perpendicular to the plane there is no linear coupling between gas and CR's nor between gas and magnetic fields. The comparison of the distributions in longitude does not yield any useful conclusion as a consequence of the complicated dependence of the observed radio emission on the unknown details (e.g. direction) of the galactic magnetic fields.

If now the 21 cm-line survey as tracer of H I is compared with the gamma-ray data (Mayer-Hasselwander *et al.*, 1982), under the hypothesis of the CR-flux density being constant throughout the Galaxy, a very instructive picture is obtained: for an emissivity of 3.4×10^{-25} ph H-atom^{-1} s^{-1} in the energy range 70 MeV–5 GeV, as derived locally

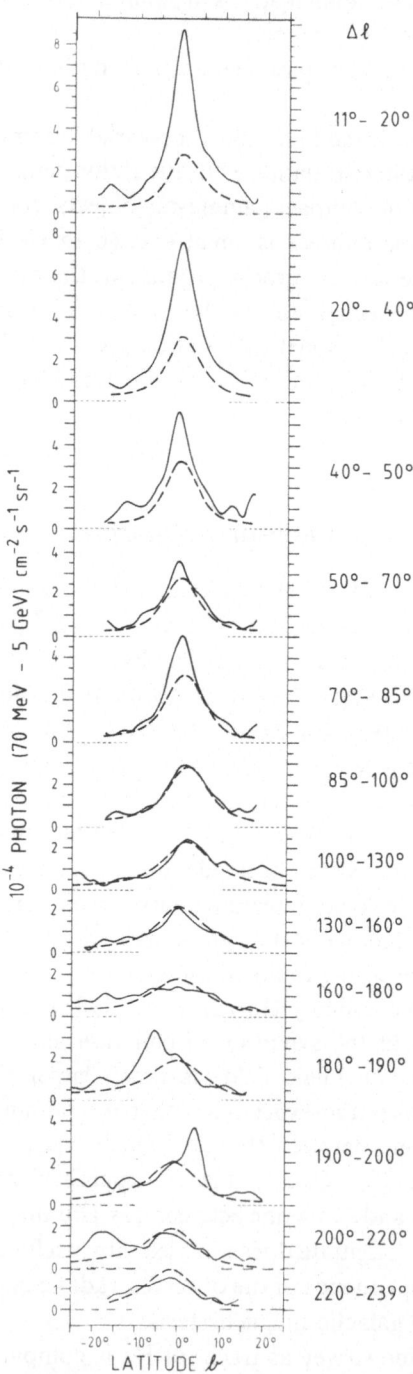

Fig. 6. Galactic latitude profiles for observed gamma-ray emission (solid line) and for emission expected from interactions between cosmic rays and H I gas (dashed line) for the energy range 70 MeV–5 GeV, under the assumption of uniformly distributed cosmic rays with a density as observed locally.

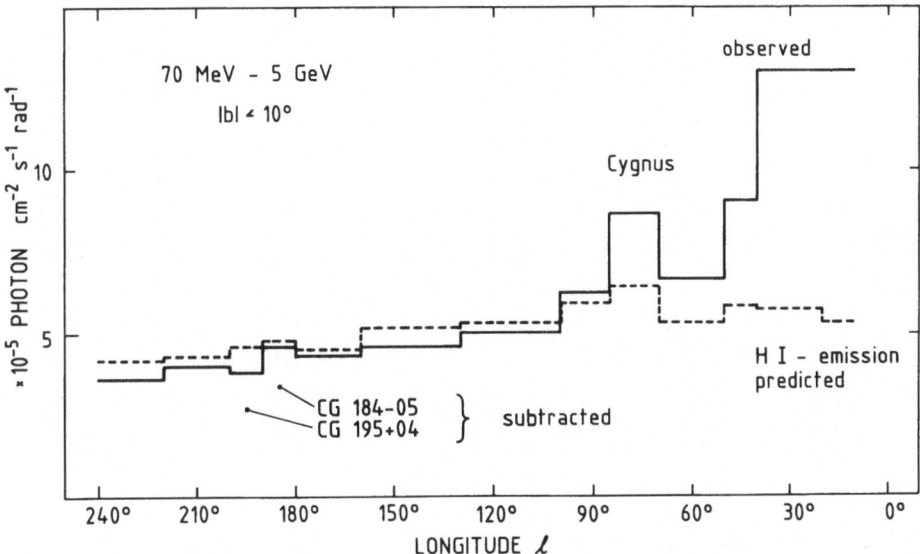

Fig. 7.   Distribution in galactic longitude of the observed gamma-ray intensity for $|b| < 10°$ (solid line) and of the expected emission from H I-gas (dashed line) in the energy range 70 MeV–5 GeV for uniformly distributed cosmic rays of an intensity as observed locally.

(Lebrun *et al.*, 1981), the latitude and longitude distributions predicted from atomic hydrogen are compared with the gamma-ray data in Figures 6 and 7.

For longitudes $l < 50°$ the predicted emission is much less than the observed emission and its latitude distribution is significantly wider. This implies that an additional source of emission is needed and that this additional component has a smaller z-scaleheight than H I. These requirements are met if a substantial fraction of the gas in the inner Galaxy is in the form of molecular hydrogen, whose scale height is known to be less than that of H I (Gordon and Burton, 1976). However, it should be noted that gamma-ray sources also have a rather small scale height (Swanenburg *et al.*, 1981), and can also very well account for most of the discepancy between the H I contribution and the observed profile. In this case, the room for molecular gas is dramatically reduced, especially if one considers that the CR density may not be uniform and could increase towards the inner Galaxy.

At $50° < l < 70°$ H I alone can account for a substantial fraction of the observed emission, indicating that in this interarm direction (between the Sagittarius and the Perseus arms) the actual amount of molecular hydrogen should certainly not exceed that of H I. This is not the case in the longitude range $70° < l < 85°$ which encompasses most of the Cygnus complex (or local arm). Detailed CO observations performed in this region have revealed the existence of several giant molecular clouds (Stark and Blitz, 1978) whose summed contributions together with the underlying H I component could account for a large fraction of the observed radiation as first shown by Protheroe *et al.* (1979). Figure 8 demonstrates the very good agreement of the spatial distribution of the

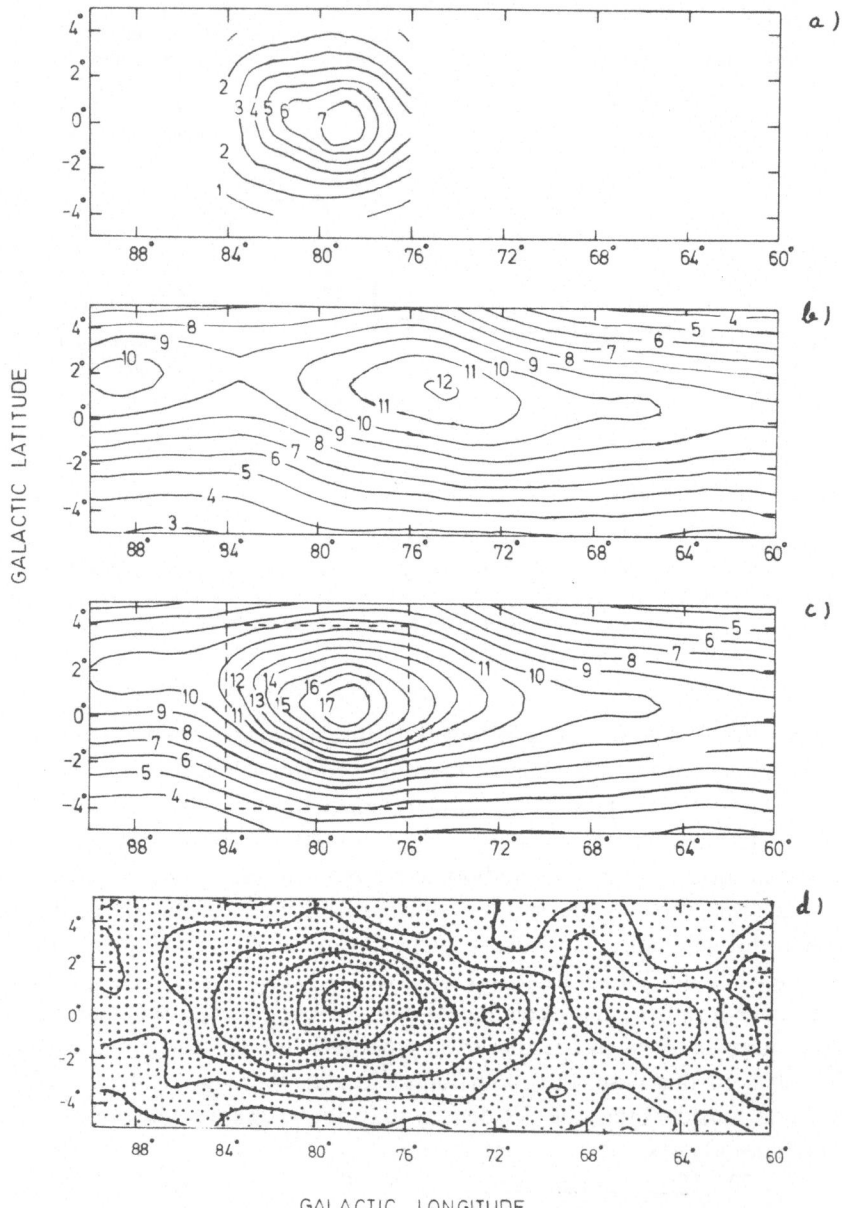

Fig. 8. Contour maps of the hydrogen column densities convoluted with the COS-B angular resolution and of the gamma-ray intensities as observed by COS-B. (a) indicates molecular hydrogen; (b) atomic hydrogen; (c) gives the sum of (a) and (b); (d) shows the observed gamma-ray intensities. Scales are arbitrary.

total predicted emission from H I and H_2 with the gamma-ray intensity contours. This analysis has the additional advantage that, as the observed region is situated near the solar circle, possible galactocentric gradients of the CR-flux density or of the ratio $[H_2]/[CO]$ would have no effect.

For $85° < l < 160°$, only a very small additional contribution can be tolerated within the observed gamma-ray profiles. However in the longitude range $105° < l < 140°$ substantial molecular hydrogen, partially related to the Perseus arm, was observed by Cohen *et al.* (1980). This would lead to excessive gamma-ray production, but one can invoke the argument first proposed by Dodds *et al.* (1975) of a marked decrease of the CR density beyond the solar circle. The same argument obviously should apply for all the other regions considered (up to $l = 220°$), where it appears that around $b = 0°$ the predicted H I emission accounts for all the observed radiation except at $180° < l < 200°$ where the contributions of two bright sources dominate.

In some regions at large distances from the galactic equator the H I contribution is not enough to match the observation. This is particularly clear at negative latitudes for $160° < l < 220°$, where local molecular complexes are known to exist (Stark and Blitz, 1978), including the Orion complex, which has been individually detected by COS-B (Caraveo *et al.*, 1980).

Similar features occur at positive latitudes at $10° < l < 40°$, where molecular complexes, related to the Gould's Belt, contribute to the observed flux. In the symmetric region at negative latitudes a strong gamma-ray excess is present as shown in Figure 6. This excess coincides well with a region of strongly enhanced galactic absorption (Strong and Lebrun, 1981).

Recently from the Columbia Telescope a new well sampled CO survey (Dame and Thaddeus, 1982) became available for the first quadrant and for the latitude range $-5.5° < b < 10.5°$. This survey makes it possible to compare directly over a large longitude range the brightness of the 2.6 mm CO line and of the 21 cm H I line with the gamma-ray intensity. This was done by Lebrun and collaborators (Lebrun, 1982; Lebrun *et al.*, 1983) by correlating the CO brightness and the H I column density (after convolution with the gamma-ray point-spread function) with the gamma-ray data in the 300 MeV–5 GeV energy range. According to Kniffen and Fichtel (1981) in this energy range an inverse-Compton component of about 20% is to be expected when the average over $|b| < 10°$ is taken. This contribution is primarily from CR-electron interactions with infrared photons which in longitude are roughly distributed like the ISM. In the analysis by Lebrun *et al.* this contribution had to be neglected, as its longitude distribution is not directly measured but derived from a model. So the emission is assumed to stem from CR interactions with the interstellar medium and from compact gamma-ray objects. In their analysis they have considered:
– CR interactions with gas atoms traced by H I (I_{HI});
– CR interactions with H_2 molecules traced by the velocity integrated brightness temperature W_{CO} (I_{H_2});
– compact source population (I_S);
– extragalactic isotropic emission (I_E);
– instrumental background (I_{BG}).
In a likelihood analysis the three free parameters A, B, and C in the expression

$$I_P = AN_{HI} + BW_{CO} + C$$

were determined so as to give the best agreement between the gamma-ray and the H I and CO surveys. Here I_P is the gamma-ray intensity predicted, $AN_{\rm H I}$ corresponds to $I_{\rm H I}$, $BW_{\rm CO}$ corresponds to the sum $I_{\rm H_2} + I_S$ (implying that these components have the same distribution), and C corresponds to the sum $I_E + I_{BG}$. This analysis requires the a priori assumption of a constant CR-flux density and of the absence of a strong galactocentric gradient in the ratio $[\rm H_2]/[\rm CO]$.

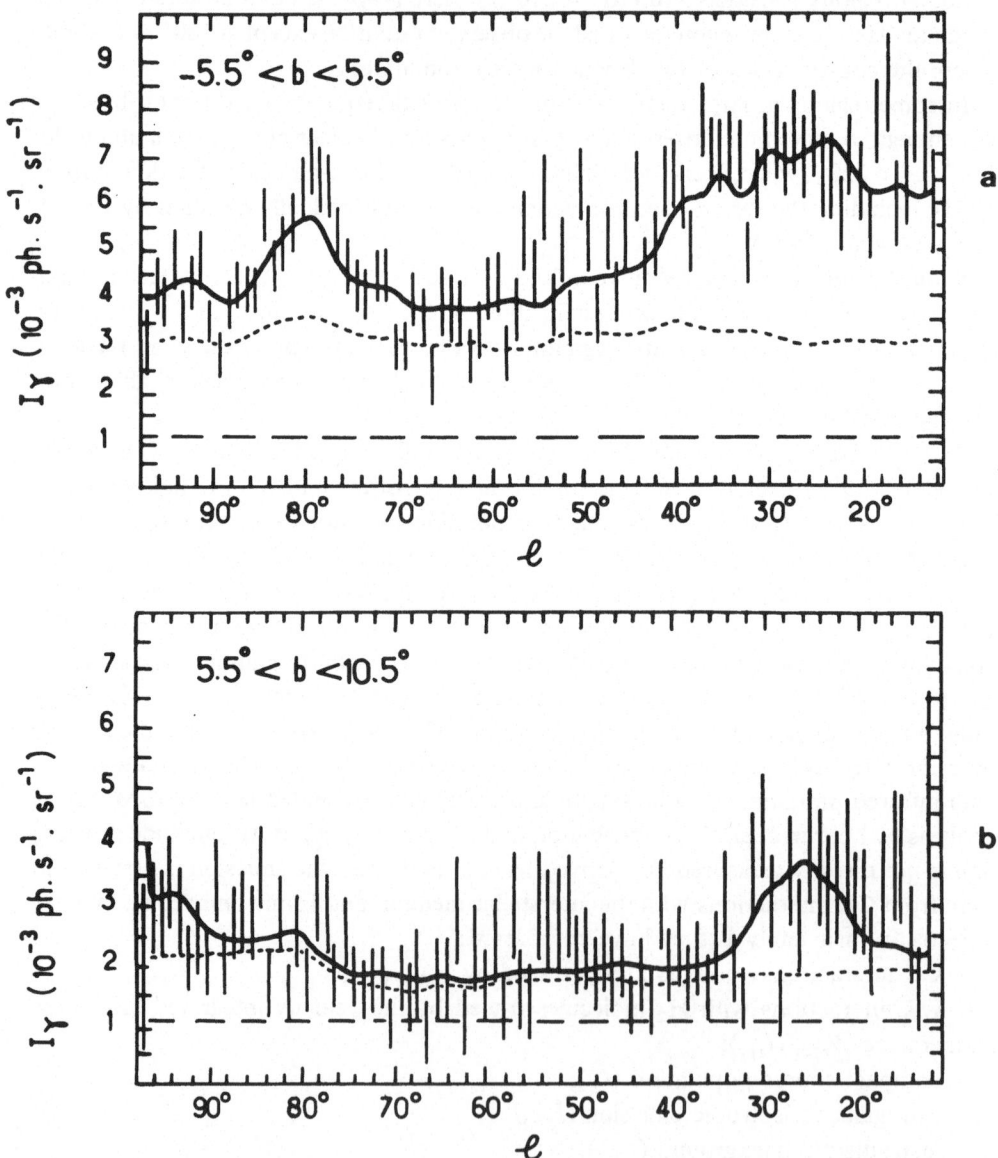

Fig. 9. Longitude profiles of the observed gamma-ray emission for $E > 300$ MeV (error bars) compared with that expected from CO plus H I (solid line) and from H I alone (dotted line). The background level B is indicated by the dashed line.

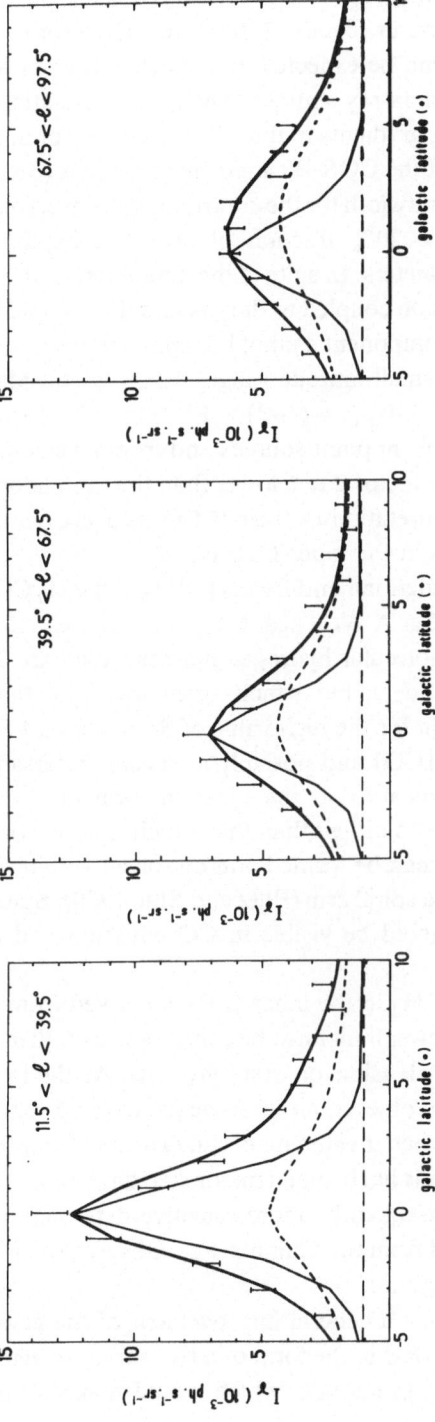

Fig. 10. Latitude profiles of the observed gamma-ray emission for $E > 300$ MeV (error bars) compared with the one expected from CO plus H I (heavy line) and from H I (dotted line). The light line shows the profile expected from a line emission at $b = 0°$ normalized to the gas tracers' predictions.

As is demonstrated by the longitude and latitude distributions in Figures 9 and 10, the correlation is overwhelming on a medium and large scale. On a fine scale, as is visible in the contour representation in Figure 11, there remain several deviations (e.g. in the Cygnus direction), which can be expected as a consequence of a (largely unresolved) population of compact gamma-ray sources. The good correlation justifies a posteriori the assumption of negligible gradients in the CR-flux density or in the ratio $[H_2]/[CO]$. Lebrun et al. on the basis of the COS-B catalogue of sources (Swanenburg et al., 1981) estimated the worst case bandwidth for the contribution from compact sources to BW_{CO} to be 0% to 70%. Also a 20% fraction of inverse-Compton gamma rays might contribute to the fitted parameters. In spite of the differences in the latitude distributions expected between the emission coupled to the gas and that coupled to the photon fields, in their analysis this latter component cannot be separated because of its relatively small amount and the relatively small latitude range, which is considered. This leads to a calibration of the ratio $N(H_2)/W_{CO} = (1-3) \times 10^{20}$ cm^{-2} K^{-1} km^{-1} s. The upper limit applies if the contributions from point sources and from inverse-Compton interactions are neglected. This ratio is a factor 2 to 6 lower than the ratio adopted by Solomon and Sanders (1980) in the interpretation of their ^{13}CO measurements, which suggested a very dominating role of H_2 in the inner Galaxy.

The recent analysis by Federman and Evans (1981) of the H_2 CO absorption between the Sun and the continuum Sgr A West near the galactic center and of the corresponding ^{13}CO emission arrives at molecular hydrogen masses which are 3–5 times lower than those of Solomon and Sanders, but which agree well with those indicated by the gamma-ray data. The reason for the high value of Solomon and Sanders primarily lies in the adopted ratio $[H_2]/[CO]$ and possibly in severe undersampling.

The gamma-ray data indicate that the mass in form of H_2 in the inner Galaxy (3 kpc $< R <$ 8 kpc) is equal to or less than that which is in the form of H I. In this case molecular cloud complexes can be formed and dissolved within several 10^7 yr i.e. well within one passage through a spiral arm (Blitz and Shu, 1980). Spiral features even more pronounced than in H I should be visible in CO emission and are indeed observed (Cohen et al., 1980).

While the upper limit for H_2 in the inner Galaxy derived from the gamma-ray data is of great significance, the lower limit must be considered as rather arbitrary and outside the possible range which is plausible on other grounds. As the lower limit depends on the contribution of compact objects, an improvement would require a better estimate of this contribution. Such a better estimate is still expected from further analysis of the available COS-B data, but it is likely that a major step has to await new data with better angular resolution and, consequently, more sensitive determination of point sources. Such new data are expected from the Gamma-Ray Observatory (GRO), planned to fly in the second half of the eighties.

Another important tool for a general improvement of the conclusions is hopefully becoming available before GRO in the form of a CO survey covering the whole galactic disc with a sufficient extent in latitude, which would make it possible to extend the correlative analysis over all longitudes.

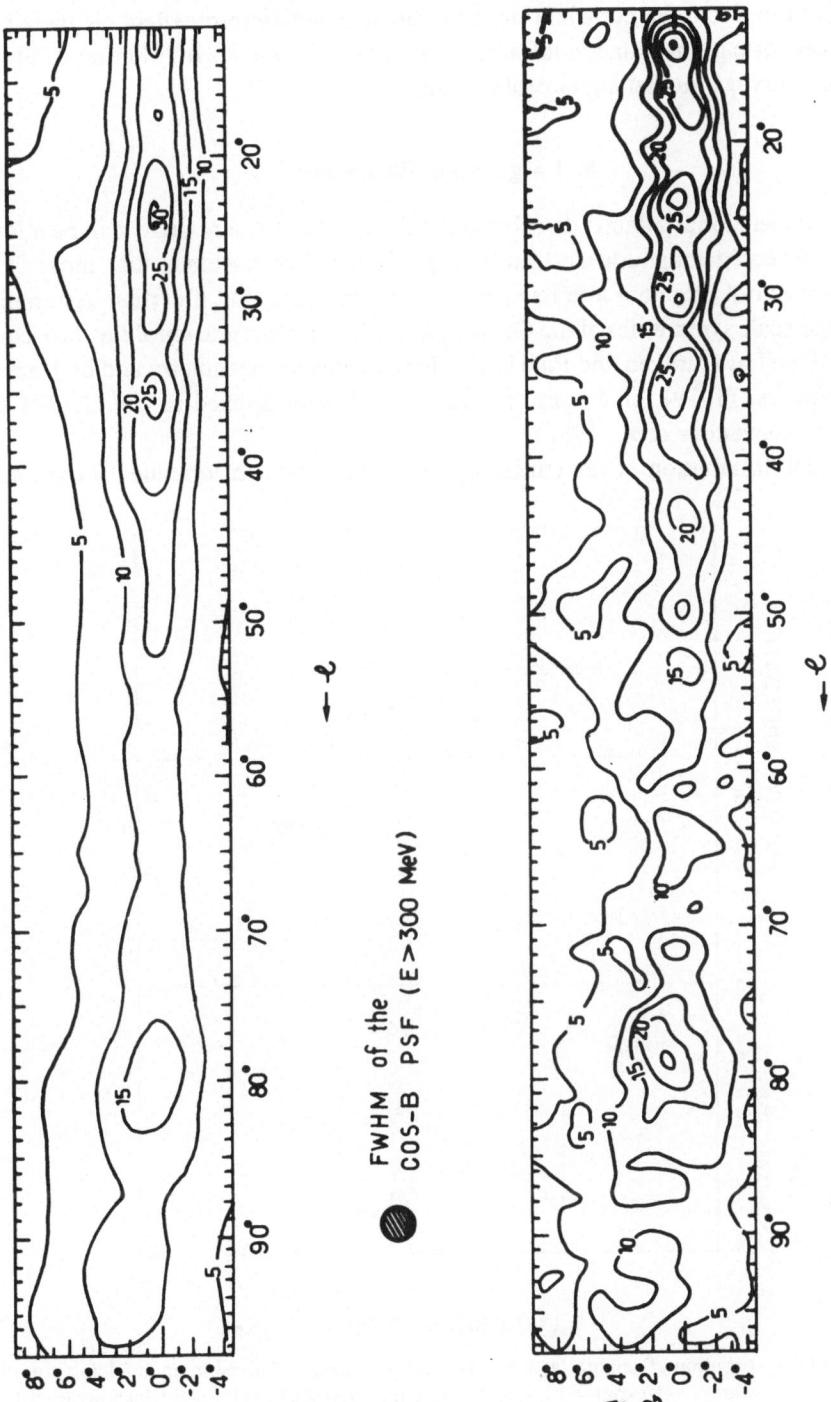

FWHM of the
COS-B PSF (E>300 MeV)

Fig. 11. Contour maps of the predicted (upper part) and observed (lower part) gamma-ray emission for $E > 300$ MeV. Contour level units are: 4.7×10^{-4} ph s^{-1} sr^{-1}. A detailed comparison of isophotes of both figures is discouraged owing to the different statistics involved. For the limitations of the gamma-ray map see Mayer-Hasselwander et al. (1982).

A problem in this type of correlation analysis is encountered when one tries to take into account in detail the contribution from the inverse Compton effect, as there is no other tracer for e.g. the galactic infrared photon field available. In addition to the use of infrared surveys, modelling is probably unavoidable.

5. Large-Scale Parameters

The three-dimensional galactic distribution of the gamma-ray emissivity can be inferred from the observed longitude and latitude profiles, if one assumes that most of the emission does not originate in a few small-scale structures and if certain assumptions on the large-scale symmetries of the Galaxy are made. Appropriate unfolding procedures were developed by Caraveo and Paul (1979) for a cylindrical symmetry and by Kanbach and Beuermann (1979) for a spiral symmetry and were applied to the COS-B data (Mayer-Hasselwander *et al.*, 1982).

The radial distribution of the emissivity per surface element obtained for the spiral

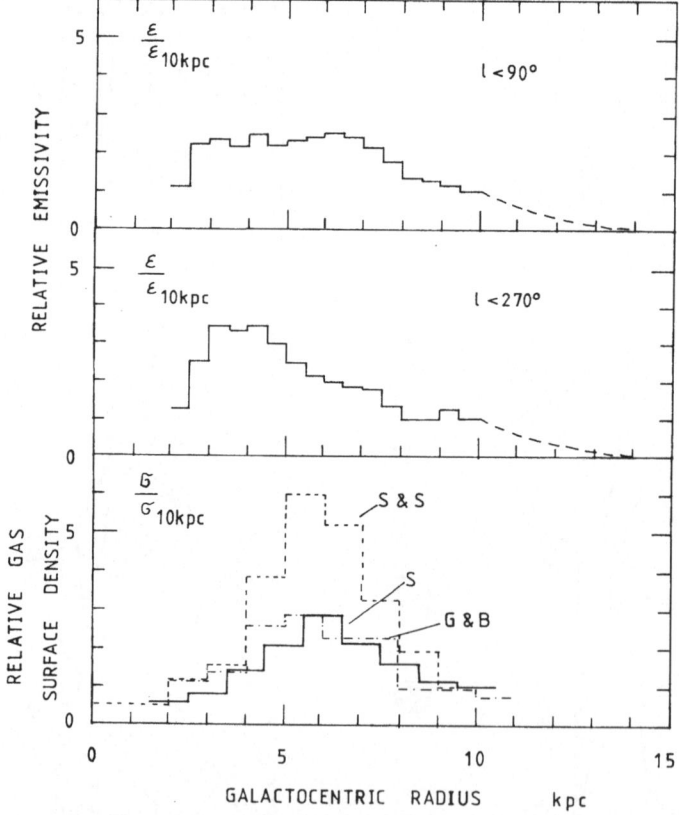

Fig. 12. Radial distribution of gamma-ray emissivity in the Galaxy is derived in an unfolding procedure. The normalisation value is $\varepsilon\,(10\,\mathrm{kpc}) = 2.1 \times 10^{-25}$ photon (> 100 MeV) cm^{-3} s^{-1} which agrees with the value for the solar neighbourhood for a gas density of 1 H-atom cm^{-3}. The first and fourth quadrant were treated separately.

pattern case (which is very similar to the result for cylindrical symmetry) is shown in Figure 12. A ring of enhanced emissivity is evident in the 3 to 6 kpc region, particularly in the fourth quadrant. The gamma-ray emission in this galactic ring is 2–3 times stronger than in the solar region. This analysis makes it possible to compare the gamma-ray emissivity at a given galactocentric radius with corresponding quantities derived e.g. from radio measurements. For comparison, the total mass distributions of the interstellar gas resulting from H I measurements and from CO-line observations (Gordon and Burton, 1976; Solomon and Sanders, 1980; and Sanders, 1982) are indicated.

The result of Solomon and Sanders must be considered too high already because of the arguments not related to gamma rays discussed in Section 4. Moreover, it cannot be reconciled with the maximum mass deducible from the gamma-ray data because of the following arguments:

– For the case of constant CR-flux density throughout the Galaxy, Figure 12 directly gives the maximum tolerable gas mass which is about a factor 2 lower. The possibility of a negative gradient of CR's is unlikely and can be neglected.

– However, one could postulate that a significant fraction of the gas is not seen by the gamma-ray-producing CR's because they cannot penetrate dense clouds. Three arguments stand against this:

– the teoretical works of Skilling and Strong (1976) and of Cesarsky and Völk (1978) show that only very low-energy protons (< 50 MeV) and electrons fail to penetrate completely a dense cloud;

– in at least one case, that of the Orion cloud complex, COS-B data (Caraveo et al., 1980) show that the totality of the gas is involved in the production of gamma rays;

– most important, if by some mechanism a fraction of the cosmic-ray spectrum, presumably the lowest-energy part of it, were excluded from the gamma-ray producing process with the gas, one should observe harder spectra from regions of dense clouds, contrary to the result presented in Section 6.

As has been mentioned before, the question of the scale height of the gamma-ray disc can now be addressed in more detail since the COS-B measurements provide a value for the intrinsic latitude extent of the gamma-ray Galaxy. The unfolding was performed with two trial scale heights of 60 pc and 130 pc and line-of-sight integrals were computed for directions out of the plane of the Galaxy. In Figure 13 the resulting model latitude profiles in terms of HWHM are compared with the intrinsic width (Mayer-Hasselwander et al., 1982) of the observed emission. Even with the uncertainties of the experimental results and with the simplifying assumptions of the model a trend is visible: dominant gamma-ray emission with a scale height of much less than 100 pc is not supported by the data. If the gas in the inner Galaxy were dominated by molecular hydrogen, one would expect the scale height of this gaseous disc to be approximately 60 pc (Cohen and Thaddeus, 1977). Therefore this result can be regarded as independent evidence for a rather small contribution of molecular gas in the inner Galaxy, at the lower end of the bandwidth of values deducible from CO data.

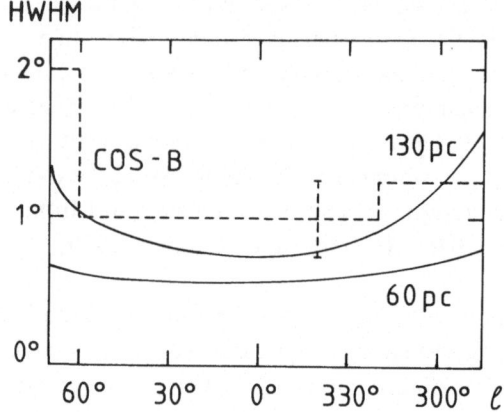

Fig. 13.   Latitude extent of the galactic gamma radiation observed (dashed line) and expected (solid lines) for two different scale heights used in the unfolding procedure.

6. Spectral Properties

The spectral ratio $R = I(70-150 \text{ MeV})/I(150 \text{ MeV}-5 \text{ GeV})$ versus galactic longitude (Mayer-Hasselwander *et al.*, 1982) is a sensitive indicator for variations in the relative contributions by the various production processes, specifically for the contribution from CR electrons relative to that from CR nucleons interacting with the ISM. This spectral ratio was first given in Mayer-Hasselwander *et al.* (1982), but as a consequence from different subtracted background values, the revised spectral ratio from Mayer-Hasselwander (1982), (which is shown in Figure 14), shows a trend with galactic longitude,

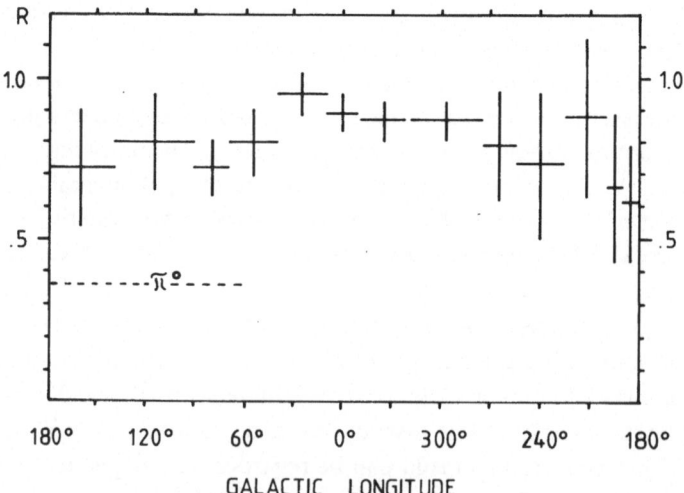

Fig. 14.   Spectral ratio R along galactic longitude for gamma emission from the galactic disc within $|b| < 10°$. R is the ratio of the intensity in the range 70–150 MeV to the intensity in the range 150 MeV–5 GeV.

such that the spectrum appears to be softer towards regions of the inner Galaxy. These regions coincide with the sites where a high concentration of dense clouds is observed. The evaluation of the properties of the COS-B instrumental background is not finalised yet, the spectral ratios presented here therefore still have to be considered with some reservations.

Basically three explanations can be offered for the trend in longitude, which seems to be indicated by the data:

– In the inner Galaxy a relatively larger point-source contribution is expected, which on average could have a softer spectrum.

– An increased relative fraction of CR electrons in the total CR flux. Such an increased flux for example could be generated by secondary electrons (Schlickeiser, 1982) which escape from dense clouds and are reaccelerated in the intercloud medium by shocks which also run from the clouds into the intercloud medium. Such a scenario has been suggested by Morfill (1982a, b). This possible gradient in the CR-electron flux density would primarily influence the 70–150 MeV energy range and disturb the linear correlation between gas cloud density and observed emission. For this energy range, the correlation between the gamma-ray emission and the H I, H_2 tracers indeed is distinctly less good as in the higher energy ranges (F. Lebrun, private communication).

– The inverse-Compton gamma rays according to the work by Hartmann et al. (1979) are estimated to contribute 20% of the emission in direction of the intense 'galactic ridge' emission ($300° < l < 60°$) but only 5% of the emission in directions outside the solar circle, when integrated over $|b| < 10°$. As this contribution would have a spectral index of 2, the observed steepening of the spectrum towards the inner Galaxy indeed could be expected as the consequence of the Compton contribution.

7. Summary

On a medium and large scale, a very close correlation is observed between the emission predicted for a constant CR flux interacting with interstellar matter (traced by atomic and molecular hydrogen). The correlation is found for gamma rays of energies above 300 MeV, where interactions of primary CR's with the gas are the dominating emission processes. The unresolved population of compact gamma-ray objects, if following the distribution of molecular hydrogen, cannot be separated at present. It causes the emissivity attributed to H_2 to be an upper limit, as it includes the source contribution. This source component may be responsible for discrepancies in the correlation at a small scale (e.g. in the direction of Cygnus). The neglected inverse-Compton component also causes the derived H I and H_2 emissivities to lie on the high side. But its contribution is apparently too small or too narrowly correlated with H I and H_2, to disturb the correlation in a recognizable way.

A coupling of the gas and CR-flux densities on the scale of galactic arms is not deducible from the data, as this correlation in the direction of individual arm tangents is not regular enough. On the other hand a general correlation with galactic spiral-arm structures as traced by H I and H_2 is clearly observed. The lack of detailed correlation with arm tangents in the inner Galaxy (Scutum, Norma, 4 kpc ring) is possibly due to the statistics introduced by a limited number of pointsources whose emission might be hiding the otherwise existing correlation with the diffuse component.

The data give no hint for the existence of a galactocentric gradient of the CR's responsible for the gamma-rays above 150 MeV inside the solar circle. Such a gradient (increase) would further reduce the available room for molecular hydrogen beyond the lower estimates presently derived from CO measurements. Outside the solar circle a decrease of CR's is required to avoid excessive gamma-ray production (> 70 MeV).

The spectral properties, showing a trend with galactic longitude, indicate that in the inner Galaxy there is either a significant pointsource component with a soft average spectrum or the CR-electron flux density at a few hundred MeV, possibly due to reaccelerated secondaries, is increased in the sites where dense clouds are concentrated. This softer spectrum, on the other hand, can be taken as a support for the view that dense clouds are completely penetrated by the gamma-ray producing CR's because, in the case of exclusion of the presumably lower-energy CR's, a harder spectrum would be expected. Also the inverse-Compton contribution would tend to make the spectrum softer towards the inner Galaxy.

Further work on existing data and the acquisition of new gamma-ray data with better angular resolution are needed, to quantify and to separate the source contribution reliably. A quantitative separation of the source contribution might be attempted when, after availability of a CO-survey covering all longitudes and after detailed accounting for the inverse Compton component, the predicted diffuse emission can be subtracted from the observed total emission.

Special emphasis has to be given to the question of time variability in the data, which for example is suggested by the comparison of the SAS-2 and COS-B data. This problem at present can be further investigated on the basis of the COS-B data. If significant time variation finally is identified, the estimate of the point-source component could be improved. The information which in such a case could be deduced concerning the diffuse components would be severely restrained yet.

Acknowledgements

Mr M. Gottwald and Dr G. Kanbach are thanked for many fruitful discussions and for reading the manuscript. Drs F. Lebrun and P. Thaddeus and their co-workers are thanked for the permission to use the figures of their paper, which was not yet published. The referee is thanked for his constructive comments.

The editors thank M. Salvati for assistance in evaluating this paper.

References

Bignami, G. F., Boella, G., Burger, J. J., Keirle, P., Mayer-Hasselwander, H. A., Paul, J. A., Pfeffermann, E., Scarsi, L., Swanenburg, B. N., Taylor, B. G., Voges, W., and Wills, R. D.: 1975, *Space Sci. Instr.* **1**, 245.
Blitz, L. and Shu, F. H.: 1980, *Astrophys. J.* **238**, 148.
Buccheri, R., *et al.*: 1983, submitted to *Astron. Astrophys.*
Caraveo, P. A. and Paul, J. A.: 1979, *Astron. Astrophys.* **75**, 340.

Caraveo, P. A., Bennett, K., Bignami, G. F., Hermsen, W., Kanbach, G., Lebrun, F., Masnou, J. L., Mayer-Hasselwander, H. A., Paul, J. A., Sacco, B., Scarsi, L., Strong, A. W., Swanenburg, B. N., and Wills, R. D.: 1980, *Astron. Astrophys.* **91**, L3.
Cesarsky, C. J. and Völk, H. J.: 1978, *Astron. Astrophys.* **70**, 367.
Cohen, R. S. and Thaddeus, P.: *Astrophys. J. Letters* **217**, L155.
Cohen, R. S., Cong, H., Dame, T. M., and Thaddeus, P.: 1980, *Astrophys. J.* **239**, L53.
Dame, T. M. and Thaddeus, P.: 1982, preprint.
Dodds, D., Strong, A. W., and Wolfendale, A. W.: 1975, *Monthly Notices Roy. Astron. Soc.* **171**, 569.
Federman, S. R. and Evans, N. J.: 1981, *Astrophys. J.* **248**, 113.
Fichtel, C. E.: 1982, 'Status and Future of High Energy Diffuse Gamma-Ray Astronomy', NASA Techn. Mem. 83957.
Fichtel, C. E. and Kniffen, D. A.: 1982, work in progress.
Fichtel, C. E. and Trombka, J. I.: 1981, 'Gamma-Ray Astrophysics', NASA SP-453.
Fichtel, C. E., Hartman, R. C., Kniffen, D. A., Thompson, D. J., Bignami, G. F., Ögelman, H., Özel, M. E., and Tümer, T.: 1975, *Astrophys. J.* **198**, 163.
Georgelin, Y. M. and Georgelin, Y. P.: 1976, *Astron. Astrophys.* **49**, 57.
Gordon, M. A. and Burton, W. B.: 1976, *Astrophys. J.* **208**, 346.
Harding, A. K. and Stecker, F. W.: 1981, *Nature* **290**, 316.
Hartman, R. C., Kniffen, D. A., Thompson, D. J., Fichtel, C. E., Ögelman, H. B., Tümer, T., and Özel, M. E.: 1979, *Astrophys. J.* **230**, 597.
Haslam, C. G. T., Kearsey, S., Osborne, J. L., Phillipps, S., and Stoffel, H.: 1981, *Nature* **289**, 470.
Kanbach, G. and Beuermann, K.: 1979, *Proc. 16th I.C.R.C. (Kyoto)* **1**, 75.
Kniffen, D. A. and Fichtel, C. E.: 1981, *Astrophys. J.* **250**, 389.
Lebrun, F., Bignami, G. F., Buccheri, R., Caraveo, P. A., Hermsen, W., Kanbach, G., Mayer-Hasselwander, H. A., Paul, J. A., Strong, A. W., and Wills, R. D.: 1982, *Astron. Astrophys.* **107**, 390.
Lebrun, F., on behalf of the Caravane Collaboration: 1982, in F. Israel and W. B. Burton (eds.), *Leiden-Workshop on Southern Galactic Surveys*, D. Reidel Publ. Co., Dordrecht, Holland.
Lebrun, F., Bennett, K., Bignami, G. F., Bloemen, J. B. G. M., Buccheri, R., Caraveo, P. A., Gottwald, M., Hermsen, W., Kanbach, G., Mayer-Hasselwander, H. A., Montmerle, T., Paul, J. A., Sacco, B., Strong, A. W., Wills, R. D., Dame, T. M., Cohen, R. S., and Thaddeus, P.: 1983, *Astrophys. J.*, is press.
Mayer-Hasselwander, H. A., Bennett, K., Bignami, G. F., Buccheri, R., D'Amico, N., Hermsen, W., Kanbach, G., Lebrun, F., Lichti, G. G., Masnou, J. L., Paul, J. A., Pinkau, K., Scarsi, L., Swanenburg, B. N., and Wills, R. D.: 1982, *Astron. Astrophys.* **105**, 164.
Mayer-Hasselwander, H. A., on behalf of the Caravane Collaboration: 1982, in W. L. H. Shuter (ed.), *Kinematics, Dynamics, and Structure of the Milky Way*, D. Reidel Publ. Co., Dordrecht, Holland.
Morfill, G. E.: 1982a, *Monthly Notices Roy. Astron. Soc.* **198**, 583.
Morfill, G. E.: 1982b, *Astrophys. J.*, in press.
Protheroe, R. J., Strong, A. W., and Wolfendale, A. W.: 1979, *Monthly Notices Roy. Astron. Soc.* **188**, 863.
Sanders, D. B.: 1981, 'NRAO Workshop on Extragal. Molecules', Greenbank, preprint.
Scarsi, L., Bennett, K., Bignami, G. F., Boella, G., Buccheri, R., Hermsen, W., Koch, L., Mayer-Hasselwander, H. A., Paul, J. A., Pfeffermann, E., Stiglitz, R., Swanenburg, B. N., Taylor, B. G., and Wills, R. D.: 1977, *Proc. 12th ESLAB Symp.*, ESA SP-124, p. 3.
Schlickeiser, R.: 1982, *Astron. Astrophys.* **106**, L5.
Simonson, C.: 1976, *Astron. Astrophys.* **46**, 261.
Solomon, P. M. and Sanders, D. B.: 1980, in P. M. Solomon and E. G. Edmunds (eds.), *Giant Molecular Clouds in the Galaxy*, Pergamon Press, p. 41.
Stark, A. A. and Blitz, L.: 1978, *Astrophys. J. Letters* **225**, L15.
Stecker, F. W.: 1971, *Cosmic Gamma Rays*, NASA SP-249.
Strong, A. W. and Lebrun, F.: 1982, *Astron. Astrophys.* **105**, 159.
Strong, A. W., Bignami, G. F., Bloemen, J. B. G. M., Buccheri, R., Caraveo, P. A., Hermsen, W., Kanbach, G., Lebrun, F., Mayer-Hasselwander, H. A., Paul, J. A., and Wills, R. D.: 1982, *Astron. Astrophys.* **115**, 404.
Skilling, J. and Strong, A. W.: 1976, *Astron. Astrophys.* **53**, 253.
Swanenburg, B. N., Bennett, K., Bignami, G. F., Buccheri, R., Caraveo, P., Hermsen, W., Kanbach, G., Lichti, G. G., Masnou, J. L., Mayer-Hasselwander, H. A., Paul, J. A., Sacco, B., Scarsi, L., and Wills, R. D.: 1981, *Astrophys. J.* **243**, L69.

ORIGIN OF THE DIFFUSE GALACTIC GAMMA-RAY EMISSION AT LOW AND MEDIUM GAMMA-RAY ENERGIES*

W. SACHER and V. SCHÖNFELDER

Max-Planck-Institut für Physik und Astrophysik, Institut für extraterrestrische Physik, 8046 Garching, F.R.G.

Abstract. Recently the galactic plane has been observed in the low and medium energy gamma-ray range in the directions towards the center and anticenter. Spectral measurements are now available at those energies, where the contribution from π°-decay gamma rays can be neglected. The high MeV-fluxes observed in both parts of the Galaxy are an indication of either a strong electron induced component or a high contribution from unresolved sources. Several interstellar cosmic-ray electron spectra have been used to calculate the contribution from electron bremsstrahlung and inverse Compton collisions with optical, infrared and 2.7 K black-body photons. From these calculations restrictions on the interstellar electron spectrum are derived.

1. Observations

The question about the origin of the diffuse galactic gamma-ray emission has widely been discussed in the literature, e.g. Clark *et al.* (1968), Kraushaar *et al.* (1972), Fichtel *et al.* (1973), Kniffen *et al.* (1973), Fichtel *et al.* (1975), Stecker (1977), Cesarsky *et al.* (1978), Fichtel *et al.* (1978), Hartmann *et al.* (1979), Hayakawa (1980), Kniffen and Fichtel (1981), Schlickeiser (1981), Stephens (1981), Mayer-Hasselwander *et al.* (1982). This discussion was mainly based on the gamma-ray measurements of the three satellite experiments OSO-3 (above 50 MeV), SAS-2 (above 35 MeV) and COS-B (above 50 MeV). The measurements have shown that in addition to the expected π°-decay component from nuclear interactions of cosmic-ray protons with interstellar matter an additional component must exist – either from interactions of cosmic-ray electrons or from unresolved gamma-ray sources or a mixture of both. It was always clear that gamma-ray measurements at low and medium gamma-ray energies below 35 MeV would be necessary in order to understand the origin of this additional component.

First measurements of this kind have now been made. The Milky Way has recently been observed at energies below 30 MeV in the center and anticenter direction, and spectral measurements are now available at those energies, where the contribution from π°-decay gamma-rays can be neglected. This progress was mainly achieved by three balloon borne experiments.

The first one was a sparkchamber experiment with a very low threshold energy of 4 MeV (Agrinier *et al.*, 1981). This experiment was directed towards the galactic center. Figure 1 shows the observed gamma-ray emission in the energy range 10 to 26 MeV (Lavigne, 1982) and clearly indicates the plane. The limited angular resolution of the telescope (9° and 14° HWHM) does not allow to resolve single sources within the excess flux.

* Proceedings of the XVIII General Assembly of the IAU: *Galactic Astrophysics and Gamma-Ray Astronomy*, held at Patras, Greece, 19 August 1982.

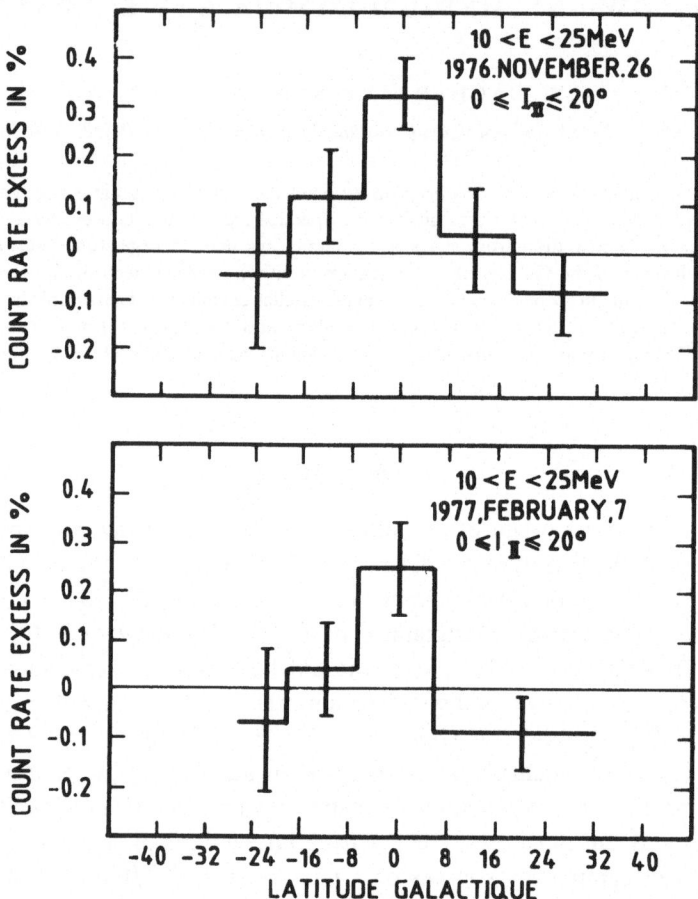

Fig. 1.   Gamma-ray observation in the direction of the galactic center above 4 MeV by Agrinier *et al.* (1981) after revision by Lavigne (1982). Here the results for the 10 to 26 MeV interval are shown, only.

The second experiment also was a sparkchamber experiment with threshold energy of 10 MeV (Bertsch and Kniffen, 1982)*. The gamma-ray emission of the galactic plane between $l = 20°$ to $45°$ was observed in the energy range 10 to 80 MeV.

The third experiment was the balloon borne Compton telescope of the Max-Planck-Institut (Graser and Schönfelder, 1982) which was directed towards the anticenter of the Galaxy. From this observation an image of the anticenter region could be reconstructed in the 1 to 10 MeV range, which is shown in Figure 2. The field of the sky observed ranged from 10° to 50° in declination and 50° to 110° in right ascension. A significantly enhanced gamma-ray emission is observed along the plane. Part of the emission is from the Crab. The angular resolution of the telescope is indicated by the circle around the Crab position. At present it cannot be decided, whether the remainder of the emission is due to further unresolved sources, or whether it is diffuse in nature.

---

* The authors became aware of these measurements only after the IAU-Meeting.

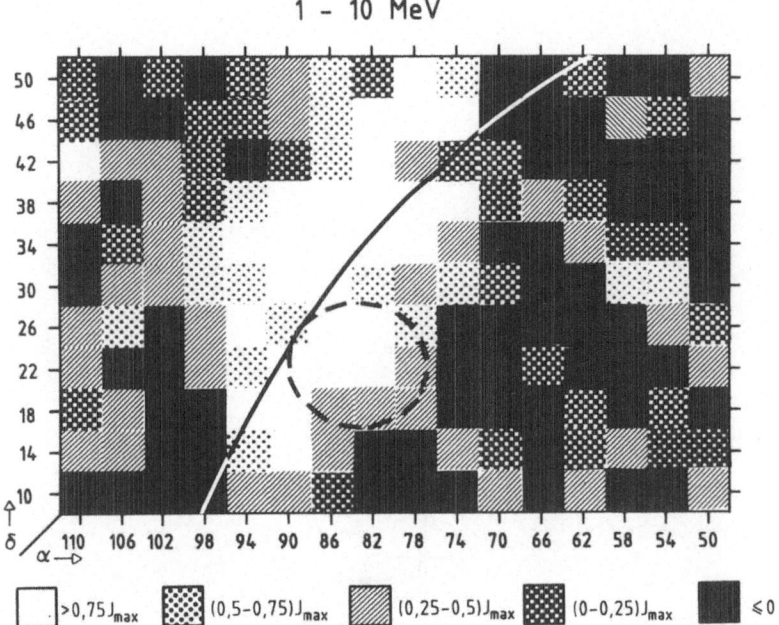

Fig. 2. Image of the anticenter of the Galaxy in the light of 1 to 10 MeV gamma rays. From Graser and Schönfelder (1982).

The lack of emission at the two ends of the plane inside the field of view may favour a contribution from localised regions rather than a diffuse origin. Therefore the observed emission can only be considered as an upper limit to the diffuse galactic gamma-ray flux. It is, however, also true that the known X-ray sources within the bright region along the plane are all very weak (Crab excluded). The strongest of these sources have intensities of only 2 to 4.5 Uhuru counts (Giacconi *et al.*, 1974). COS-B has detected two gamma-ray sources within the sky field of Figure 2, namely, Crab ($\alpha = 82.88°$, $\delta = 21.98°$) and Geminga (2CG 195 + 04) ($\alpha = 97.95°$, $\delta = 17.95°$); see Swanenburg *et al.* (1981).

In Figure 3 the spectral informations for both parts of the Galaxy – the center and anticenter region – are compared with each other.

For the center region X-ray and low energy gamma-ray measurements exist up to 1 MeV; between 4 to 26 MeV there are the new measurements of Agrinier *et al.* (1981), between 10 to 80 MeV the new measurements of Bertsch and Kniffen (1982), above 35 MeV the SAS-2 (Kniffen and Fichtel, 1981) and above 50 MeV the COS-B measurements (Paul *et al.*, 1978). Very recently the results of Agrinier *et al.* (1981) have been revised and are now lower by a factor of about two (Lavigne, 1982). Only these revised fluxes are shown for the energy-range 4 to 26 MeV in Figure 3. Between 1 to 4 MeV at present only upper limits exist. Below 100 keV the spectrum steepens rapidly due to the presence of discrete X-ray sources, which certainly contribute to the spectrum above 100 keV as well (Wheaton, 1976; Matteson, 1982). Some additional measurements,

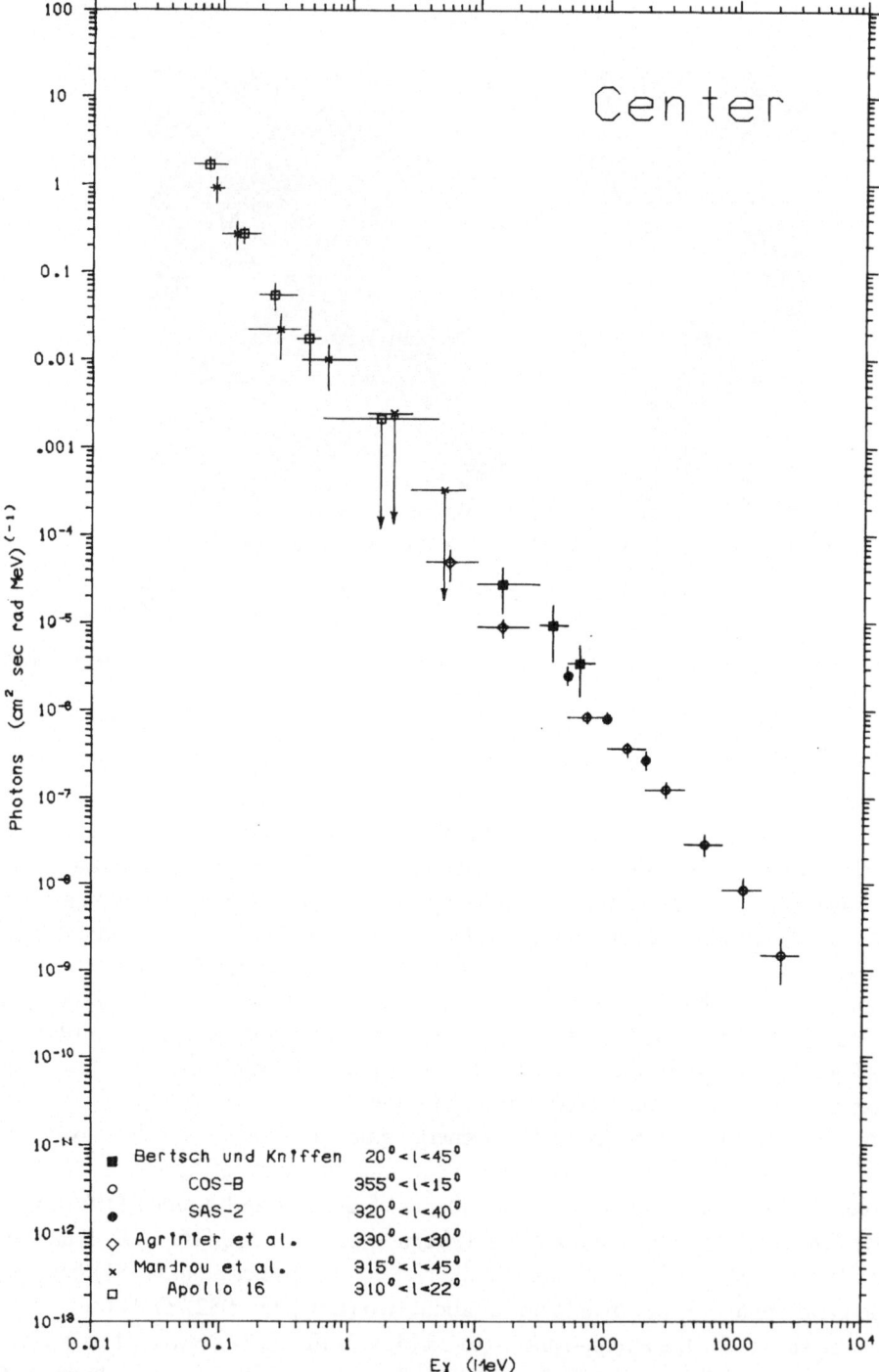

Fig. 3.   Comparison of the energy spectra from the galactic center and anticenter region. COS-B data are from Paul *et al.* (1978), SAS-2 data from Kniffen and Fichtel (1981), Apollo-16 data from Gilman *et al.* (1976), the results of Mandrou *et al.* (1980) are from a balloon experiment. The contribution of the Crab was substracted from the 1 to 10 MeV flux measured by the Compton telescope (Graser and Schönfelder, 1982).

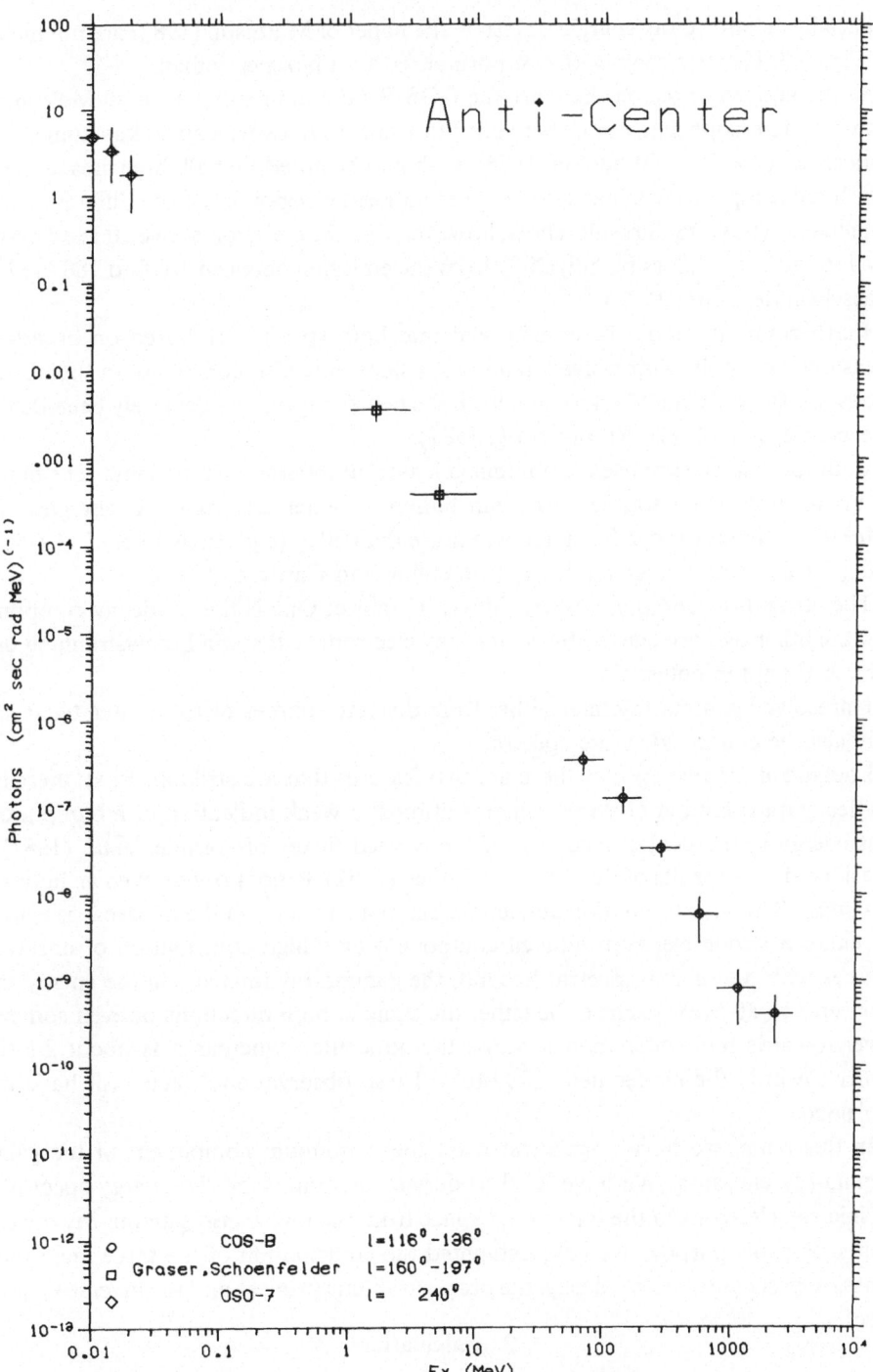

Fig. 3.   (continued).

especially at hard X-ray energies, given in the paper of Matteson (1982) are not included in Figure 3, because they were not normalised to a flux per radian.

In the anticenter region there are the COS-B measurements above 50 MeV and the Compton telescope data points between 1 to 10 MeV. Between 6 to 40 keV some OSO-7 data exist at $l = 240°$ (Wheaton, 1976). It should be noted that all three measurements, which are compared with each other in the anticenter region, are from different galactic longitudes. The COS-B results show, however, that the emission above 70 MeV between $l = 116°$ to $136°$ differs by only 20% from the emission between $160°$ to $180°$ (Mayer-Hasselwander et al., 1982).

Furthermore it should be emphasized that both spectra are based on broad-band measurements only. Unresolved gamma-ray lines may also contribute to the observed fluxes. So far, only the 511-keV line from the center region has definitely been detected (Leventhal et al., 1982; Riegler et al., 1981).

Both spectra are supposed to contain at least four different components. The first one is due to unresolved sources. Its contribution is rather uncertain. At energies above 100 MeV estimates range from a few to more than 50% (e.g. Bignami et al., 1978; Coe et al., 1978; Protheroe et al., 1979; Rothenflug and Caraveo, 1980).

The other three components are diffuse in nature: One is the $\pi°$-decay component, and the other two are caused by cosmic-ray electrons, either via bremsstrahlung or via inverse Compton collisions.

Unresolved gamma-ray lines either from discrete sources or from interstellar space may also be contained in the spectra.

Looking at the two spectra there are two features that are striking: First, there is no $\pi°$-decay maximum at 68 MeV; there is at most a weak indication of a bump around that energy in the center spectrum, if the revised fluxes of Agrinier et al. (1981) are considered. The results of Bertsch and Kniffen (1982) do not provide even an indication of a bump. The strong emission outside the gamma-ray range of the $\pi°$-decay is evidence for either a strong electron induced component or a high contribution of unresolved sources with power law spectra. Second, the gamma-ray intensity in the energy range between 1 to 30 MeV seems to be either the same in both directions or even somewhat lower towards the center than towards the anticenter, whereas it is about 2.5-times higher towards the center near 100 MeV. These observational facts will have to be explained.

In this paper we have concentrated on the continuum component of the galactic gamma-ray emission. We have tried to derive constraints on the energy spectrum of cosmic-ray electrons in the interstellar space from the low energy gamma-ray measurements. For this purpose we have estimated the contribution of the three main diffuse emission processes, the $\pi°$-decay, the bremsstrahlung process and the inverse Compton effect.

2. Calculations

The calculation of the $\pi°$-decay component is rather straight forward. Its contribution has been calculated by many groups in the past (e.g. Stecker, 1977; Fichtel et al., 1976; Strong et al., 1978; Stephens and Badwhar, 1981; Schlickeiser, 1982).

The distribution of interstellar matter was derived from the paper of Gordon and Burton (1976). Their molecular hydrogen distribution was scaled down by a factor of about 1.7 according to Blitz and Shu (1980), and corrected for the isotopic abundance ratio $(C^{12})/(C^{13})$. The resulting density distribution is shown in Figure 4. The scale heights of the neutral and molecular hydrogen disks and their variation with galacto-centric distance are indicated on the figure.

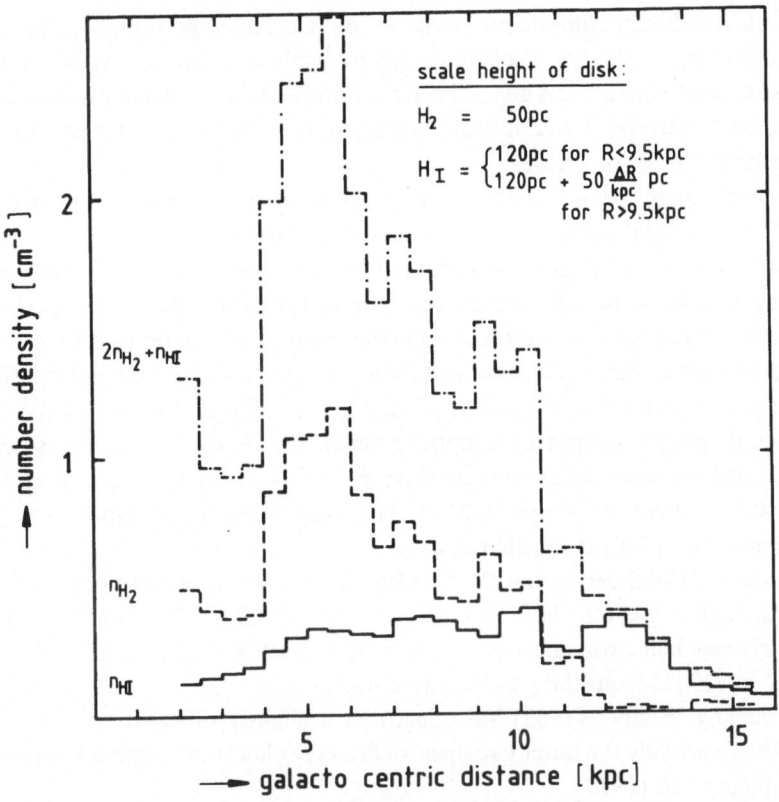

Fig. 4.   Density distribution of interstellar hydrogen as a function of galactocentric distance used in this paper (for references see text).

The π°-decay production spectrum was assumed to be that of Stephens and Badwhar (1981). There the influence of heavier nuclei in interstellar matter as well as in the cosmic radiation is taken into account.

The estimate of the electron induced component is much more complicated and contains more uncertainties. The greatest uncertainty is the interstellar electron spectrum itself.

In Table I the mean electron energies are listed that are required for the various production processes:

TABLE I

E_γ	Bremsstrahlung	Inverse Compton collisions		
		on 2.7 K	on infrared (500 K)	on optical (5000 K)
1 MeV	$E_e = 3$ MeV	$E_e = 17$ GeV	$E_e = 1.3$ GeV	$E_e = 0.4$ GeV
10 MeV	$E_e = 30$ MeV	$E_e = 50$ GeV	$E_e = 4$ GeV	$E_e = 1.2$ GeV

The bremsstrahlung component at energy E_γ is on average produced by relativistic electrons of energy $3E_\gamma$. The electron energy to produce a gamma-ray of – say 10 MeV by inverse Compton collisions depends on the nature of the scattered photon, which may be an optical, infrared or 2.7 K blackbody photon. In all three cases the required electron energy is near 1 GeV or greater.

Our knowledge of the interstellar electron spectrum over this wide energy range from a few MeV to several 100 GeV is summarised in Figure 5.

At low energies the electron spectrum measured near Earth is several orders of magnitude smaller than the interstellar electron spectrum due to the effect of solar modulation. At energies above several GeV the solar modulation no longer exist and there the various electron measurements tend to agree. Three different possible interstellar electron spectra have been drawn into Figure 5. They all are within the upper and lower bound spectra derived by Cummings et al. (1973) from radio measurements.

Kniffen and Fichtel (1981) have used an $E^{-2.14}$ electron spectrum to calculate the bremsstrahlung spectrum above 10 MeV. This electron spectrum which was suggested by Webber et al. (1980) is labelled K & F.

Lebrun et al. (1982) derived an $E^{-2.8}$ electron spectrum when interpreting the COS-B data from higher galactic latitudes betwen 10° to 20°. This spectrum is labelled L. Similarly steep electron spectra have also been derived by Cesarski et al. (1978) and Strong et al. (1981) from the gamma-ray data.

Very recently Webber (1982) has suggested a slightly different spectrum which is labelled W. Especially the latter two spectra are very close to the upper bound spectrum of Cummings et al. (1973).

These three electron spectra (K & F, L, and W) have been used to calculate the gamma-ray production at low energies by bremsstrahlung and inverse Compton collisions. They are assumed to describe the combined intensity of primary and secondary electrons in interstellar space. The shape of the electron source function is assumed to be the same throughout the Galaxy. This certainly is a simplifying assumption.

As can be seen from Figure 5 the electron spectrum is rather well defined above 1 GeV, which is the range responsible for inverse Compton collisions. The largest uncertainties exist at energies below – say 100 MeV: this is the range, which is responsible for bremsstrahlung production of low energy gamma rays.

For the calculation of the bremsstrahlung yield it is important to recognize that the electron spectra cannot be extrapolated into the MeV-range with constant slope. At these energies ionisation loss becomes the dominant energy loss process, as is illustrated

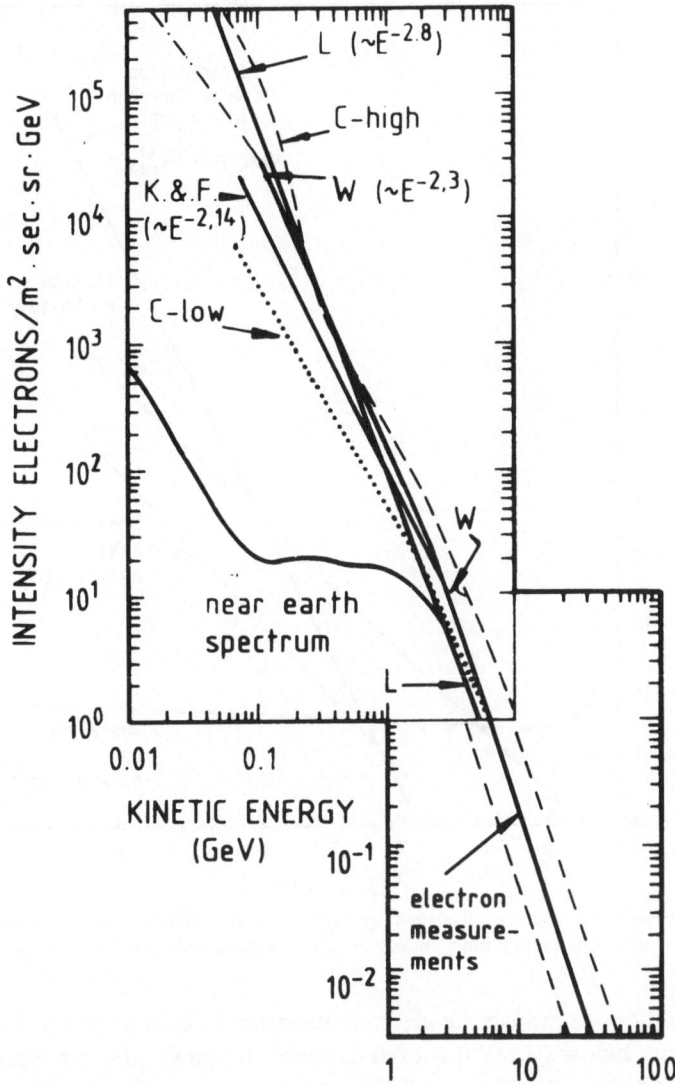

Fig. 5. Three possible interstellar electron spectra (K & F, L, and W) which all are within the upper and lower bound spectra (C-high and C-low) derived by Cummings *et al.* (1973) from radio measurements are compared with the observed near earth spectrum (detailed description in the text). The measurements above 10 GeV are summarised by Taira *et al.* (1979).

in Figure 6, where the various loss processes are compared with each other. Above several GeV the energy loss due to synchrotron radiation and inverse Compton collisions is highest. At MeV-energies the ionisation loss rate is dominant and is higher than the effective loss rate through escape.

Figure 7 illustrates how the various loss processes influence the equilibrium spectrum: In the dominant synchrotron and inverse Compton range the electron spectrum is by one power steeper than the source spectrum which is assumed to go like $E^{-\alpha}$. In the

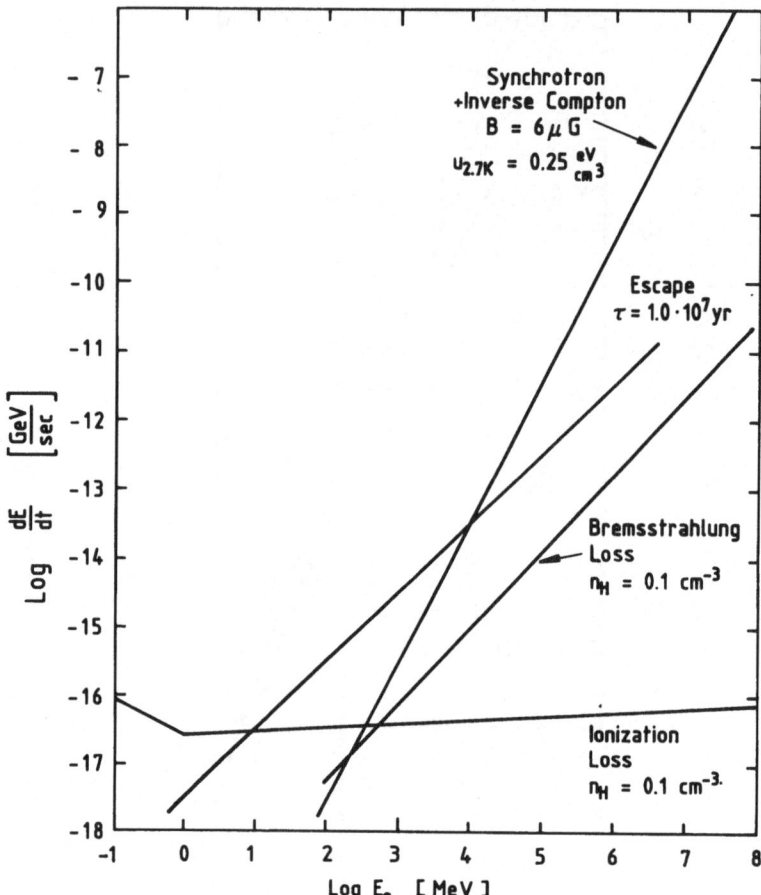

Fig. 6.   Various energy loss processes of cosmic ray electrons in interstellar space. For simplicity the escape
time is assumed to be energy independent in this schematic drawing. From Ramaty (1974).

diminant ionisation loss range the electron spectrum is by one power flatter than the
source spectrum. Below 1 MeV it is even flatter by 1.5 power. For an electron confinement time of 10^7 yr the onset of the dominating ionisation loss range is at
$E_e = 94$ MeV n_H ($\alpha - 1$), where n_H is the mean hydrogen density within the electron
confinement region. If the electron confinement disk is assumed to have a scale height
of 1 kpc, the mean local hydrogen density within this confinement region is about
0.1 cm^{-3}. The scale height of 1 kpc was derived by Kniffen and Fichtel (1981) from
radio measurements of Baldwin (1976). Values near 750 pc for the scaleheight have been
suggested by Ormes and Freier (1978) and Hayakawa (1980).

Let us now assume that the electron escape time is the same everywhere in the Galaxy
– say 10^7 yr, other values will be consider later. In this case the break point in the energy
spectrum due to ionisation losses depends on n_H only. Therefore, in the center region
of the Galaxy, where the hydrogen density is higher, the break point is shifted towards
higher energies and the intensity of electrons on the low energy end of the spectrum is

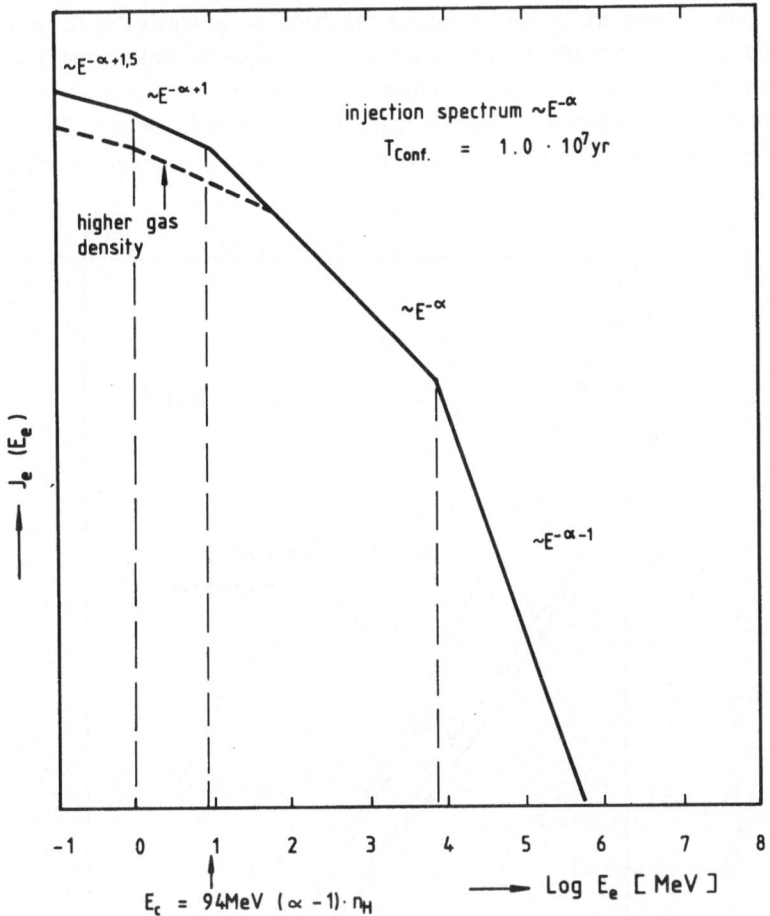

Fig. 7. Equilibrium spectrum of cosmic-ray electrons in interstellar space that have a differential source spectrum going like $E^{-\alpha}$. The break point due to ionisation losses depends on the mean interstellar hydrogen density within the confinement region of the electrons, the exponent of the electron source spectrum, and the escape time of cosmic-ray electrons.

reduced. This results in a lower bremsstrahlung yield of MeV-photons per hydrogen atom.

If bremsstrahlung would be the main production process at MeV energies, it would then be easy to understand, why the observed MeV-gamma-ray flux is about the same from the direction of the anticenter as of the center (which is true for the Bertsch and Kniffen results (Bertsch and Kniffen, 1982)), whereas it is about 2.5 times higher at 100 MeV in the direction of the center. In this case the gamma-ray intensity produced within the next 2 to 3 kpc around the Sun would be practically the same because of comparable interstellar mass column densities in both directions. Towards the direction of the center an additional gamma-ray intensity would be produced in the high density region near 5 kpc from the center. Whereas this additional component is important at 100 MeV, it is considerably reduced at MeV-energies due to the higher ionisation losses of the electrons. Therefore, at MeV-energies the direction towards the center would not be expected to be much brighter than the direction towards the anticenter. The presence of gamma-ray sources may, however, change this picture.

In calculating the bremsstrahlung spectrum down to MeV – or even X-ray energies, it is important to use the exact bremsstrahlung cross section formulas (Koch and Motz,

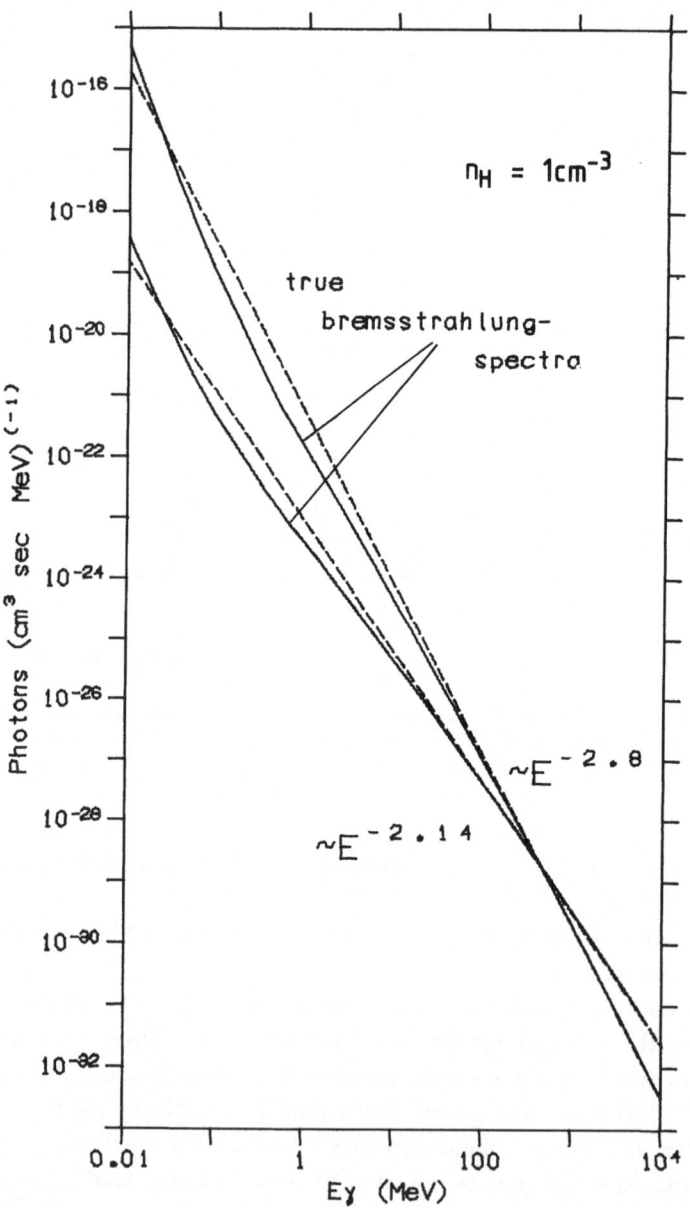

Fig. 8.   Bremsstrahlung spectra calculated for $E^{-2.14}$ and $E^{-2.8}$ electron spectra using the production formulas of Blumenthal and Gould (1970) and Koch and Motz (1959).

1959; Blumenthal and Gould, 1970). At these energies the resulting bremsstrahlung spectrum no longer has the same spectral index as the electron spectrum as is illustrated in Figure 8. Here, the bremsstrahlungs spectra for an $E^{-2.14}$ and an $E^{-2.8}$ electron spectrum are shown. Below 10 MeV the true bremsstrahlung spectra more and more deviate from the simple power law dependence. For energies greater than some 100 MeV, our results are consistent with the ones calculated by the formula for highly relativistic electrons, given for example by Stecker (1977).

Figure 9 summarises the results of our calculations of the three gamma-ray emission processes in the Galaxy. The measured spectra from Figure 3 are compared with calculations of the π°-decay-, inverse Compton-, and bremsstrahlung component.

The bremsstrahlung component was calculated for the three different interstellar electron spectra shown in Figure 5, which were labelled K & F, L, and W. For the L-spectrum two different electron escape times were considered. For L1 the rather low value of 10^7 yr was assumed, whereas for all other spectra (L2, K & F, W) $\tau = 2.5 \times 10^7$ yr was taken. These values together with the mean hydrogen density determine the break point due to ionisation losses. (The breakpoint of the W-spectrum was not assumed to be at 100 MeV as suggested by Webber (1982).) The scaleheight of cosmic-ray electrons was assumed to be 750 pc.

For the calculation of the inverse Compton component the energy density distribution of photons in the Galaxy from the paper of Kniffen and Fichtel (1981) was used. With the sourcefunctions of Piccinotti and Bignami (1976) and with the corrections to these functions of Schlickeiser (1979) we get an inverse Compton spectrum, which is shown for the K & F electron spectrum only. The deviation of the spectra derived from the other electron spectra are less than a factor of two in the energy-range 1 to 100 MeV.

For all three gamma-ray components the cosmic ray surface density was assumed to be proportional to the matter surface density. This leads to a slight increase of the cosmic-ray density towards the center and a decrease beyond 12 kpc. A correlation between cosmic ray and matter density was already suggested by Hayakawa and Tanaka (1970) in order to interpret the first cosmic gamma-ray measurements of OSO-3 (Clark *et al.*, 1968). Since that time various correlations have been used in the papers on the origin of the diffuse galactic gamma-ray emission summarized in the introduction.

3. Discussion

By comparing the calculated spectra with the measurements the following conclusions can be derived:

3.1. CENTER REGION

The interpretation of the gamma-ray measurements towards the center at low gamma-ray energies (say 1 to 30 MeV) at present suffers from the fact that between 1 to 4 MeV only upper limits exist and that between 10 to 30 MeV the revised results of Agrinier *et al.* (1981) and those of Bertsch and Kniffen (1982) differ by a factor of about three.

Whereas the revised measurements of Agrinier *et al.* (1981) are significantly below the

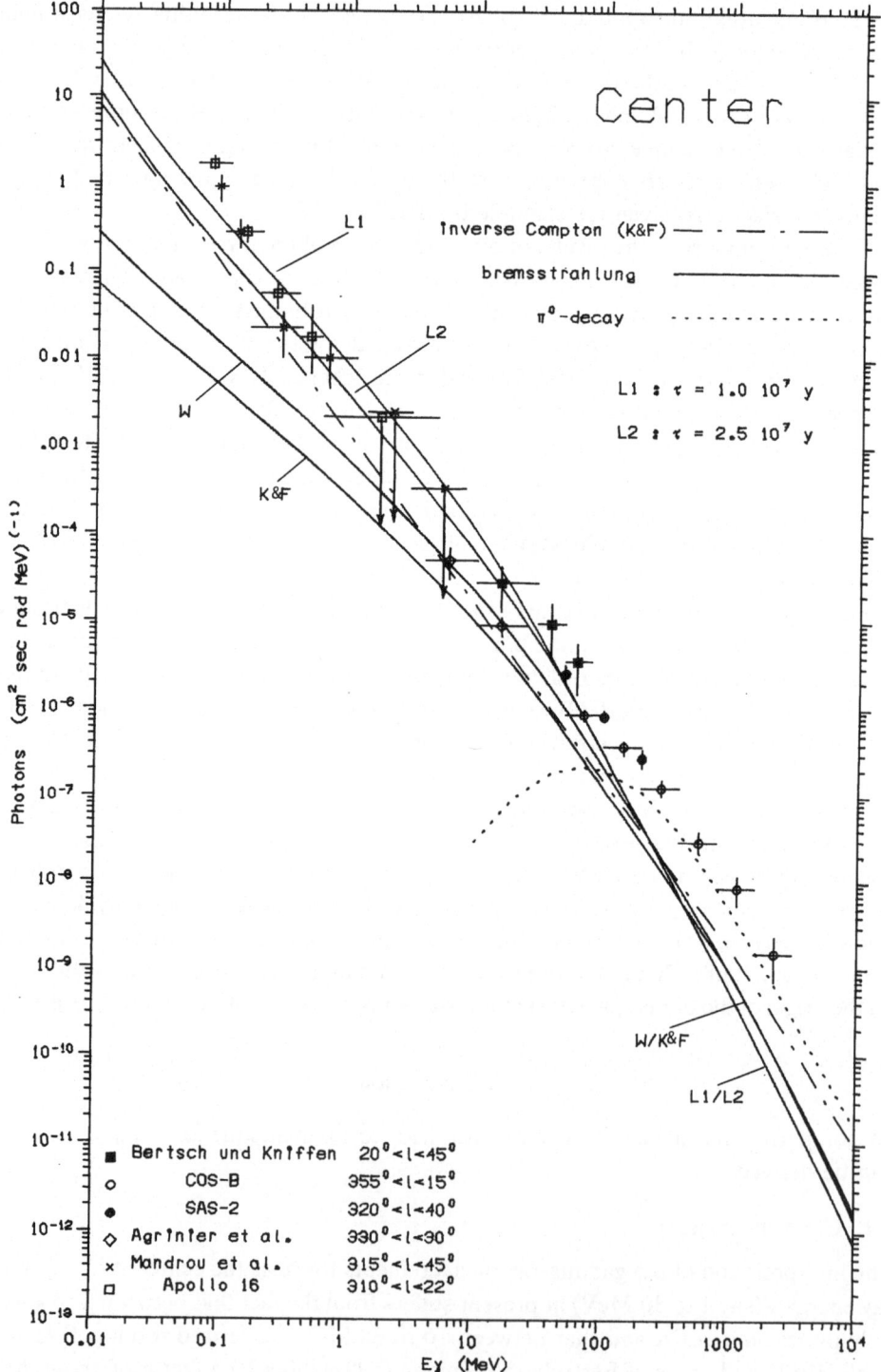

Fig. 9. Comparison between the calculations of the diffuse galactic gamma ray emission from $|b| < 10°$ due to $\pi°$-decay, inverse Compton collisions and bremsstrahlung with the observed spectra of Figure 3. Details are described in the text.

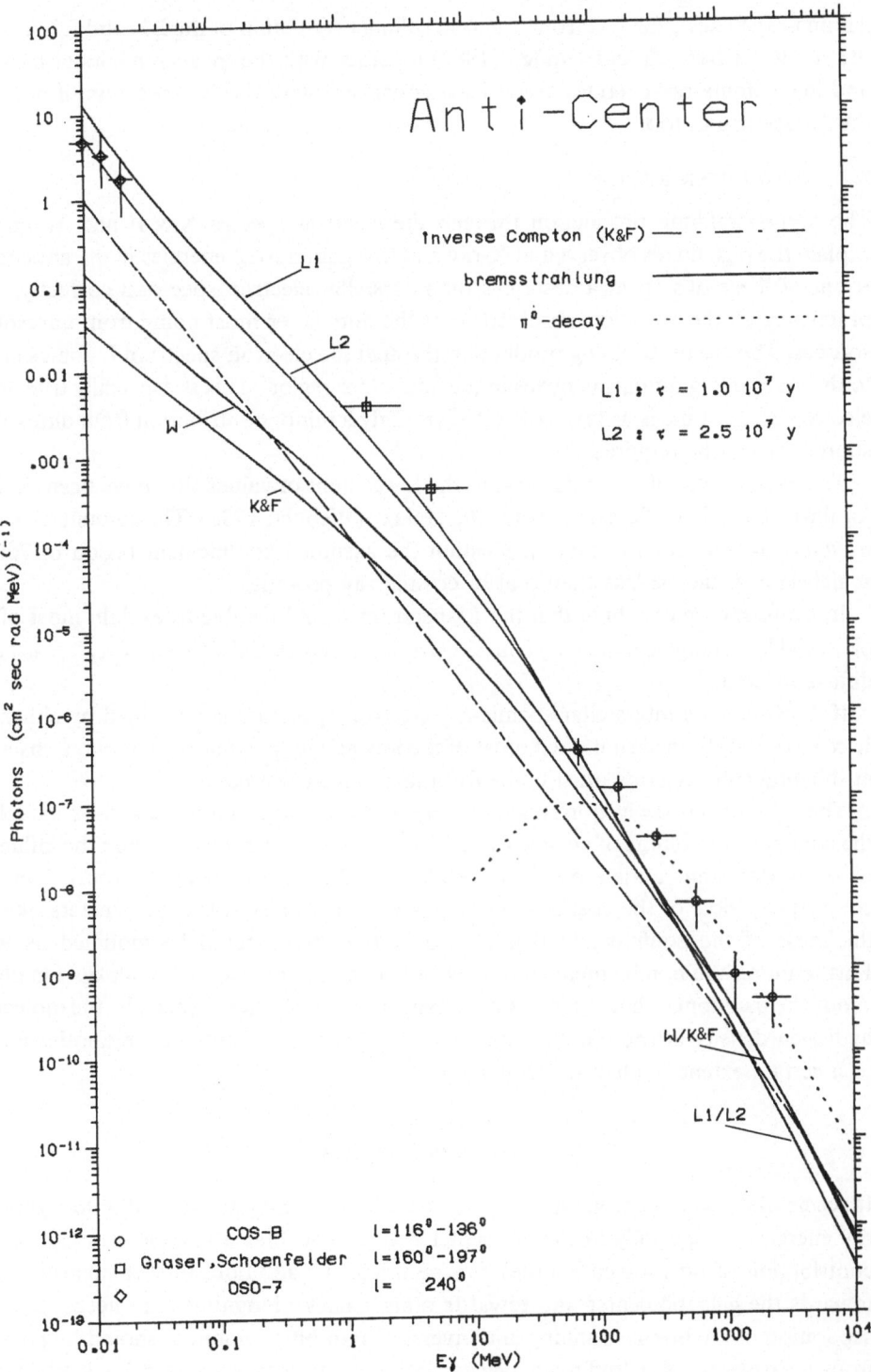

Fig. 9. (continued).

gamma-ray fluxes expected from bremsstrahlung production of the L1- and L2-spectra, the results of Bertsch and Kniffen (1982) together with the measurements at hard X- and low gamma-ray energies are at least consistent with the L2-spectrum, if not with the L1-spectrum, too.

3.2. ANTICENTER REGION

The bremsstrahlung production through the electron spectra K & F and W cannot explain the high fluxes observed at X-ray and low gamma-ray energies in the anticenter. If one of these two spectra describes the interstellar electron spectrum correctly, then practically all the emission observed from the anticenter must come from unresolved sources. The bremsstrahlung production through the electron spectrum L comes closer to the measured gamma-ray fluxes in the anticenter region. This is especially true, if the electron escape time is as low as 1×10^7 yr. Only a minor contribution from unresolved sources would be required.

The escape time of 10^7 yr is close to the lower limit of values that have been derived for the leakage time of cosmic rays in the Galaxy (Webber, 1982). The amount of matter traversed by electrons in this time within the assumed confinement region of 750 pc scaleheight would be less than that of cosmic ray protons.

In summary we conclude that the L-spectrum would be able to explain most of the observed low energy gamma-ray emission from the center and anticenter region as being diffuse in origin.

If, however, the interstellar cosmic-ray electron spectrum is described by either the K & F- or the W-spectrum, then most of the low energy gamma-ray emission observed in the anticenter regions must come from unresolved sources.

The conclusions derived in this paper are all based on a simple leaky box model for the confinement volume of cosmic rays. This model is a simplification and the influence of other diffusion models on these results should be investigated. Apart from the confinement time of the cosmic-ray electrons in the Galaxy other parameters like the thickness of the confinement region to a certain exent could be modified as well. Furthermore, the confinement time may not be constant over the Galaxy. Finally, it should be mentioned that the interstellar hydrogen distribution, especially the molecular hydrogen density at present still contains uncertainties. All these uncertainties may – to a certain extent – affect our conclusions.

4. Future Aspects

It seems that much more gamma-ray source measurements are needed at low gamma-ray energies – especially in the anticenter region – so that a reliable estimate on the contribution of unresolved sources can be made. In addition, further measurements towards the galactic center and towards other galactic longitudes are necessary. The separation of the bremsstrahlung and inverse Compton component should be possible from measurements at different galactic latitudes as was suggested by Kniffen and Fichtel (1981). The shape of the bremsstrahlung component may then provide a good means to determine the confinement time of cosmic-ray electrons in the galaxy. All these measurements should be made by GRO at the end of our decade.

Acknowledgements

The authors would like to thank Dr. P. Caraveo for several helpful comments. The editors thank Dr. P. Caraveo for assistance in evaluating this paper.

References

Agrinier, B., et al.: 1981, Proc. of 17th Int. Cosmic Ray Conf. 9, 72.
Baldwin, J. E.: 1976, 'The structure and Content of the Galaxy and Galactic Gamma-Rays', NASA/GSFC CP002.
Bertsch, D. L. and Kniffen, D. A.: 1982, preliminary results in Fichtel and Kniffen, NASA, TM 83992.
Bignami, G. F., Caraveo, P., and Maraschi, L.: 1978, Astron. Astrophys. 67, 149.
Blitz, L. and Shu, F. H.: 1980, Astrophys. J. 238, 148.
Blumenthal, G. R. and Gould, R. J.: 1970, Rev. Mod. Phys. 42, 237.
Burton, W. B. and Gordon, M. A.: 1976, Astrophys. J. 207, 189.
Cesarsky, C. J., Paul, J. A., and Shukla, P. G.: 1978, Astrophys. Space Sci. 59, 73.
Clark, G. W., Garmire, G. P., and Kraushaar, W. L.: 1968, Astrophys. J. Letters 153, L203.
Coe, M. J., Quenby, J. J., and Engel, A. R.: 1978, Nature 274, 343.
Cummings, A. C., Stone, E. C., and Vogt, R. E.: 1973, Proc. 13th Int. Cosmic Ray Conf. 1, 335.
Fichtel, C. E., Kniffen, D. A., and Hartmann, R. C.: 1973, Astrophys. J. Letters 186, L99.
Fichtel, C. E., et al.: 1975, Astrophys. J. 198, 163.
Fichtel, C. E., et al.: 1976, Astrophys. J. 208, 211.
Fichtel, C. E., Simpson, G. A., and Thompson, D. J.: 1978, Astrophys. J. 222, 823.
Giacconi, R., et al.: 1974, Astrophys. J. (Suppl. Ser. 237) 237, 37.
Gilman, D.: 1976, NASA TM 79619, p. 190.
Gordon, M. A., and Burton, W. B.: 1976, Astrophys. J. 208, 346.
Graser, U. and Schönfelder, V.: 1982, Astrophys. J. 263, 677.
Hartmann, R. C.,et al.: 1979, Astrophys. J. 230, 597.
Hayakawa, S. and Tanaka, Y.: 1970, in L. Gratton (ed.), 'Non-Solar, X-, and Gamma-Ray Astronomy', IAU Symp. 37, 374.
Hayakawa, S.: 1980, in R. Cowsik and R. D. Wills (eds.), 'Non-Solar Gamma-Rays', Cospar Advances in Space Exploration 7, 175.
Koch, H. W. and Motz, J. W.: 1959, Rev. Mod. Phys. 31, 921.
Kniffen, D. A., et al.: 1973, Astrophys. J. Letters 186, L105.
Kniffen, D. A. and Fichtel, C. E.: 1981, Astrophys. J. 250, 389.
Kraushaar, W. L., et al.: 1972, Astrophys. J. 177, 341.
Lavigne, J. M.: 1982, paper presented at the 8th Europ. Cosmic-Ray Conf., Rome.
Lebrun, F., et al.: 1982, Astron. Astrophys. 107, 390.
Leventhal, M., et al.: 1982, Astrophys. J. Letters 260, L1.
Mandrou, P., et al.: 1980, Astrophys. J. 237, 424.
Mayer-Hasselwander, L. A., et al.: 1982, Astron. Astrophys. 105, 164.
Matteson, J. L.: 1982, in G. R. Riegeler and R. D. Blanford (eds.), The Galactic Center, American Inst. Phys., New York (to be published).
Ormes, J. and Freier, P.: 1978, Astrophys. J. 222, 471.
Paul, J. A., et al.: 1978, Astron. Astrophys. 68, L31.
Piccinotti, G. and Bignami, G. F.: 1976, Astron. Astrophys. 52, 69.
Protheroe, R. J., et al.: 1979, Nature 277, 542.
Ramaty, R.: 1974, in F. B. McDonald and C. E. Fichtel (eds.), High Energy Particles and Quanta in Astrophysics, MIT Press, Cambridge, p. 122.
Riegler, G. R., et al.: 1981, Astrophys. J. Letters 248, L13.

Rothenflug, R. and Caraveo, P.: 1980, *Astron. Astrophys.* **81**, 218.
Schlickeiser, R.: 1982, *Astron. Astrophys. Letters* **100**, L5.
Schlickeiser, R.: 1981, *Fortschritte der Physik* **29**, 95.
Schlickeiser, R.: 1979, *Astrophys. J.* **233**, 294.
Stecker, F. W.: 1977, *Astrophys. J.* **212**, 60.
Stephens, S. A. and Badwhar, G. D.: 1981, *Astrophys. Space Sci.* **76**, 213.
Stephens, S. A.: 1981, *Astrophys. Space Sci.* **79**, 419.
Strong, A. W., *et al.*: 1981, *Proc. of 17th Int. Cosmic Ray Conf., Paris* **1**, 146.
Strong, A. W., *et al.*: 1978, *Monthly Notices Roy. Astron. Soc.* **182**, 751.
Swanenburg, B. N., *et al.*: 1981, *Astrophys. J. Letters* **243**, L69.
Taira, T., *et al.*: 1979, *Proc. of the 16th Int. Cosmic Ray Conf.* **1**, 478.
Webber, W. R.: 1982, presented at the *Intern. School of Cosmic Ray Astrophys.*, Erice, Sicily.
Webber, W. R., Simpson, G. A., and Cane, H. V.: 1980, *Astrophys. J.* **236**, 448.
Wheaton, W. A.: 1976, Thesis, UCSD, U.S.A.

THE USE OF GAMMA RAYS TO TRACE THE
LOCAL INTERSTELLAR HYDROGEN*

A. W. STRONG

Istituto di Fisica Cosmica del CNR, Milano, Italy

Abstract. COS-B data in the latitude range $11° < |b| < 19°$ show the gamma-ray intensity to be closely correlated with the total line-of-sight absorption derived from galaxy counts. It is found that to a good approximation the gamma-ray intensity is proportional to the total gas column density, and on this basis a map of the angular distribution of local molecular hydrogen at intermediate latitudes is presented. Comparing this with a simular map produced from galaxy counts, several structures appear in both maps in regions related to Gould's Belt and elsewhere. Some regions in the Southern celestial hemisphere not accessible in galaxy counts show high molecular hydrogen column densities.

1. The Use of Gamma Rays as a Local Gas Tracer

At present there is no direct method to determine the distribution of total hydrogen column density for the local interstellar medium. The only direct measurement are via UV absorption towards nearby stars (Bohlin *et al.*, 1978); the H_2 column densities are limited to about 100 directions. Indirect tracing of total hydrogen via interstellar reddening of stars (e.g. Fitzgerald, 1968; Lucke, 1978) severely undersamples the interstellar medium at intermediate latitudes and is more suited to studying the three-dimensional distribution of dust at low latitudes. Galaxy counts provide uniform sampling of total absorption in the Northern celestial hemisphere, and their use has been studied in detail (Heiles, 1976; Burstein and Heiles, 1978; Lebrun, 1979; Strong and Lebrun, 1982; Strong, 1983). This method is limited by statistics when the absorption is large, as well as by galaxy clustering and variations in the gas-to-dust ratio. In addition the response of the counts to patchy absorption on scales less than a degree is uncertain.

Gamma rays produced in cosmic-ray-gas interaction (via pion-decay and bremsstrahlung) give in principle a linear and fully sampled tracer of the total hydrogen, provided the cosmic-ray density can be assumed uniform and the contribution from other components such as gamma-ray discrete sources can be ignored. All gas regardless of scale is completely sampled. One conclusion of the work reported here (see Strong *et al.*, 1982a, for a detailed account) is that these assumptions are justified to a good approximation, and that the traditional idea of gamma rays as a diagnostic for cosmic-ray fluxes may perhaps be replaced with advantage by their use for delineating local gas structure.

Lebrun and Paul (1979, 1983) used the SAS-2 database to study correlations with galaxy counts. Lebrun *et al.* (1982) extended this work to the COS-B database, and showed that the absorption deduced from galaxy counts is much more successful than

* Proceedings of the XVIII General Assembly of the IAU: *Galactic Astrophysics and Gamma-Ray Astronomy*, held at Patras, Greece, 19 August 1982.

atomic hydrogen as a predictor of the gamma-ray intensities at intermediate latitudes. Strong *et al.* (1982a) developed this study to consider the spatial distribution of the gas and its interpretation.

2. Comparison of Gamma-Ray Intensities with Reddening from Galaxy Counts

A description of the galactic emission measured by COS-B is given in Mayer-Hassel-wander *et al.* (1982) and the reader is referred to this work for details of the gamma-ray observations. The present work uses additional observation periods to supplement the database. This study is restricted to $11° < |b| < 19°$ where both gamma ray and galaxy count data can be used. No attempt was made to correct for the presence of point sources which in this latitude range should in any case give a negligible contribution to the intensity.

To compare gamma-ray intensity I_γ and total gas column density we assume

$$I_\gamma = (q/4\pi)\tilde{N}_{HT} + I_b,\tag{2.1}$$

where q is the emissivity per hydrogen atom, N_{HT} is the total gas column density in atoms cm^{-2} and I_b is a residual background. The tilde denotes convolution with the point-spread-function of the instrument for the energy range considered.

To estimate N_{HT} we use the digitized Shane–Wirtanen galaxy count data (Shane and Wirtanen, 1967; Seldner *et al.*, 1977). Following Strong and Lebrun (1982) and Lebrun *et al.* (1982) we adopt

$$N_{HT} = 2 \times 10^{21} \log(50/N_g)\,\mathrm{cm}^{-2},\tag{2.2}$$

where N_g = no. galaxies per square degree (here averaged over $3° \times 3°$ cells). Fits to Equation (2.1) gave $q/4\pi = (1.4, 0.53, 0.59) \times 10^{-26}\,\mathrm{s}^{-1}\,\mathrm{sr}^{-1}$ for energy ranges 70–150, 150–300, and 300–5000 MeV, respectively. The longitude range 0–40° was excluded from this fit since it appeared to lie above the linear relation of the rest of the data; the choice of area for the fitting is the main source of uncertainty in the q values. At present an uncertainty of 25% should be taken for the above values.

Figure 1 shows the longitude distribution of expected and observed gamma-ray intensity (70–5000 MeV) in $5° \times 8°$ bins. The contribution from atomic hydrogen for the same emissivity is also shown, using H I data from Heiles and Habing (1974), Heiles and Clearly (1979), Weaver and Williams (1973), and Strong *et al.* (1982b). The similarity of the gamma-ray and galaxy count distributions leads one to the conclusion that they are measuring the same quantity: the total gas column density. The effects of variations in cosmic-ray intensity, point sources, gas-to-dust fluctuations, etc., are not large enough to upset the general correlation.

The deviation in $l < 40°$ can be partly explained by the failure of the logarithmic relation of absorption to galaxy counts when the absorption is large (see Strong, 1983); combined with the uncertainty in the q values there is no reason at present to consider this region abnormal, although the problem deserves further study.

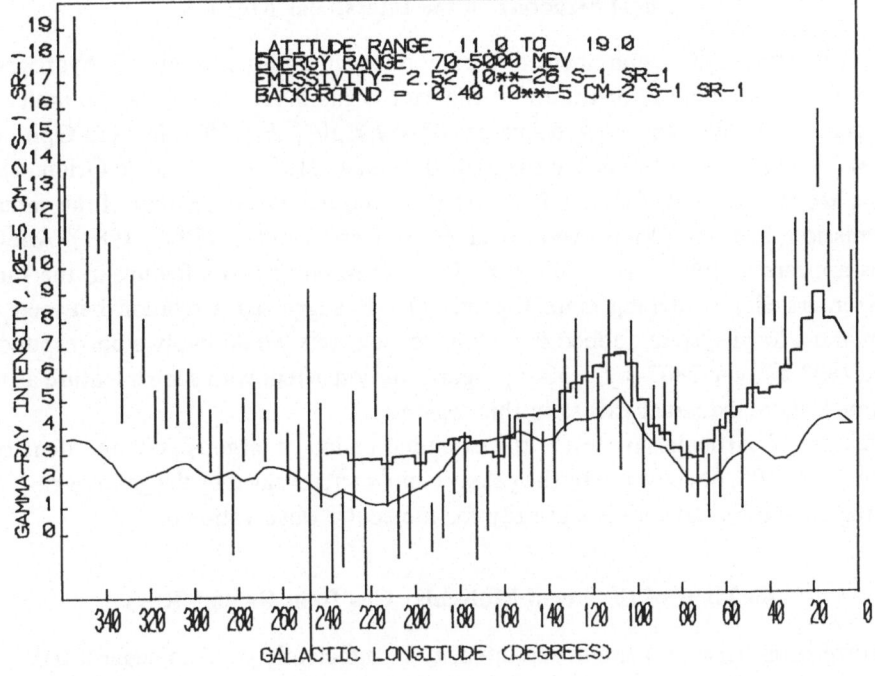

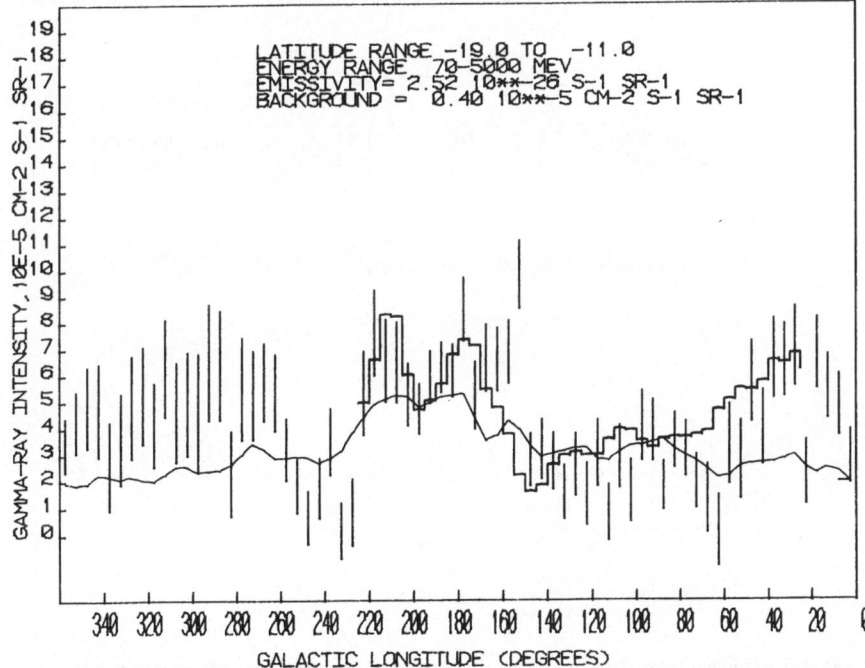

Fig. 1.   Comparison of measured and predicted gamma-ray intensities in the energy range 70–5000 MeV.
Bars: average measured intensity with statistical errors based on total counts in the bin. *Thick line*: predicted
intensity based on fit to Equations (2.1) and (2.2) with the parameters of Section 2. *Thin line*: predicted
intensity for atomic hydrogen alone, using the same parameters.

3. Local Structure in the Interstellar Medium

Figure 1 illustrates the requirement of a component in addition to atomic hydrogen to reproduce the gamma-ray distribution; the excess gas must presumably be molecular hydrogen. Particular excesses occur in $0° < l < 30°$, $b > 10°$, in Sco-Oph, and $160° < l < 180°$, $b < -10°$ in Taurus and $205° < l < 215°$, $b < -10°$ in Orion. These regions are related to the Gould's Belt system. A more detailed account of the gamma-ray emission from the Orion region is given by Caraveo *et al.* (1980, 1981). Another excess appears at $100° < l < 120°$, $b > 10°$, corresponding to a feature in the Lucke (1978) interstellar reddening map. Bignami (1981) suggested a relation between this region and a local system dubbed the 'Dolidze' belt; this would imply a corresponding excess 180° away at $280° < l < 300°$; Figure 1 is consistent with such a feature as part of a much more extended excess in this region.

A feature of particular interest is the extended region of large H_2 column density at $l > 300°$, $b > 10°$, not covered by the galaxy counts but revealed by the gamma-ray data. This region should have a high priority for molecular observations.

4. Maps of the Local Molecular Gas from Gamma Rays

The correlation between gamma-ray intensity is sufficiently good to suggest using the gamma-ray data to map the molecular gas column density. Figure 2 shows the result

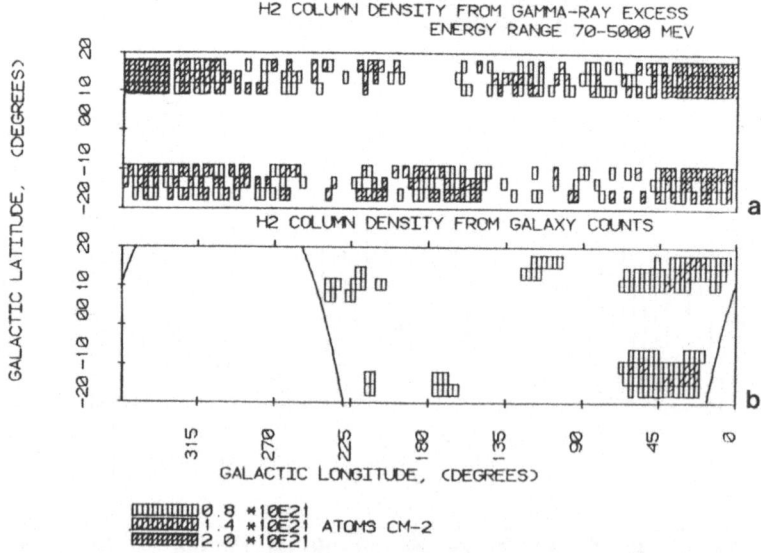

Fig. 2. (a) Map of molecular-hydrogen column density based on gamma-ray excess relative to expectation for atomic hydrogen. Binsize is $3° \times 3°$. Greyscale levels are 0.8–1.4, 1.4–2.0, and 2.0×10^{21} atom cm^{-2}. (b) Map of molecular hydrogen column densities derived from galaxy counts. Presentation is identical to (a). The continuous line corresponds to declination $-23°$, the Southern limit of the Lick counts. Since the convolution is as for (a), cells within $5°$ of this limit are excluded.

of inverting Equation (2.1) and subtracting the atomic hydrogen column density derived from 21-cm surveys. The map is the average of values from the three energy ranges. An identical presentation of a map derived from galaxy counts is also shown, allowing a critical comparison of the two techniques. Features common to both maps have a good chance of being real gas enhancements; those occuring in only one map are more likely of different origin (statistical, gamma-ray sources, etc.). The overall agreement is encouraging, and adds credibility to the claim that gamma rays trace mainly the total hydrogen. However in view of the uncertainty in the q-values (Section 2) the absolute scale of Figure 2 for the gamma-ray map should be taken as provisional at this stage.

5. Conclusions

Gamma rays are a useful quantitative tracer of total hydrogen at intermediate latitudes on scales greater than a few degrees. Unlike other tracers, they provide a linear measure of gas in all phases. It may in future be possible to calibrate other surveys (such as CO lines) using the gamma-ray data, provided the sampling and sky coverage is sufficient. On large scales the error in the gamma-ray estimate should eventually be less than uncertainties in deducing H_2 densities from molecular observation. In future a significant tightening of the gamma-ray estimates is expected.

The overview of local structures revealed in the gamma rays provides useful clues as to what to expect and look for at other wavelengths. In particular, in the Southern celestial hemisphere, the longitude range $l > 300°$ is expected to contain much molecular gas at intermediate latitudes.

Acknowledgement

The editors thank A. Dean for assitance in evaluating this paper.

References

Bignami, G. F.: 1981, *Phil. Trans. Roy. Soc. London* **A301**, 555.
Bignami, G. F., Barbareschi, L., Bloemen, J. B. G. M., Buccheri, R., Caraveo, P. A., Hermsen, W., Kanbach, G., Lebrun, F., Mayer-Hasselwander, H. A., Paul, J. A., Strong, A. W., and Wills, R. D.: 1981, *Proc. 17th Int. Cosmic Roy. Conf., Paris* **1**, 182.
Bohlin, R. C., Savage, B. D., and Drake, J. F.: 1978, *Astrophys. J.* **224**, 132.
Burstein, D. and Heiles, C.: 1978, *Astrophys. J.* **225**, 40.
Caraveo, P. A., Bennett, K., Bignami, G. F., Hermsen, W., Lebrun, F., Masnou, J. L., Mayer-Hasselwander, H. A., Paul, J. A., Sacco, B., Scarsi, L., Strong, A. W., Swanenburg, B. N., and Wills, R. D.: 1980, *Astron. Astrophys.* **91**, L3.
Caraveo, P. A., Barbareschi, L., Bennett, K., Bignami, G. F., Hermsen, W., Kanbach, G., Lebrun, F., Masnou, J. L., Mayer-Hasselwander, H. A., Sacco, B., Strong, A. W., and Wills, R. D.: 1981, *Proc. 17th Int. Conf. Cosmic Rays, Paris* **1**, 139.
Fitzgerald, M. P.: 1968, *Astron. J.* **73**, 983.
Heiles, C.: 1976, *Astrophys. J.* **204**, 379.
Heiles, C. and Clearly, M. N.: 1979, *Australian J. Phys. Astrophys. Suppl.* **47**, 1.
Heiles, C. and Habing, H. J.: 1974, *Astron. Astrophys. Suppl.* **14**, 1.
Lebrun, F. and Paul, J. A.: 1979, *Proc. 16th Int. Cosmic Ray Conf.* **12**, 13.
Lebrun, F. and Paul, J. A.: 1983, *Astrophys. J.* **266**, 276.

Lebrun, F., Bignami, G. F., Buccheri, R., Caraveo, P. A., Hermsen, W., Kanbach, G., Mayer-Hasselwander, H. A., Paul, J. A., Strong, A. W., and Wills, R. D.: 1982, *Astron. Astrophys.* **107**, 390.

Lucke, P. B.: 1978, *Astron. Astrophys.* **64**, 367.

Mayer-Hasselwander, H. A., Bennett, K., Bignami, G. F., Buccheri, R., Caraveo, P. A., Hermsen, W., Kanbach, G., Lebrun, F., Lichti, G. G., Masnou, J. L., Paul, J. A., Pinkau, K., Sacco, B., Scarsi, L., Swanenburg, B. N., and Wills, R. D.: 1982, *Astron. Astrophys.* **105**, 164.

Seldner, M., Siebers, B., Groth, E. J., and Peebles, P. J. E.: 1977, *Astron. J.* **82**, 249.

Shane, C. D. and Wirtanen, C. A.: 1967, *Publ. Lick. Obs.* **22**, 1.

Strong, A. W.: 1983, *Monthly Notices Roy. Astron. Soc.* **202**, 1015.

Strong, A. W. and Lebrun, F.: 1982, *Astron. Astrophys.* **105**, 159.

Strong, A. W., Bignami, G. F., Bloemen, J. B. G. M., Buccheri, R., Caraveo, P. A., Hermsen, W., Kanbach, G., Lebrun, F., Mayer-Hasselwander, H. A., Paul, J. A., and Wills, R. D.: 1982a, *Astron. Astrophys.* **115**, 404.

Weaver, H. and Williams, R. W.: 1973, *Astron. Astrophys. Suppl.* **8**, 1.

NONTHERMAL RADIO EMISSION FROM THE GALAXY*

G. KANBACH

Max-Planck-Institut für Physik und Astrophysik, Institut für extraterrestrische Physik, 8046 Garching, F.R.G.

Abstract. Synchrotron radio emission from interstellar space has long been recognized as a useful tool to probe into the galactic distribution of high energy electrons and magnetic fields. We first review the results obtained from the local (<2 kpc distant) region of the Galaxy and conclude that the observed local synchrotron emissivity is consistently explained by the measured cosmic ray electron spectrum and the interstellar magnetic field if some reasonable assumptions are allowed. The large scale distribution of radio emissivity shows evidence for spiral structure and is likely to originate in two distinct disk systems: a thin disk (thickness 250 pc in the inner Galaxy) formed by population I objects which emits about 10% of the galactic radio luminosity and a thick disk (2.5 kpc thick in the inner Galaxy) which constitutes the truly diffuse emission and produces 90% of the total luminosity.

1. Introduction

Our knowledge of the large-scale spatial and spectral distribution of cosmic-ray electrons in the Galaxy derives either from theoretical concepts of the generation and propagation of these particles (e.g. connected to the supernova rate in the Galaxy) or from the observation of synchrotron and bremsstrahlung radiation from the galactic disk. Synchrotron emission (also called magnetobremsstrahlung) originates from the movement of cosmic-ray electrons in galactic magnetic fields. Electron bremsstrahlung results from interactions with the interstellar medium and is thought to contribute the major part of the observed diffuse galactic gamma-ray emission at energies below 100 MeV. Unlike the case of cosmic-ray nucleons, where the gamma radiation produced in collisions with the interstellar medium is the only information measureable from far away locations in the Galaxy, we therefore have for the cosmic-ray electrons two channels of information, magnetobremsstrahlung and matter bremsstrahlung, which can be used to complement our understanding of these particles. This review summarises recent results on the interpretation of synchrotron emission from local and large scale regions of the Galaxy and is based mainly on the works of Webber (1982), Webber *et al.* (1980), Phillipps *et al.* (1981a, b), and Beuermann *et al.* (1982, hereafter referred to as BKB).

2. Basic Relations of Synchrotron Emission

Diffuse radio emission from the Galaxy is caused both by thermal and nonthermal processes. The observed thermal radiation which is the result of a black body emission spectrum ($S(v) \propto v^2$, Rayleigh-Jeans law) modified in the optically thin case by an absorption coefficient ($\tau \propto v^{-2.1}$) exhibits a nearly constant spectrum as a function of frequency. Nonthermal radiation from the synchrotron process however typically mirrors the form of the relativistic electron spectrum and therefore steeper power law forms are expected. As a consequence the relative strength of thermal and nonthermal

* Proceedings of the XVIII General Assembly of the IAU: *Galactic Astrophysics and Gamma-Ray Astronomy*, held at Patras, Greece, 19 August 1982.

emission is a strong function of frequency: lower frequencies ($\lesssim 1$ GHz) have to be used if predominantly nonthermal emission is to be observed. At lower frequencies (= longer wavelengths) however the beamsize of a radiotelescope increases and is finally not anymore useful for the observation of fine spatial details in the Galaxy. Below about 5 MHz also the effect of free-free absorption in the interstellar medium attenuates the radiation substantially. The new 408 MHz All-Sky Survey published by Haslam *et al.* (1981a, b) with an angular resolution of $0.8°$ can therefore be considered a good compromise between the mentioned effects and gives a suitable data base for studies of galactic structure.

An isotropic ensemble of relativistic electrons with a differential spectrum $KE^{-\gamma}$ moving in a homogeneous magnetic field ($B_\perp$: field component perpendicular to line of sight $B_\perp = |B| \sin \psi$) produces synchrotron radiation with a spectral distribution $\propto \nu^{(1-\gamma)/2}$ and source strength (or emissivity per unit volume) $\propto KB_\perp^{(1+\gamma)/2}$. The typical frequency of radiation emitted by an electron of energy E is given by $\nu \approx 20$ MHz $B_\perp (\mu G) E(\text{GeV})^2$, i.e. in a representative interstellar field of 5 μG a 1 GeV electron emits 100 MHz radiation. The observed intensity in a given galactic direction is obtained by the line of sight integration over the emissivity, which is considered to be a function of galactic location. It is important to keep in mind that the average emissivity over the line of sight is defined as $\langle \varepsilon \rangle \propto \langle KB_\perp^{(1+\gamma)/2} \rangle$ which takes into account possible correlated variations of electron spectrum and magnetic field (Cowsik and Mitteldorf, 1974). Only if restrictive further assumptions are made different averages become meaningful: if K = const. and only magnetic field fluctuations are considered $\langle \varepsilon \rangle \propto \langle B_\perp^{(1+\gamma)/2} \rangle$ (Rockstroh and Webber, 1978) and only for $B_\perp$ = const. is $\langle \varepsilon \rangle \propto \langle K \rangle B_\perp^{(1+\gamma)/2}$ a correct description. If observations of local synchrotron emissivity (on a galactic scale) are analysed such assumptions appear reasonable. For a large scale study however one should always qualify the conclusions one draws on electron spectrum and magnetic field with the assumptions on their constancy and the size of the emitting volume.

3. Synchrotron Emission from the Local Region

We have seen that three ingredients determine the synchrotron intensity observed from a given direction: the electron spectrum, the magnetic field and the size of the emitting volume (the 'line of sight'). For the interpretation of synchrotron emission from the galactic vicinity of the Sun one can measure or put limits on these three quantities and ask for the consistency of the result. First the *line of sight* can be limited by considering our position near the edge of the Galaxy. Observations towards the galactic anticenter extend most probably over equivalent dimensions not in excess of 4 kpc (Badhwar *et al.*, 1977). In the direction of the galactic poles the line of sight is limited by the extent of a possible halo or thick disk distribution of emissivity. Arguments derived from the large-scale latitudinal distribution of synchrotron brightness can be used to limit the equivalent extent of the emission at $b = 90°$ to ~ 2 kpc (Beuermann *et al.*, 1982). Another way to limit the line of sight very accurately is afforded by the fact that

H II-regions are opaque to low frequency radiation ($\lesssim 10$ MHz). The continuum bright-ness observed towards a H II region of well known distance must then originate in the foreground region. This method to determine the synchrotron emissivity is however only applicable at very low frequencies and corresponding low electron energies. If the spectral form of the synchrotron source function is independent of galactic location (van der Kruit, 1978) one can apply the result of the H II-region study to the distribution of higher frequency emission as well. Specifically for the anticenter direction Rockstroh and Webber (1978) compared the intensity towards several H II-regions with the background intensity in the same region and conclude that the line of sight extends over 2–3 kpc. The same authors have compiled a list of H II regions at various galactic locations in front of which low frequency synchrotron emission has been measured. We can sort these data according to galactocentric radius and use the derived emissivities as indicators of the average line of sight emissivity in the direction of the H II region. Figure 1 shows these emissivities at a frequency of 10 MHz. The galactic gradient of the synchrotron strength is quite evident. The typical radial scale length is about 2.5 kpc in the neighbourhood of the Sun which gives independent support to the above derived extent of the radio Galaxy in the anticenter direction.

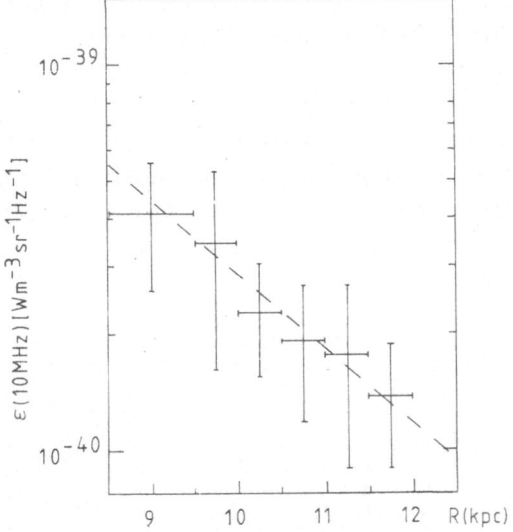

Fig. 1.   10 MHz synchrotron emissivity as a function of galactocentric radius. The data are derived from the measurement of foreground radio continuum in the direction of H II regions.

The interstellar *magnetic field* is a difficult quantity to measure. The Faraday rotation of polarized radiation in the interstellar plasma has been the principal method to measure the large scale fields. Manchester (1974) derived from pulsar rotation measures that the directed field has a magnitude of 2.2 ± 0.4 µG out to a distance of more than 1 kpc and that irregular field components of similar magnitude are present in regions of spatial dimensions of ~ 100 pc. On a galactic scale the field cannot be measured in such a direct way. However one can rather safely infer from observations of external galaxies that the direction of the ordered galactic field generally follows the spiral arms (Beck, 1982).

The local *electron spectrum* at energies below a few GeV is severely affected by solar modulation and planetary electron sources like Jupiter. At high energies however the interstellar spectrum should be directly measurable and recent results above 10 GeV (compiled by Webber, 1982) have converged to a consistent description of the spectrum of (700 ± 100) (electrons $m^{-2} s^{-1} sr^{-1} GeV^{-1}) \times E^{-3.3}$. Previous electron measurements at these high energies differed by as much as a factor of 10 and made a comparison with the synchrotron intensity ambigous (see e.g. Badhwar *et al.*, 1977). The present accurate electron spectrum however allows for an absolute calculation of the synchrotron intensity at frequencies $\gtrsim 10$ GHz as a function of transverse magnetic field and line of sight. In order to explain the observed radio brightness at 10 GHz in the galactic anticenter of 10^{-22} W m^{-2} sr^{-1} Hz^{-1} with the high energy electron spectrum a relation between the field and the extent of the emission is obtained as shown in Figure 2. We find that the effective magnetic field strength $[\langle B_\perp^{(1+\gamma)/2} \rangle]^{2/(1+\gamma)}$ should be between 5 and 7 µG if the above limits on the line of sight (2–4 kpc) are applied. These effective B fields can be satisfactorily explained by postulating appropriate distributions of the interstellar field strength (Rockstroh and Webber, 1978) or as Cowsik and Mitteldorf (1974) have shown by considering the correlated enhancements of electron spectrum and field fluctuations which result from turbulent motion of the interstellar medium.

In conclusion one can say that the local synchrotron emissivity, the observed cosmic ray electron spectrum and the interstellar magnetic fields fit together in a consistent picture if certain reasonable assumptions on local galactic structure and magnetic field distributions are admitted.

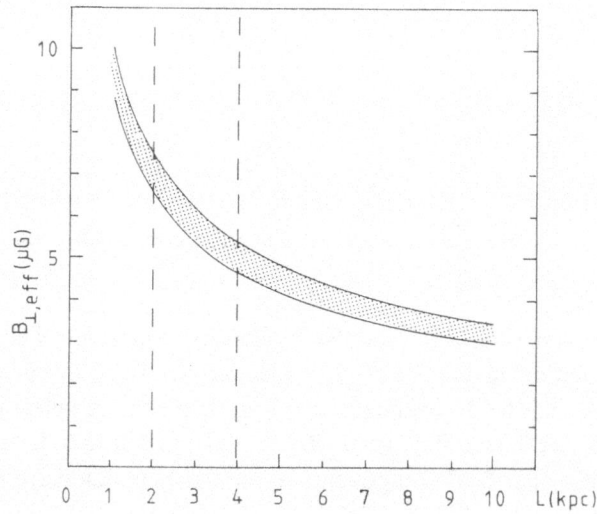

Fig. 2. Combinations of effective field strength and extent of line of sight to explain the synchrotron intensity at 10 GHz in the galactic anticenter direction.

4. Large Scale Distribution of Synchrotron Emission

Recently a new all sky survey at 408 MHz has been published by Haslam *et al.* (1981a, b). This survey is particularily suited for work on the large scale structure of nonthermal radio emission from the Galaxy because its angular resolution of 51 arc min, which is equivalent to a spatial resolution element of 150 pc at a distance of 10 kpc, corresponds to the size of the objects of investigation. There are two different ways to interprete the measurements in terms of the large scale distribution of emissivity: Firstly the 'model building' approach, which tries to construct plausible distributions of synchrotron emissivity and compares the resulting line of sight integrals with the measured map. Secondly the attempt to deconvolve or unfold the line of sight integrals directly. The latter method has the advantage that it derives its results more directly from the measurement and recent studies (Kanbach and Beuermann, 1979; Phillipps *et al.*, 1981a, b; Beuermann *et al.*, 1982) are all based on this approach.

However, the deconvolution method necessitates a number of critical assumptions on the large scale geometry of the galactic disc. Essentially one has to make up for the lack of distance information in the observed line of sight integrals by assuming a geometrical structure (n.b. not a source strength distribution) within the galactic plane and perpendicular to this plane in the z-direction. In the case of spiral structures one furthermore has to impose a form for the radial dependence of the emissivity, including the overall size of the Galaxy, which has to be derived in an iterative procedure to be consistent with the measurements. The structure of the magnetic field also has to be assumed in some plausible model.

The justification for the necessary assumptions can either come from galactic structure studies at other wavelengths, from theoretical concepts of spiral arm structure and the physics of the emission regions, or from comparison with observations of external galaxies.

The assumptions and results in the works of Beuermann *et al.* (1982, BKB) and Phillipps *et al.* (1981a, b), shall now be reviewed in more detail.

The underlying geometrical pattern of the Galaxy is assumed to be of spiral form. BKB use a logarithmic spiral structure with an inclination angle of 13° while Phillipps *et al.* (1981) found 12° to give the sharpest features in their model. The resulting morphology of the four armed spiral arm pattern in both deconvolutions agrees quite well with the model found by Georgelin and Georgelin (1976) on the basis of a H II region study.

The magnetic field structure which determines the directional distribution of synchrotron emission is initially assumed as a free parameter. As was shown in the previous section a galactic field composed of regular, i.e. directed along the spiral structure and irregular components is observed locally. BKB model the large scale field as a vector evenly distributed in a conical solid angle around the preferred spiral direction. The opening angle of this cone defines the limits of the angular deviation a 'wandering field line' can have with respect to the principal direction. This angle is equivalent to a parameter that describes the relative strengths of the effective parallel and transverse field components. A similar separation of the synchrotron emission coming from a directed and from an irregular field component has been adopted by Phillipps *et al.* (1981). Both studies conclude that the absence of a substantial enhancement in synchrotron brightness in directions perpendicular to the arm structure (at small longitudes) with respect to the tangential directions precludes a highly directed magnetic field structure. Moderate alignment with the directed and the irregular field components of equal magnitude is the preferred large scale field structure in both models.

The treatment of the distribution of synchrotron emissivity perpendicular to the galactic plane is different in the two studies. Phillipps *et al.* (1981b) start from the result of their deconvolution in the galactic plane ($z = 0$) and fit the observed latitude profiles by adjusting the z-dependence in the form of a sixth order polynominal with a linear fall off at larger values of z. Their main conclusions are that all models require a non-spherical halo of emission extending to distances of about 10 kpc from the plane and the scale height of the disk emission is a function of galactocentric radius. The thickness of the radio disk increases in the outer galaxy and mirrors the structure of the gaseous disk. These authors basically treat the radio galaxy as one coherent entity extending from the intense galactic disk at low latitudes to the outer reaches of the galactic halo.

BKB, on the other side, derive a three dimensional model that is based on a two disk concept of the Galaxy. The observation of some external galaxies, seen edge-on, shows that the z-distribution of emissivity is strongly peaked at $z = 0$ and exhibits wings that extend to several kpc above the plane of the system (Allen *et al.*, 1978). The assumption made by BKB implies that two distinct physical systems, namely a narrow and a broad disk, form the galactic radio background and have to be evaluated separately. Since the two disk systems have grossly different latitude extents one can use two longitude profiles to deconvolve both systems simultaneously: a central one at $b = 0$ where narrow and broad disk contribute about equally and one at $b = 3°$ where the narrow disk has fallen off to a small fraction of the broad disk. The functional form of the z-dependences of the two disks is based on an approximate description of the hydrostatic equilibrium of a cosmic ray, magnetic field gas system in the field of the dominating mass of the Galaxy in the form of the stellar disk (Fuchs *et al.*, 1976). The width of the disk systems corresponds to their respective temperatures: the narrow disk is a few hundred parsecs wide and is based in a relatively cool medium of less than 10^4 K temperature while the several kiloparsec of the broad disk indicate that it is supported in the hot (about 10^5 K) phase of the interstellar medium. Figure 3 shows the separation of the two galactic disk systems in latitude profiles centered on the galactic longitude 322°. At this longitude no apparent localised sources perturb the galactic background radiation. The excellent fit of the resulting model depends directly only on two parameters describing the thickness of the disks while the scale height of the stellar base disk is an indirect parameter that is deduced from independent observations. Similar to previous conclusions in other studies, BKB find it necessary to increase the thickness of the radio galaxy with larger galactocentric radius in order to fit the latitude distributions in all galactic quadrants. Table I summarises the disk thickness parameters used in the model. The result of the simultaneous deconvolution of the two disk components is a two-

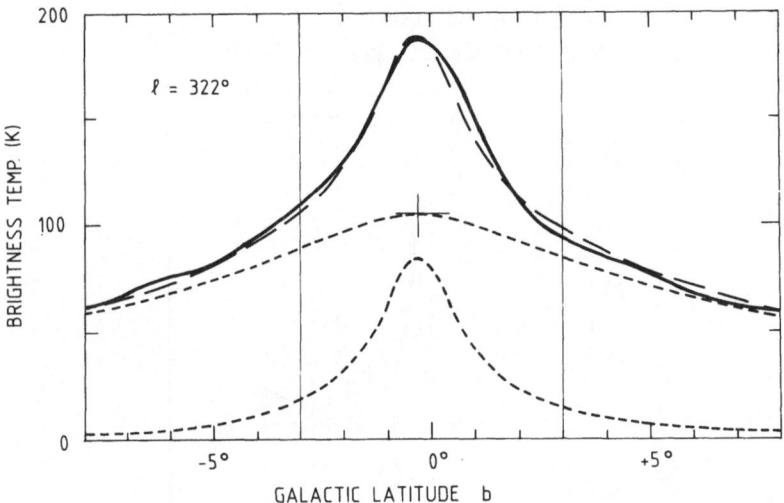

Fig. 3.   Latitude distribution of nonthermal radio brightness at $l = 322°$. *Solid line*: Observation of Haslam *et al.* (1981). *Dashed lines*: Contribution of the thin and thick disk systems and their sum in the model of Beuermann *et al.* (1982).

TABLE I

Parameters of the thin and thick radio disk according to the model of Beuermann *et al.* (1982)

	Radial range	Thin disk	Thick disk
Full equivalent width	0–8 kpc	250 pc	2.3 kpc
	8–12 kpc	370 pc	3.6 kpc
	12–20 kpc	690 pc	6.3 kpc
Radial scale length of emissivity at $z = 0$ and $R > 5$ kpc		3.1 kpc	4.0 kpc
Monochromatic power at 408 MHz		8×10^{20} W Hz^{-1}	8×10^{21} W Hz^{-1}

dimensional emissivity map in the galactic plane. As was mentioned before the derived spiral arm pattern agrees quite well with the four armed spiral structure described by Georgelin and Georgelin (1976). This is, however, not too surprising because the Georgelin study was also based in part on nonthermal radio observations of the so called tangential direction to spiral arms. The coincidence of the nonthermal radio arms with the basic H II region distribution is nevertheless illustrating the close connection that exists between parts of the synchrotron emissivity and young galactic objects. The quantitative result of the unfolding procedure is given in Figure 4. Here the emissivity values of the spiral segments are given at the tangential points, i.e. those points on the spiral that are viewed tangentially from the Sun. The peaks visible in this result from

280   G. KANBACH

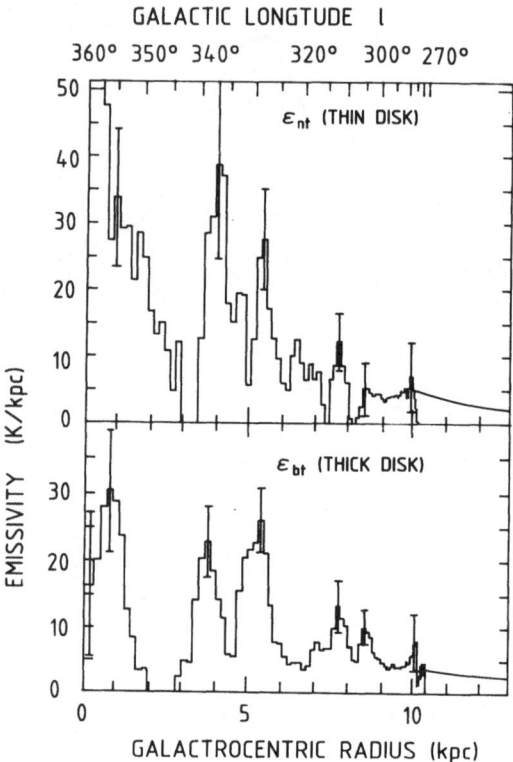

Fig. 4.   Result of the unfolding of the thin and thick disk (Beuermann *et al.*, 1982). Given are the emissivities at the tangential points of the spiral segments in the fourth galactic quadrant.

the fourth galactic quadrant correspond to well known spiralarm features: Carina at $l = 285°$, Crux-Centaurus at $l = 310°$, Norma at $l = 329°$, and the 3 kpc arm at $l = 338°$. The arm-interarm modulation extends well into the thick disk which indicates that the galactic arm shocks reach large distances from the gaseous plane. The average arm contrast of the radio galaxy is about 2. External systems like M51 or M81 (Segalovitz, 1976, 1977) show similar contrast values. On the scale of spiral arms the correlations between gas density, magnetic field strength and cosmic-ray population may be quite complex as we have seen in the discussion of the local synchrotron emissivity. It is therefore not possible to uniquely attribute the emissivity enhancement to variations in either magnetic field or cosmic-ray electron density alone. Theoretical work on galactic spiral arm shocks (Soukup and Yuan, 1981) may however provide some arguments for a compression of gas and magnetic field that extends to large z heights and has the required shock strength to explain the synchrotron emission enhancement even with an overall smooth electron distribution.

An average of the detailed spiral arm emissivity maps for the two disk systems performed over galactocentric circles allows one to investigate the overall distribution of the synchrotron source strength. Figure 5 shows this radial distribution. We will first consider the result for the *thick radio disk*. The wide distribution and the apparent

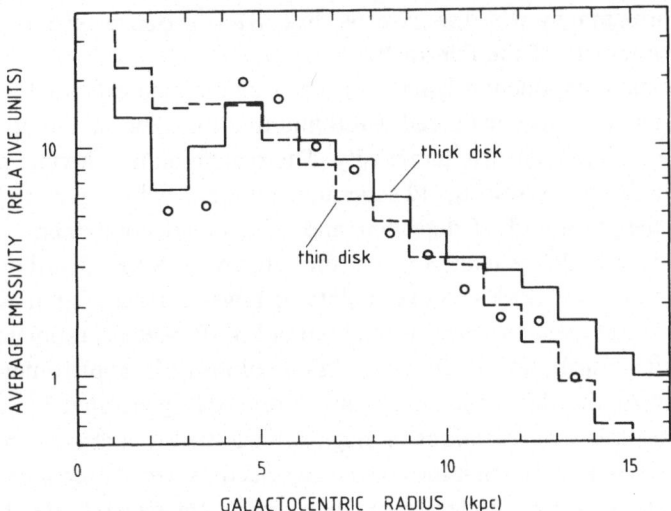

Fig. 5. Radial distribution of the average emissivity at 408 MHz in the thick (full line) and thin (dashed line) disk systems. The open circles show the sum of the proposed constituents of the thin radio disk.

absence of small scale localised structure in the broad disk suggest that the origin of this radio radiation is truly diffuse and not connected with a hypothetical unresolved source population. The radial decrease of the thick disk volume emissivity starts from a maximum around 5 kpc galactocentric radius and shows a scale length of about 4 kpc (see Table I). This corresponds to a relative enhancement of a factor 3.5 at 5 kpc with respect to the solar neighbourhood. If equipartition between cosmic rays and magnetic field energy density is applicable on a galactic scale (i.e. $n_e \propto H^2$) we can deduce that the average field strength at $R = 5$ kpc is about 40% higher than locally and the electron density is enhanced by about 90%. Equivalent decreases of H and n_e would apply to the outer galaxy around $R = 15$ kpc. The total power radiated by the thick disk at 408 MHz is about 8×10^{21} W Hz^{-1}. If we integrate this monochromatic power over a typical radio spectrum (Webber, 1982) the total luminosity of the thick disk is about 2×10^{31} W. It is illustrative to compare this radio luminosity with the total energy released by a supernova explosion of about 10^{51} erg or 10^{44} W. If the rate of supernovae is of the order of once in 100 years then only a modest amount of about 10^{-3} of the total energy released would be required to maintain the high energy electron population of the thick radio disk. BKB point out that the radial distribution of the thick disk agrees with that of supernova remnants and the total mass surface density of the Galaxy. The distribution of population I objects like extended low density H II regions indicates a much steeper fall off with radius. This result would place the origin of the cosmic-ray electrons in the general disk population rather than in the extremely young objects.

The *thin radio disk* which dominates the galactic radio map due to its narrow and sharp distribution plays nevertheless only a minor role in the luminosity budget of the radio galaxy. As shown in Table I only about 9% of the total unresolved galactic emission at 408 MHz originate from the thin disk. Since the latitude extent of this radio disk coincides approximately with the gaseous disk, BKB propose several constituents to explain the luminosity of the thin disk.

First a thermal component originating in the well mapped extended low density H II regions (Mezger, 1978) is considered. BKB find that this thermal radiation contributes about one third of the thin-disk luminosity. The remaining nonthermal fraction of this disk is made up from unresolved, old supernova remnants whose surface brightness has fallen below the threshold of detection and of a cloud component. The old SNR contribution is readily estimated from the observed SNR distribution using an evolutionary model for SNRs to extrapolate to lower surface brightness values. The volume fill factor of these probably broken-up old SNR shells is estimated to be about 20% and their contribution to the thin disk luminosity is approximately 15%. The remaining half of the thin disk luminosity is probably generated by enhanced synchrotron emission in condensations of magnetic field tied to a cloud component in the cool interstellar medium. Even if the cosmic-ray electron density remains unchanged by the cloud condensations a moderate enhancement of the general interstellar magnetic field by factors of 4 to 8 is sufficient to provide the necessary synchrotron luminosity. The volume fill factor of the hydrogen clouds is estimated at about 20%. In Figure 5 the summed contribution of ELD H II regions, old SNRs and clouds is shown and one concludes that the strength and structure of the thin radio disk as derived from the deconvolution agrees very well with the proposed explanation.

5. Conclusions

The large-scale nonthermal radio emission of the Galaxy according to the model of BKB originates in two distinct disk systems: About 90% of the total radiation is emitted by a thick disk of a full equivalent width that varies between 2 and 6 kpc depending on galactocentric distance. It is proposed by BKB that the electrons and magnetic fields in this thick disk are coupled to the hot (10^5 K) and tenuous (10^{-2} cm^{-3}) phase of the interstellar medium. Embedded in this system is the gaseous disk of the Galaxy (thickness between 250 and 700 pc) that contains the constituens for the thin radio disk. Since the luminosity of the thin disk was found to be rather small (about 10% of the total luminosity) a satisfactory explanation for the origin of this radiation can be proposed. Thermal radiation from extended low density H II regions, old SNRs with low surface brightness and magnetic field condensations within clouds form the thin, fragmentary radio disk with a volume fill factor of about 40%. Both disk systems were found to have a spiral structure. The four armed pattern of radio spiral arms agrees well with models based on other spiral arm tracers. It could be suggested that the spiral structure of the thin radio disk is due to the distribution of the mentioned localised although unresolved components, whereas the pattern in the thick diffuse disk is produced by large scale spiral arm shocks which compress the medium. The large scale distribution of cosmic-ray electrons as deduced from the thick disk shows only a moderate (factor of two) increase of the electron density toward the 5 kpc region of the Galaxy. The consequences of such an increase on the source function of gamma-ray bremsstrahlung in the Galaxy remains to be determined.

Acknowledgement

The author would like to thank K. Beuermann for the long collaboration on this subject. The editors thank M. Morini for assistance in evaluating this paper.

References

Allen, R. J., Baldwin, J. E., and Sancisi, R.: 1978, *Astron. Astrophys.* **62**, 397.

Badhwar, G. D., Daniel, R. R., and Stephens, S. A.: 1977, *Nature* **265**, 424.

Beck, R.: in 'Magnetic Field and Spiral Structure', *IAU Symp.* **100**.

Beuermann, K., Kanbach, G., and Berkhuijsen, E. M.: 1982, submitted to *Astron. Astrophys.*

Cowsik, R. and Mitteldorf, J.: 1974, *Astrophys. J.* **189**, 51.

Georgelin, Y. M. and Georgelin, Y. P.: 1976, *Astron. Astrophys.* **49**, 57.

Fuchs, B., Schlickeiser, R., and Thielheim, K. O.: 1976, *Astrophys. J.* **206**, 589.

Haslam, C. G. T., Klein, U., Salter, C. J., Stoffel, H., Wilson, W. E., Clearly, M. N., Cooke, D. J., Thomasson, P.: 1981a, *Astron. Astrophys.* **100**, 209.

Haslam, C. G. T., Salter, C. J., Stoffel, H., and Wilson, W. E.: 1981b, *Astron. Astrophys. Suppl. Ser.* **47**, 1.

Kanbach, G. and Beuermann, K.: 1979, *16th Int. Conf. Cosmic Rays, Kyoto* **1**, 75.

Manchester, R. N.: 1974, *Astrophys. J.* **188**, 637.

Mezger, P. G.: 1978, *Astron. Astrophys.* **70**, 565.

Phillipps, S., Kearsay, S., Osborne, J. L., Haslam, C. G. T., and Stoffel, H.: 1981a, *Astron. Astrophys.* **98**, 286.

Phillipps, S., Kearsay, S., Osborne, J. L., Haslam, C. G. T., and Stoffel, H.: 1981b, *Astron. Astrophys.* **103**, 405.

Rockstroh, J. M. and Webber, W. R.: 1978, *Astrophys. J.* **224**, 677.

Segalovitz, A.: 1976, *Astron. Astrophys.* **54**, 703.

Segalovitz, A.: 1977, *Astron. Astrophys.* **55**, 203.

Soukup, J. E. and Yuan, C.: 1981, *Astrophys. J.* **256**, 376.

van der Kruit, P. C.: 1978, in E. M. Berkhuijsen and R. Wielebinski (eds.), 'Structure and Properties of nearby Galaxies', *IAU Symp.* **77**, 33.

Webber, W. R.: 1982, *Cosmic Ray Electrons and Positrons*, Review at Int. School of Cosmic Ray Astrophysics, Erice, June 1982.

Webber, W. R., Simpson, G. A., and Cane, H. V.: 1980, *Astrophys. J.* **236**, 448.

COMPARISON OF FAR-INFRARED RADIATION WITH GAMMA RADIATION*

(Large-Scale Structure of the Galaxy)

NAOKI ODA

Max-Planck-Institut für extraterrestrische Physik, D-8046 Garching bei München, F.R.G.

Abstract. The gamma-ray emissivity for the narrow component (FWHM = 2°) at the 0.3–5 GeV range is derived as a function of the galactocentric distance. The narrow component might result from the interaction between cosmic rays and H_2 gas. The mass of gas in the Galactic Center is not large enough to produce the gamma-ray peak, but enough to produce the far-infrared peak. The relation of far-infrared dip and near-infrared hump near $l = 356°$ to gamma-ray hump is discussed.

1. Introduction

For the investigation of the large-scale structure of the Galaxy, i.e. the distributions of stars and interstellar matter, radio, infrared (IR) and gamma observations are quite powerful, because these radiations can easily reach us from the far side of the Galaxy. Recently, farinfrared (FIR) observations have been made by Maihara *et al.* (1979), Nishimura *et al.* (1980), and Gispert *et al.* (1982), mid-IR observations by Price (1981) and near-IR observations by Oda *et al.* (1979), and Hayakawa *et al.* (1981). They are listed in Table I. The large-scale IR surveys are reviewed by Okuda (1981). The distribution of gamma radiation has been obtained by the COS-B satellite in three energy ranges (Mayer-Hasselwander *et al.*, 1982).

TABLE I

Large-scale IR surveys and COS-B experiment

Observers (year)	L (deg)	B (deg)	λ (µm)	Beam
Maihara *et al.* (1979)	340–32	± 5	100–300	0.7° × 1°
Nishimura *et al.* (1980)	350–45	± 2	100–300	0.5°
Gispert *et al.* (1982)	0–90	± 4	114 – 196	0.37°
			73–94	0.4°
Price (1981)	0–30	± 5	4	
	0–320	± 5	11	10.5′ × 5′ or
	0–320	± 5	20	10.5′ × 3.35′
	40–85, 280–320	± 5	27	
Oda *et al.* (1979)	345–32	± 10	2.4	0.6°
Hayakawa *et al.* (1981)	288–65	± 10	2.4	0.4°, 0.8°, 1.7°
			3.4	2°
Mayer-Hasselwander *et al.* (1982)	0–360	± 20	0.3–5 GeV	2.2°
			150–300 MeV	4.4°
			70–150 MeV	7°

* Proceedings of the XVIII General Assembly of the IAU: *Galactic Astrophysics and Gamma-Ray Astronomy*, held at Patras, Greece, 19 August 1982.

In general, the FIR radiation is emitted by dust grains which are heated up to several 10 K by stars. There have been only few explanations so far about the observed diffuse FIR emission from the galactic plane, which has been reviewed by Drapatz (1981). Mezger (1978) has explained the FIR emission with the model that the radiation of mainly early-type stars is absorbed by dust grains in the extended low-density (ELD) H II-regions and is converted to FIR emission. Drapatz (1979) has taken the viewpoint that FIR radiation originates from the dust grains located inside dense clouds (especially the outer parts of the clouds) which are illuminated mainly by external sources (OB stars, AF stars, and late-type stars). His estimate is in quite good agreement with observations. He also has pointed out that protostellar clouds (cocoons) contribute to the diffuse FIR radiation field by at most 10–20%.

Kniffen and Fichtel (1981) have shown that the dominant mechanism to produce gamma-ray photons with energies higher than 100 MeV in the latitude $|b| \leq 1°$ is cosmic-ray (CR)-matter interaction and that the contribution to the gamma-ray flux due to the inverse Compton process over the $|b| \leq 10°$ region is at most 25% for $E_\gamma > 100$ MeV. With the spectrum of CR electrons ($E_e^{-2.8}$ for $E_e \geq 2.5$ GeV), the inverse Compton contribution can be reduced to 9% at $E_\gamma \geq 300$ MeV. Therefore COS-B data in the 0.3–5 GeV range can be used to study the matter distribution in the Galaxy. Further, the distribution of FIR emission will be compared with that of gamma radiation at the 0.3–5 GeV range to obtain information on the distribution of interstellar matter in the Galaxy.

2. Basic Data and their Characteristics

2.1. BASIC DATA

Figure 1 shows the gamma-ray skymap in the 0.3–5 GeV range, reproduced from Mayer-Hasselwander *et al.* (1982) and the FIR map which is synthesized from three maps (Maihara *et al.*, 1979; Nishimura *et al.*, 1980; and Gispert *et al.*, 1982). The flux-level of FIR emission in the map is not adjusted among these observations (see each paper for the absolute fluxes).

The longitude profiles of gamma and FIR radiations are shown in Figures 2a and 2b, respectively. The gamma radiation is divided into the wide component and the narrow component (Mayer-Hasselwander *et al.*, 1982). The narrow component with an intrinsic FWHM of 2° comes from remote regions and the wide component with the FWHM of about 10° is largely produced locally (within about 1.5 kpc). The longitude profile of the narrow and wide components (dashed curve in Figure 2a) is obtained by averaging the COS-B data (Figure 5 in Mayer-Hasselwander *et al.*) in the 0.3–5 GeV range within $|b| \leq 1°$, i.e. the beam size. Then, I subtract the wide component from the longitude profile to extract the structure of the inner galaxy. From Figures 7 and 8 in Mayer-Hasselwander *et al.* (1982), the flux of the wide component is estimated to about 4×10^{-5} photon cm^{-2} s^{-1} sr^{-1}. Of course, more detailed treatment to estimate this flux would be desirable. The longitude profile of the narrow component thus obtained

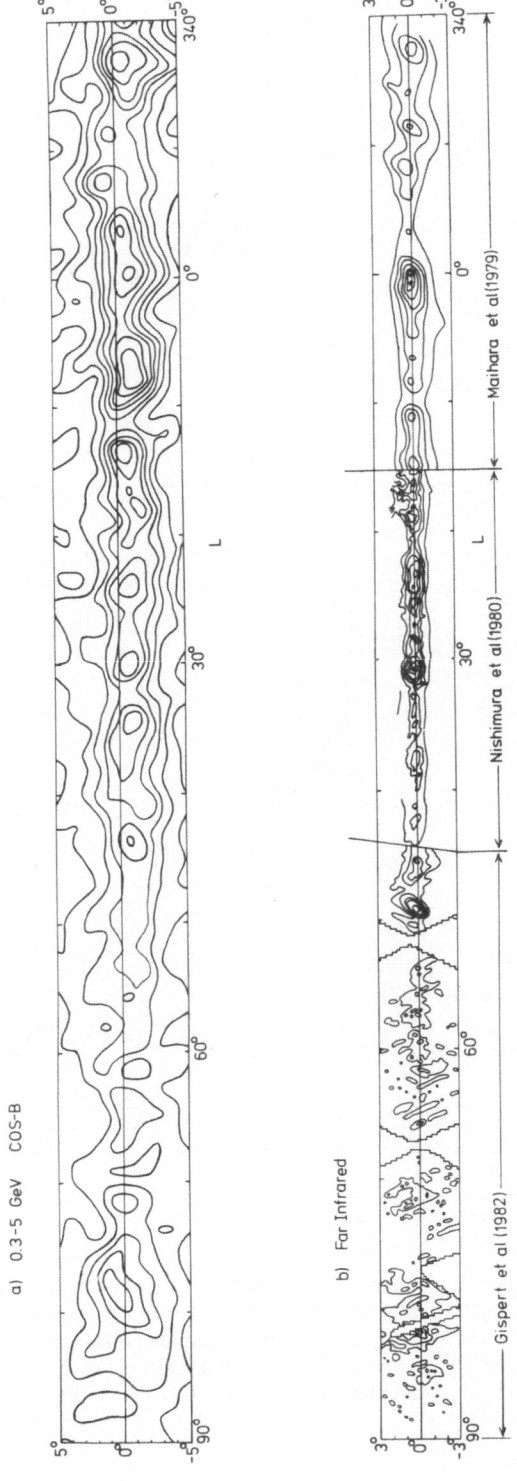

Fig. 1.   (a) Gamma-ray skymap at the 0.3–5 GeV range in the galactic coordinate frame. (b) Far-infrared map produced from Maihara *et al.* (1979) in the $l = 340°–15°$ region, Nishimura *et al.* (1980) in the $l = 15°–45°$ region and Gispert *et al.* (1982) in the $l > 45°$ region.

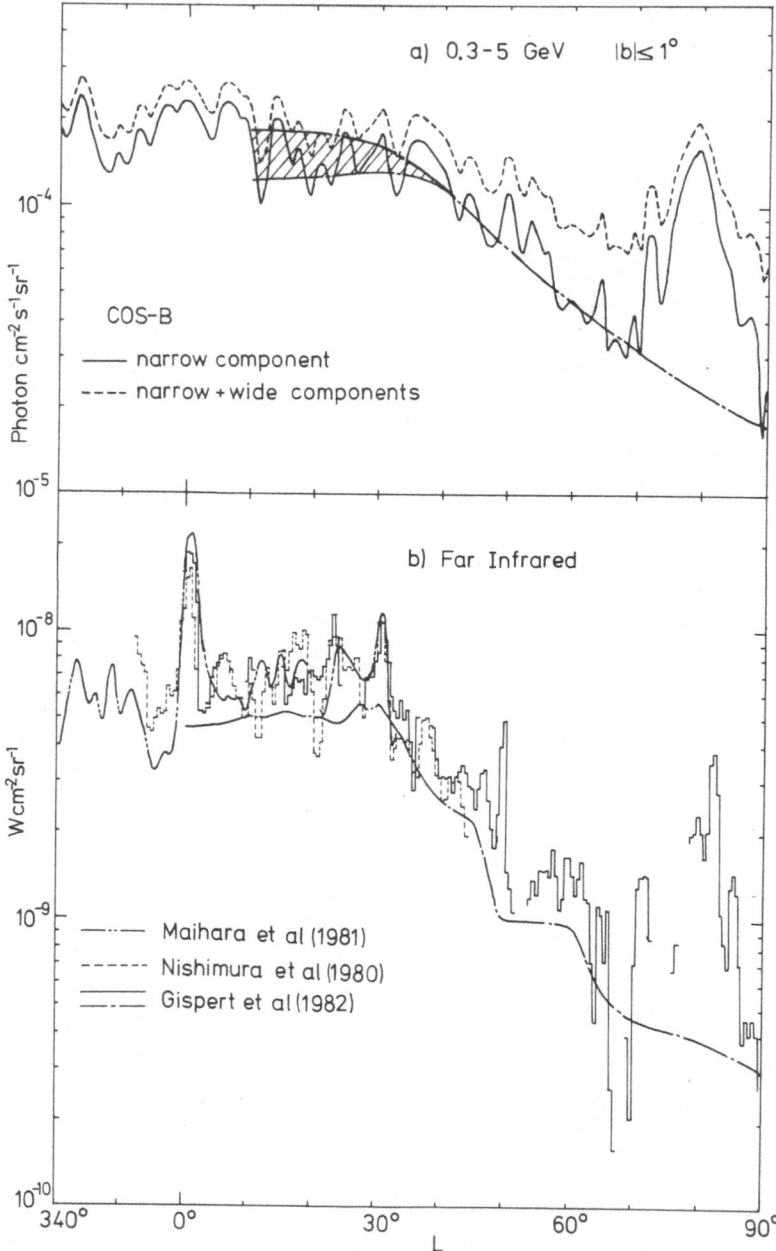

Fig. 2.. (a) Longitude profile of gamma radiation; dashed curve: the profile of the narrow and wide components; solid curve: the profile of the narrow component. The dash-dot curve with shaded area relates to the dashed curve with shaded area in Figure 3 (see text). (b) Longitude profile of far-infrared radiation. Dash-two dots curve: Maihara *et al.* (1981); dashed curve: Nishimura *et al.* (1980); solid curve and dash-dot curve: Gispert *et al.* (1982). The dash-dot curve relates to the solid line in Figure 3 (see text).

is shown by the solid curve in Figure 2a. It can be clearly seen that the intensity-gradient from $l = 30°$ to $70°$ is steeper in the narrow component than in the sum of narrow and wide components. The overall gross profile of the narrow component (the dash-dot curve in Figure 2a), except for the Cygnus region around $l = 80°$, will be discussed later, taking into account the effect of the point spread function of the COS-B experiment.

Since the three above mentioned FIR observations differ in beam-size and observed wavelength region, the following procedure is taken to determine the flux-level: Gispert *et al.* have found an excellent agreement of fluxes at $l = 30.8°$ between their data and Nishimura *et al.* According to Maihara *et al.* (1981), there is also good agreement of the fluxes of diffuse emission between their data and Nishimura *et al.* The data are combined in Figure 2b. The FIR emission is divided into two components; one comes from discrete H II-regions which can easily be seen as sharp spikes, and the other is called a continuous component (dash-dotted curve in Figure 2b) which is either a really diffuse component or the sum of unresolved distant sources.

2.2. CHARACTERISTICS

The distribution of FIR emission has the following characteristics:
(i) The emission is concentrated in a very thin plane of the Galaxy (FWHM of about 1.6° in latitude).
(ii) The brightness is fairly constant up to $l = 30°$ and decreases abruptly at $l \gtrsim 30°$.
(iii) A strong concentration of emission at the Galactic Center can be seen.
(iv) Most of the emission peaks are identified with discrete H II-regions.
(v) The surface brightness near $l = 356°$ is fairly low.
The characteristics of the gamma radiation are the following:
(i) The longitude profile is fairly flat up to $l = 35°$ and the emission decreases moderately at $l \gtrsim 35°$.
(ii) There is no prominent peak at the Galactic Center.
(iii) Displacement of the ridge is about 0.5° below the galactic plane.
In both FIR- and gamma-ray distributions, the Cygnus region is outstanding.

3. Emissivities

To fit the observations, I take the following simple expressions for the gamma-ray emissivity in the 0.3–5 GeV range as a function of the galactocentric distance R, assuming the scale-height of 130 pc and axisymmetry:

$$\varepsilon = A \exp\left(-\frac{R - R_c}{R_1}\right) \quad \text{for} \quad R \gtrsim R_c,$$

$$\varepsilon = A \exp\left(\frac{R - R_c}{R_2}\right) \quad \text{for} \quad R < R_c.$$

The following two data sets are taken to fit the upper and lower limit in the longitude distribution (Figure 2a): $A = 1.4 \times 10^{-25}$ or 1.1×10^{-25} photon s^{-1} cm^{-3}; $R_c = 6$ or 6.5 kpc, $R_1 = 2$ kpc, and $R_2 = 20$ or 5 kpc, respectively. As the gamma-ray spectrum is proportional to E^{-2} (Mayer-Hasselwander *et al.*, 1982), one can estimate an 'effective' energy of 0.39 GeV and calculate the gamma-ray surface emissivity in $L_\odot$ pc^{-2} (dashed lines in Figure 3). The dashed lines in Figure 3 reproduce the overall profile of the narrow component shown by the dash-dot curves in Figure 2a. To produce the longitude profile of the narrow and wide component, a scale-length of $R_1 = 3$ kpc should be taken instead of 2 kpc.

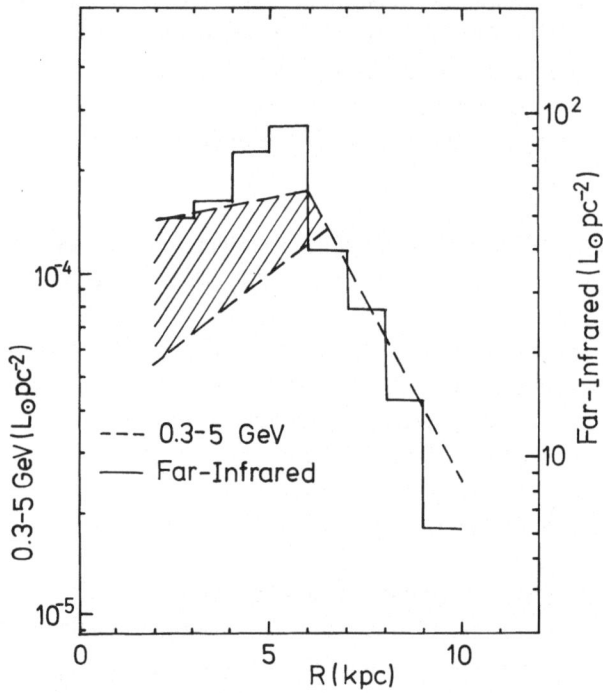

Fig. 3. Dashed curve with shaded area: distribution of surface emissivity for the narrow component of gamma radiation at the 0.3–5 GeV range; solid line: distribution of surface emissivity of far-infrared radiation (Gispert *et al.*, 1982).

The solid line in Figure 3 shows the radial distribution of the total wavelength-integrated FIR emissivity for the continuous component which is estimated from the emission rate in the 114–196 μm region (Gispert *et al.*, 1982). One finds that both the FIR- and gamma-ray distributions have peaks in the 5–6 kpc region.

4. Discussion

The gamma-ray emissivity in the 0.3–5 GeV range could be due to interactions between CR and gas, which means that the gamma-rays trace the product of CR density and gas density. According to Gordon and Burton (1976) and Solomon and Sanders (1979), the ratio of H_2 gas density in the 5–6 kpc region to that in the solar neighborhood is 5 and the H I gas density ratio is 0.8, while the gamma-ray emissivity ratio obtained here for the narrow component is 6. This suggests that the narrow component of gamma radiation results from the interaction between H_2 gas and CR, while the wide component results from the interaction between H I gas and CR and that CR is uniformly distributed in the Galaxy. One can estimate the gamma-ray production rate at the 0.3–5 GeV range, using the distributions of both H_2 gas and volume emissivity of gamma radiation derived in the previous section. The volume emissivity of gamma radiation and H_2 gas density in the 5–6 kpc region are 1.4×10^{-25} photon s^{-1} cm^{-3} and $n(H_2) = 2$ cm^{-3} (Gordon and Burton, 1976), respectively, from which one can obtain the gamma-ray production rate of 7×10^{-26} photon s^{-1} H_2^{-1}, equal to the gamma-ray production rate in the solar neighborhood (Lebrun $et\ al.$, 1982). Taking into account the hydrogen gas density re-analyzed by Blitz and Shu (1980) in molecular as well as atomic form, the gamma-ray production rate becomes 2 times higher than the local value. This indicates that there might be CR enhancement in th 5–6 kpc region which seems plausible as the distribution of young supernova remnants, which can produce and accelerate CR particles, has a peak in the same region (Kodaira, 1974).

The total FIR emissivity is determined by the product of stellar radiation and the influence of the dust distribution. Drapatz (1979) has taken the following spatial distributions of stars: the distribution of OB stars is proportional to that of giant H II-regions which has a peak in the 5–6 kpc region. All other stars (AF stars and late-type stars) follow Schmidt's model. Assuming that the outer parts of cloud material, which is illuminated by the external radiation field, contribute to the FIR emission, he found that the mean dust density averaged over clouds and intercloud regions is of moderate influence (intensity $\sim \sqrt{\text{density}}$). Additionally, if the enhancement of late-type stars in the 5–6 kpc region is taken into account (Maihara $et\ al.$, 1978; Hayakawa $et\ al.$, 1981), the FIR emissivity distribution can be explained.

There is a prominent FIR peak at the Galactic Center, while no gamma-ray peak can be seen. Audouze $et\ al.$ (1979) have shown that the amount of gas of $1.3 \times 10^8\ M_\odot$ they obtain for the nuclear disk (235 pc in radius, 115 pc in thickness) is not large enough to make a prominent gamma-ray peak and that the CR density at the Galactic Center is not significantly higher than the local density. This value might even be lower. Blitz and Shu (1980) and Oda $et\ al.$ (1979) have given the value of $N(^{13}CO)/A_v = 1.2 \times 10^{16}$ cm^{-2} mag^{-1} and $A_v = 10$ mag for the nuclear disk, respectively. If one takes $N(H_2)/N(^{13}CO) = 3.7 \times 10^5$ (Audouze $et\ al.$, 1979) and also takes into account the contribution of He gas to the total mass of gas (40%), a value of $4 \times 10^7\ M_\odot$ is obtained for the mass of gas within the nuclear disk. From the FIR flux from the Galactic Center, the mass of dust grains is estimated to be $3 \times 10^5\ M_\odot$ (Maihara $et\ al.$, 1981) which is consistent with the mass of gas, taking into account a reasonable mass ratio of gas to dust of approximately 100.

Near $l = 356°$, the FIR brightness is fairly low. This is a region where an anomalous emission feature has been found in the 2.4 μm map (Okuda $et\ al.$, 1977; Ito $et\ al.$, 1977).

This can be understood in terms of a deficiency of interstellar dust in this direction. This interpretation is supported by Hamajima *et al.* (1981). They investigated the spatial distribution of M-type stars near the Galactic Center with objective-prism plates in the 6800–8000 Å wavelength region on a Schmidt telescope. They found that the surface number density of these stars is enhanced and concluded that the enhancement is due to the low interstellar extinction in this direction. This fact supports the idea that the gamma-ray enhancement near $l = 356°$ is not caused by diffuse emission but by a localized source, i.e. 2CG 356 + 00 which has been seen only in one of four observations (Swanenburg *et al.*, 1981).

Acknowledgements

The author thanks Drs S. Drapatz, H. Zinnecker, G. Kanbach, H. A. Mayer-Hasselwander, and M. Gottwald for useful discussions and reading a manuscript.
The editors thank W. Brinkmann for assistance in evaluating this paper.

References

Audouze, J., Lequeux, J., Masnou, J.-L., and Puget, J.-L.: 1979, *Astron. Astrophys.* **80**, 276.
Blitz, L. and Shu, F. H.: 1980, *Astrophys. J.* **238**, 148.
Drapatz, S.: 1979, *Astron. Astrophys.* **75**, 26.
Drapatz, S.: 1981, in C. G. Wynn-Williams and D. R. Cruikshank (eds.), 'Infrared Astronomy', *IAU Symp.* **96**, 261.
Gispert, R., Puget, J. L., and Serra, G.: 1982, *Astron. Astrophys.* **106**, 293.
Gordon, M. A. and Burton, W. B.: 1976, *Astrophys. J.* **208**, 346.
Hamajima, K., Ichikawa, T., Ishida, K., Hidayat, B., and Raharto, M.: 1981, *Publ. Astron. Soc. Japan* **33**, 591.
Hayakawa, S., Matsumoto, T., Murakami, H., Uyama, K., Thomas, J. A., and Yamagami, T.: 1981, *Astron. Astrophys.* **100**, 116.
Ito, K., Matsumoto, T., and Uyama, K.: 1977, *Nature* **265**, 517.
Kniffen, D. A. and Fichtel, C. E.: 1981, *Astrophys. J.* **250**, 389.
Kodaira, K.: 1974, *Publ. Astron. Soc. Japan* **26**, 255.
Lebrun, F., Bignami, G. F., Buccheri, R., Caraveo, P. A., Hermsen, W., Kanbach, G., Mayer-Hasselwander, H. A., Paul, J. A., Strong, A. W., and Wills, R. D.: 1982, *Astron. Astrophys.* **107**, 390.
Maihara, T., Oda, N., Sugiyama, T., and Okuda, H.: 1978, *Publ. Astron. Soc. Japan* **30**, 1.
Maihara, T., Oda, N., and Okuda, H.: 1979, *Astrophys. J. Letters* **227**, L129.
Maihara, T., Oda, N., Shibai, H., and Okuda, H.: 1981, *Astron. Astrophys.* **97**, 139.
Mayer-Hasselwander, H. A., Bennett, K., Bignami, G. F., Buccheri, R., Caraveo, P. A., Hermsen, W., Kanbach, G., Lebrun, F., Lichti, G. G., Masnou, J. L., Paul, J. A., Pinkau, K., Sacco, B., Scarsi, L., Swanenburg, B. N., and Wills, R. D.: 1982, *Astron. Astrophys.* **105**, 164.
Mezger, P. G.: 1978, *Astron. Astrophys.* **70**, 565.
Nishimura, T., Low, F. J., and Kurtz, R. F.: 1980, *Astrophys. J. Letters* **239**, L101.
Oda, N., Maihara, T., Sugiyama, F., and Okuda, H.: 1979, *Astron. Astrophys.* **72**, 309.
Okuda, H.: 1981, in C. G. Wynn-Williams and D. P. Cruikshank (eds.), 'Infrared Astronomy', *IAU Symp.* **96**, 247.
Okuda, H., Maihara, T., Oda, N., and Sugiyama, T.: 1977, *Nature* **265**, 515.
Price, S. D.: 1981, *Astron. J.* **86**, 193.
Solomon, P. M. and Sanders, D. B.: 1979, in P. M. Solomon and M. G. Edmunds (eds.), *Giant Molecular Clouds in the Galalxy*, Pergamon Press, Oxford, p. 41.
Swanenburg, B. N., Bennett, K., Bignami, G. F., Buccheri, R., Caraveo, P. A., Hermsen, W., Kanbach, G., Lichti, G. G., Masnou, J. L., Mayer-Hasselwander, H. A., Paul, J. A., Sacco, B., Scarsi, L., and Wills, R. D.: 1981, *Astrophys. J. Letters* **243**, L69.

OBSERVATIONS OF SUPERNOVA EXPLOSIONS
AND COSMIC RAYS*

G. G. C. PALUMBO

Istituto TE.S.R.E.-CNR, Via De' Castagnoli, 1, 40126 Bologna, Italy

Abstract. A summary of coordinated observations at various frequencies of two type II and three type I SNe is given. Since type II events emit radio and X-ray radiation in the early phases it is shown how one has data to estimate the Cosmic Ray output from these SNe.

1. Introduction

For at least fifty years now the origin of the Cosmic Radiation, i.e. particles, either protons or nucleons, whose energy may range from a few GeV up to 10^{20} eV has been associated, one way or another, with Supernova (SN) explosions. To my knowledge it was Professor F. Zwicky, the promotor of the systematic optical study of SNe, who, in 1934, first associated SNe and Cosmic Rays (CR) (see Figure 1 in Kirshner, 1981). Since SN explosions are so spectacular and energetic, involving a whole star of a few solar masses (Type I SNe) or of several (> 10) solar masses (Type II SNe) with a release of some 10^{50} erg, it is not surprising that high energy photons, i.e. X and gamma rays may also be produced. Actually a whole zoo of strange phenomena such as neutrino bursts, gravitational waves, gamma-ray bursts, X-ray flashes, radio wave pulses have also been predicted and from time to time experiments have hinted at their presence but not as yet confirmed it. Because of the specific theme of this paper however, we shall not concern ourselves with any of these predictions but rather concentrate on high energy particles and photons.

The SN as such lasts only a few months. The ejecta, which expand for some 10^4 yr in interstellar space, quickly become optically very faint but brighten up at radio and X-ray frequencies. These are the so called Supernova Remnants (SNR). The SN explosion, by its very nature, is typically an event rather than an astronomical object (Arnett, 1982) and the time behaviour of such an event contains more information than a steady source. On the contrary, the SN progenitor star as well as the SNR are typical astronomical objects.

On a human scale SNe are rather rare events. In our Galaxy the last SN observed was Kepler's SN which occurred in 1604 AD and the most recent one is probably Cassiopea A which appears to have exploded around 1700 AD but no historical report of it has been found.

Although the frequency of SNe (f_{SN}) varies from galaxy to galaxy according to mass and morphological type and is not the same for type I and II events (see Tammann,

---

* Proceedings of the XVIII General Assembly of the IAU: *Galactic Astrophysics and Gamma-Ray Astronomy*, held at Paras, Greece, 19 August 1982.

Space Science Reviews **36** (1983) 293–304.   0038–6308/83/0363–0293$01.80.

1978, and references therein) frequencies generally assumed are either one SN every 50 or one SN every 100 years per galaxy. We must point out that it has been shown that the error associated to the best estimate of f_{SN} is 50% or larger (Craven et al., 1978). Because of their rarity the study of SN in modern times inevitably has been based on events which occurred in nearby galaxies. One therefore is forced to extrapolate what one observes in SNRs in our own Galaxy to those far away objects whose remnants are generally too faint to be observed even one month after the explosion.

The basic reasoning behind any CR production from SNe is, as I mentioned, at least 50 years old.

The rate Q_{CR} of CR production in our Galaxy is:

$$Q_{CR} = \frac{N_G V_G}{\tau_G} = 10^{43} \text{ s}^{-1},$$

where $N_G = 10^{-10}$ is the number of CR particles per cm^{-3} observed near Earth, $V_G = 6 \times 10^{68}$ cm^3 is the volume of our Galaxy, assuming there is a halo surrounding a disk and τ_G is the travel time of CR particles if trapped in the Galaxy.

$$\tau_G = \frac{x}{c\rho_G} = 3 \times 10^{15} \text{ s},$$

with $x = 3$ g cm^{-2} thickness of matter to be traversed to account for the measured intensity of relativistic light nuclei (Li, Be, B) and $\rho_G = 3 \times 10^{-26}$ g cm^{-3} is the mean density of the Galaxy; c is the speed of light.

From the study of SNRs (to be precise mainly the Crab Nebula) it is known that the number of relativistic electrons of energy above 100 MeV is $Ne = 2 \times 10^{49}$. A round figure universally adopted for SNe however is $Ne(\text{SN}) = 10^{50}$.

From the above figure one can estimate the number n_e of relativistic electrons in the Galaxy supplied by SNe. One obtains

$$n_e = \frac{N_e \tau_G f_{SN}}{V_G} = 10^{-12} \text{ cm}^{-3},$$

where all symbols are explained above (the classical calculation assumes one SN every 50 years in our Galaxy, as we have done here). The value obtained coincides with the number of electrons required to justify the radio emission observed from our Galaxy.

A further assumption, based on the similarity of the radio spectral indeces of SNRs (~ 0.7) which imply integral power exponents of their electron spectra of ~ 1.4 and the spectrum of galactic electrons, is that the ratio k between relativistic protons (N_p) and relativistic electrons (N_e) in the Galaxy is 100.

With the numbers derived above at hand we can estimate the rate of CRs from SNe $Q_{CR}(\text{SN}) = KN_e f_{SN} = 10^{43}$ s^{-1} which coincides precisely with the value Q_{CR} derived from observations of the CR flux.

This argument, changed in the details, amplified and occasionally turned around is still the main point on which the statement 'SNe are the origin of CRs' is based. Detailed theoretical studies of SN explosions substantiate this statement and it has been shown (Colgate, 1975) that assuming a given model not only the energy requirements are met but the CR spectrum and composition are justified. From the observational point of view however, the only available data on SNe have been optical light curves and spectra.

In the past four years SNe in external galaxies have been intensively studied with a new approach. Observations were coordinated among observatories to enable astronomers to follow each event with any available instrument for as long as possible. A wealth of new informamtion has therefore been collected and now we have ultraviolet spectra as well as optical spectra of SNe. We know that some SNe become radio emitters shortly after the explosion and that some also emit X-ray photons. Additional infrared fluxes and *UBV* photometry are now available to give a detailed set of physical parameters concerning each event. In what follows I shall concentrate mainly on these new results. I shall try to review the new picture which is emerging from the data laboriously obtained pertaining to: the nature of the progenitor star, the physical processes observed at work during the explosion, and the processes involved in the formation of the young SNRs. With such detailed data it should be possible to compare the copious existing theories with observations.

It should also be possible to say something more quantitative about the claimed connection between SNe and CRs. For studies on SNe and SNRs previous to this coordinated effort there are a number of recent, excellent review articles (Meyerott and Gillespie, 1980; Chevalier, 1981; Wheeler, 1980, 1981; Wefel, 1981; Kirshner, 1981) to which the reader may refer.

2. The Data

Since April 1979 five SNe have been studied using a different approach from the traditional one, i.e. taking spectra of the event with an optical telescope. Besides optical spectra coordinated observations at all available frequencies of the electromagnetic spectrum were organized. A followup of the events was also assured as far as was feasible. In Table I the observations are summarized and references to the papers already published are given.

From all these observations the following new information has emerged:

(a) From the ultraviolet spectra, taken by the International Ultraviolet Explorer (IUE) satellite between 1150 and 3500 Å, various emission and absorption lines are present. This is the first time that such information has been acquired on SNe in the UV portion of the spectrum. As we shall see later one can derive from it the physical conditions of the material surrounding the SN star, stringent requirements on the nature of the stellar progenitor and data on abundances relevant to the problem of nucleosynthesis. A comprehensive review of UV data was given by Panagia (1980, 1982). We shall quote, in what follows, only the main results of relevance to our discussion.

(b) The two type II SNe are now known to be radio emitters. From the point of view of particle production radio information is essential. The radio light curves at different frequencies and spectral data obtained at the Very Large Array (VLA) will be used and

TABLE I

Data on the five SNe observed at many frequencies

SN	Ty	mag. (GAL)	GAL (NGC)	R.A. (1950)	Decl. (1950)	mag. (SN)	v (km s^{-1})	Other SNe	OPT. Sp + Ph	UV LW	UV SW	VLA	X	IR
1979c	II	12.23	4321 (M100)	12°20'22".96	+16°05'56".5	10.6	1620	1901b 1914a 1959e	yes (a)	18	11 (c)	det. (e)	not (f)	yes
1980k	II	11.78	6946	20°34'26".7	+59°55'56".5	10.5	46	1917a 1939c 1948b 1968d 1969p	yes (b)	24	13 (d)	det. (e)	det. (g)	yes (h)
1980n	I	12.62	1316 Fornax A	03°20'47"	−37°23'06"	10.1	1774	1981d	yes	7	1	Not Obs.	Not Obs.	yes (h)
1981b	I	12.00	4536	12°31'54"	+02°27'42"	11.2	1927		yes	4	2	Not Det.	Not Obs.	yes (h)
1982	I	13.50	2268	07°00'48"	+84°27'48"	12.2	2261		yes	2	0	Not Obs.	Not Obs.	Not Obs.

a Panagia et al. (1980)
Barbon et al. (1982a)
b Barbon et al. (1982b)

c Panagia et al. (1980)
Fransson et al. (1982)
Benvenuti et al. (1982)

d Fransson (1982)
e Weiler et al. (1981a)
Weiler et al. (1981b)

f Palumbo et al. (1981)
g Canizares et al. (1982)
h Elias et al. (1981)

discussed in what follows. For details about radio measurements and their interpretations see the original papers (Weiler *et al.*, 1981a, b).

(c) At least one type II SN is an X-ray emitter. Canizares *et al.* (1982) detected SN 1980k with the Einstein Observatory some 50 days after maximum light. X-ray emission from SN events has been established beyond any doubt for the first time. The relevance of this finding will be discussed later in connection to the fact that SN 1979c (also a type II) apparently was not an X-emitter. For a review of UV radio and X-ray data from SNe see also Palumbo (1982).

(d) Infrared fluxes have been carefully measured for one of the type I SNe, (Elias *et al.*, 1981). These data turned out to be of paramount importance in interpreting the UV and spectral continuum (Palumbo, 1982). With all these new data at hand it has been possible to tie all information together and derive a coherent picture of the physical conditions of the observed events.

(e) No gamma rays were observed by COS-B from SN 1980k (Cavallo *et al.*, 1981b).

For convenience of presentation I shall deal with the SN types separately.

2.1. TYPE I SNe

Three events have so far been observed. While optical and UV spectra are available, no radio detection has, so far, been announced; however, observations at the VLA on regular one month intervals are continuing. Because the Einstein Observatory operation has ended, it has been impossible to search for X-ray emission from any of these objects. Furthermore, since there were various pointing and brightness constraints, the UV information acquired is mostly around the period of peak luminosity. From limited statistics based on three events we can confirm the well known claim from old optical data i.e. type I SNe are identical events and therefore we conclude that they are standard candles. The behaviour is identical in the UV (Figure 1) (Panagia, 1982) as well as in the IR (Elias *et al.*, 1981). Since the UV and the IR radiation is lower than the black-body extrapolation of the optical continuum (9000 K) Panagia has suggested that the optical features are emission bands above a general continuum at 6000 K which fit both UV and IR reasonably well.

While no definite line identification has yet been completed for the UV spectra obtained, a number of 'bands' are apparent on all three events with multiplets of Fe I, Fe II, and Mg II.

2.2. TYPE II SNe

The two events studied SN 1979c and SN 1980k, were type II events of subclass L i.e. displaying a linear decline in their light curve (Barbon *et al.*, 1982a, b). The data on SN 1979c have been extensively presented in Panagia *et al.* (1980) and discussed in Fransson *et al.* (1982).

The data on SN 1980k have not yet appeared in detail. Properties common to both events are:

(i) Both SNe show a UV excess continuum shortward of 2000 Å.

(ii) Both SNe became radio emitters shortly after the explosion; SN 1979c one year

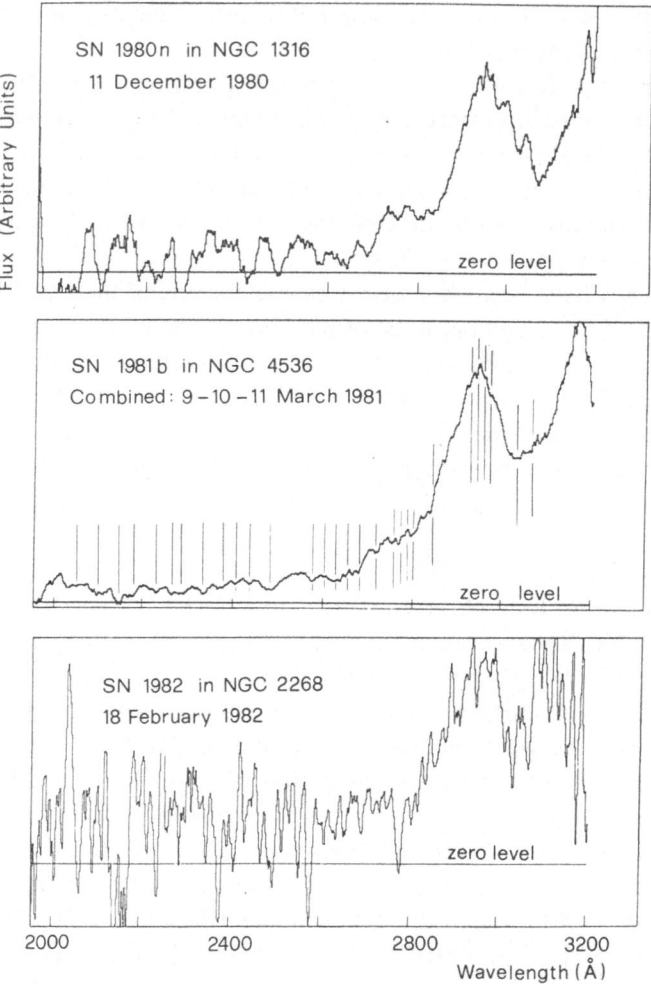

Fig. 1.   Ultraviolet spectra taken with IUE of SN 1980n, SN 1981b, and SN 1982 in NGC 2268.

later, SN 1980k one month later. From the so far best studied event SN 1979c, the following physical parameters have been derived (Panagia, 1982; Fransson *et al.*, 1982).

(a) The ultraviolet emitting region is a shell of material surrounding the photosphere extending from it to 1.3 photospheric radii.

(b) In this region the electron density varies from 2×10^9 electrons cm^{-3} to 10^{10} electron cm^{-3} and their temperature ranges from 10^4 to 2×10^4 K correspondingly.

(c) The shell expands at a velocity $v \sim 8400$ km s^{-1} comparable to the velocity of the photosphere of ~ 9200 km s^{-1} measured from Hα and Mg II at 2800 K.

Although the case of SN 1980k is not as clear because its spectra are severely contaminated by interstellar lines, a preliminary analysis of the data (Fransson, private communication) reveals that this second type II SN appears to have similar properties to SN 1979c.

Let us now look in some detail at the radio emission from both SNe. To put the radio SNe in perspective, Weiler *et al.* (1981) compare peak fluxes at 6 cm to the flux of the known strong radio source; SNR Cas A. SN 1979c was 300 times brighter than Cas A while SN 1980k was 9 times brighter. For sake of comparison the Crab now is one tenth of Cas A at 6 cm. Weiler *et al.* also present these new radio sources in a different fashion; i.e. they compare the flux in Jy of various objects as if they all were at a distance of 10 kpc. It turns out that while Cas A is of the order of 70 Jy and the Crab 26 Jy, SN 1980k is 650 Jy and SN 1979c above 2×10^4 Jy. This comparison alone provides two important pieces of information: (a) SNe, at least type II SNe, are, at the outburst, extremely powerful radio sources; (b) it is unlikely a SN has occurred in our Galaxy and been missed since radio telescopes have been in operation (i.e. 20–30 yr). This is a hard argument to fight for those who like to keep the SN frequency in our Galaxy very high and claim SNe are frequent but missed because obscured.

Weiler *et al.* (1981b) have presented radio light curves of the two events. Making use of the same data plus some additional points (Weiler, private communication) we have obtained the hand fitted radio light curves given in Figure 2. Minor bumps and humps were ignored since they might not have real physical significance. For comparison I have also plotted on the same scale the optical light curve. One should note in Figure 2 that the flux scale is linear and expanded for SN 1980k, intrinsically a weaker source. No upper limits have been included although they were estimated before the events began to be detected. The 15 GHz ($\lambda = 2$ cm) curve is based on only three data points affected by large errors and therefore little significance should be attached to it.

The high brightness temperature ($T > 9 \times 10^9$ K) and rather steep spectrum indicate that radio emission takes its origin from non thermal processes. A complete theory to account for the observed radio emission has not yet been given; however a simple model consisting of a uniformly filled sphere of fully ionized hydrogen surrounding a central radio source fits the data reasonably well if $10^{-3} M_\odot$ of hydrogen are present. This model accounts for both the observed time delay between 6 and 20 cm data as well as the amount of material surrounding the SN ejected by the pre SN star while in the super giant stage as stellar wind, a value obtained from the study of the UV spectrum i.e. $10^{-2} M_\odot$ (Panagia *et al.*, 1980).

This model apparently does not fit SN 1980k very well. The time delay between 6 and 20 cm cannot be accounted for. Weiler *et al.*(1981b) point out that because of the early turn-on only $\sim 10^{-4} M_\odot$ of ejecta are accounted for. Fransson (1982) is able to reconcile the apparent absence of lines in the UV spectra; the early radio turn-on and strong X-ray emission from SN 1980k with the presence of UV lines, late radio turn on and no X-rays from SN 1979c. His argument is based on the fact that if the mass loss from the stellar SN progenitor varies from $10^{-5} M_\odot$ yr^{-1} to $10^{-4} M_\odot$ yr^{-1} dramatic effects are expected.

At high mass loss a sharp cut-off at 3 keV is expected in the X-ray flux whereas the wind is almost transparent to X-ray photons for $\dot{M} = 10^{-5} M_\odot$ yr^{-1}. Transparency is achieved via the increased degree of ionization of the wind material as $\dot{M}$ decreases. Fransson computes the number of ionizing photons needed to completely ionize He in

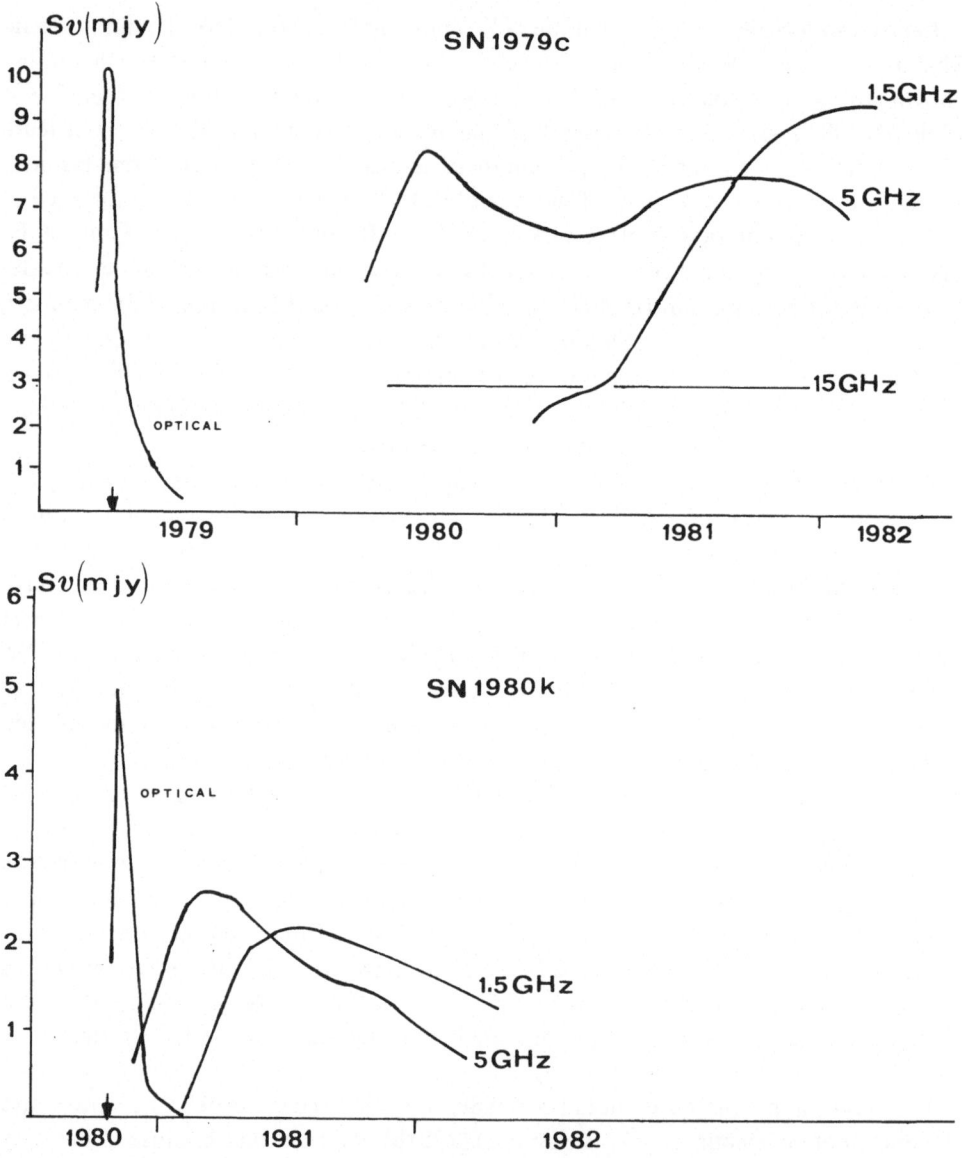

Fig. 2.   Hand fitted radio light curves of the type II SN 1979c and SN 1980k. Extrapolation of data by
Weiler *et al.* (1981b) and Weiler private communication. For comparison a sketch of the optical light curves
is also drawn. Note: the flux scale for SN 1980k is expanded. (1.5 GHz = 20 cm; 5 GHz = 6 cm;
15 GHz = 2 cm.)

the wind. In the case of SN 1980k the number of ionizing photons exceeds the number
of recombinations, He^+ becomes ionized and a He^{++} zone is formed.

These arguments nicely account for stronger UV lines, late radio turn on (~ 400 days
at $\lambda = 6$ cm) and absence of X-ray photons in SN 1979c as well as for weak or absent
UV lines, early radio turn on (~ 38 days at $\lambda = 6$ cm) and X-ray detection in SN 1980k
within one single theoretical framework.

It follows that the mass loss rate of SN 1980k must have been 50 times less than that of SN 1979c.

So far most of the data collected, other than optical and UV, is about type II SN events. This seems unfortunate since, as far as CR production is concerned type I SNe have always been the theorist's favourite. Nonetheless type II events have a total luminosity $L_T \sim 2.4 \times 10^{43}$ erg s^{-1} and a radiation output $E_{rad} \sim 7 \times 10^{49}$ erg. Furthermore, when it comes to CR energy production everyone seems to forget about SN types and simply multiplies a reasonable estimate by a reasonable rate. It appears therefore worth while to evaluate what the CR output might be for type II events using the available information.

3. Cosmic Rays from Type II SNe

The radio detection of SNe in the very early phases of the expansion is of particular interest for the CR production problem. Estimate can be made of the magnetic field strength and the total relativistic particle energy content (see Ginzburg and Syrovatskii, 1964). For the young remnant this has been done by Palumbo and Cavallo (1981). For SN 1979c for instance $B \sim 0.5$ G and $W_{CR} \sim 10^{48}$ erg if the ratio $k = W_p/W_e = 100$ and $B = 0.12$ G and $W_{CR} \sim 7.5 \times 10^{46}$ erg if $k = 1$. From the simple estimate it appears that CRs are already present in the young remnant but about two orders of magnitude below the canonical requirement of 10^{50} erg per SN. Under the assumption that the turnover in the radio spectrum at 5 GHz where the measured maximum flux is 8.3 mJy (see Figure 2) is due to syncrotron reabsorption, from the well known formula from syncrotron theory giving the frequency at which the optical depth τ equals one i.e.

$$v_{\tau=1} \sim 16 B^{1/5} \theta^{-4/5} F_v^{2/5},$$

where v is given in MHz, B in gauss, θ, the diameter of the radio source in arc sec, and F_v its flux in flux units, one can again estimate the magnetic field.

One can also estimate the total energy in magnetic field

$$W_B = 8.86 \times 10^{15} v^{10} \theta^8 F_v^{-4} R^3$$

which, in a regime of equipartition equals the cosmic-ray energy content.

Here v is the frequency of observation in GHz, θ the source angular diameter in arc sec, F_v the flux at the frequency v in flux units, and R the radius of the young SNR in pc.

Assuming equipartition one therefore derives $W_{CR} \sim 6 \times 10^{50}$ erg when $v = 5$ GHz, $F_v = 8.3$ mJy and the SNR at ~ 16 Mpc had a radius of $\sim 6 \times 10^{16}$ cm, i.e. after one year. This value however depends on high power exponents of poorly known parameters. A more realistic, less parameter dependent estimate comes from standard formulae (Ginzburg and Syrovatskii, 1964)

$$W_{CR}(k) = 0.19[kA(\Gamma, v)F_v D^2]^{4/7} d^{9/7}.$$

Here $A(\Gamma, v)$ is a function depending on the radio spectral index $\Gamma = \frac{1}{2}(\gamma - 1)$ and on the frequency of observation v, F_v as above, D the source distance and $d = 2R$ its diameter.

For SN 1980k ($k = 1$) one obtains $W_{CR} \sim 7 \times 10^{44}$ erg at about 140 days after maximum.

In order to compare these young SNRs to the old, well studied, galactic SNRs Cavallo *et al.* (1981a) have made use of a well known empirical relationship derived from SNRs observations in our Galaxy.

It appears in fact that the surface brightness Σ of a SNR is proportional to the SNR diameter d ($\Sigma d^{-\alpha}$) and not to its age. Since

$$\Sigma \propto \frac{F_v}{\theta^2} = \frac{F_v}{4R^2} D^2 \, ,$$

where R is the radius and D the distance of SNR as above it follows $F_v D^2 \propto R^{2-\alpha}$.

Combining this relationship with the W_{CR} above for a given radio source they are able to show that

$$W_{CR}(k = 1) \propto d^{(17 - 4\alpha)/7} \, ,$$

where, again empirically, $3 < \alpha < 4$ (Milne, 1979; Carswell and Lerche, 1979). The CR energy content, therefore, is $W_{CR} = k^{4/7} W_{CR}$ ($k = 1$).

To fulfil the CR energy requirement for our Galaxy, i.e.

$$W_{CR} \sim 10^{48} f_{SN}^{-1} \text{ erg yr}^{-1}$$

one requires $k = (0.2 f_{SN}^{-1})^{7/4}$.

The SN frequency, although uncertain, is $30 < f_{SN}^{-1} < 100$ yr.

Therefore k ranges between 20 and 200. The value $k = 100$ generally assumed seems to be well justified. The corresponding final relativistic energy is $\sim 10^{50}$ erg or about 10% of the initial average injection energy estimated to input a SN explosion. This indicates that equipartition is never achieved.

Another point worth noticing is that $W_{CR} \propto d^{0.4}$.

In a diameter versus CR energy content plot (see Cavallo *et al.*, 1981a; Cavallo, 1982) this is a straight line on which, within the uncertainties, at the lower end lie the young radio SNRs (SN 1979c, SN 1980k, and SN 1970q in M101, the only other young SNR detected as radio emitter (Gottesman *et al.*, 1972; Allen *et al.*, 1976)). On the upper end the old SNRs for which distances are known populate the line region. This implies that CR injection occurs throughout the expansion of the remnant but the largest energy output occurs towards the end of the expansion phase.

If π° decay is the mechanism by which gamma rays are produced the same reasoning applies to gamma photons which accounts for the non detection of SN 1980k by COS-B (Cavallo *et al.*, 1981b).

A further implication from the d versus W_{CR} plot is that because of continuous injection there seems to be no need for a pulsar for accelerating particles. As Cavallo (1982) has shown there is ground to believe that the dominant acceleration mechanism for relativistic particles operates by shock fronts as theoretically predicted by Bell (1978).

4. Concluding Remarks

Most of the arguments so far put forward by various authors are quite simple and not conclusive. However no manipulation of well known figures has been attempted and a consistent set of parameters has already been obtained from the data about the two type II SNe studied.

The results seem encouraging enough to stimulate further, more detailed work. The point I want to stress here is that thanks to continuous coordinated observations of relatively few objects covering as much of the electromagnetic spectrum as possible for as long as the event is visible one can obtain sufficient data to reach a consistent set of parameters. With these one can eventually choose models, identify radiation production mechanisms and finally sort out the SN phenomenon.

The effort to observe one event well rather than several poorly has to be continued to confirm the findings on the two type II SNe (and eventually to catch a type II without linear decline) and provide sufficient data about type I. The latter may still provide surprises also for CR particle production.

Acknowledgement

The editors thank G. Lichti for assistance in evaluating this paper.

References

Allen, R. J., Goss, W. M., Ekers, R. D., and De Bruyn, A. G.: 1976, *Astron. Astrophys.* **48**, 253.
Arnett, W. D.: 1982, *Astrophys. J.* **254**, 1.
Barbon, R., Ciatti, F., Rosino, L., Ortolani, S., and Rafanelli, P.: 1982a, *Astron. Astrophys.* **116**, 43.
Barbon, R., Ciatti, F., and Rosino, L.: 1982b, *Astron. Astrophys.* **116**, 35.
Bell, A. R.: 1978, *Monthly Notices Roy. Astron. Soc.* **182**, 147 and 443.
Benvenuti, P., Sanz Fernandez De Cordoba, L., Wamsteker, W., Macchetto, F., Palumbo, G. G. C., and Panagia, N.: 1982, ESA Special Report, 1046.
Canizares, C. R., Kriss, G. A., and Feigelson, E. D.: 1982, *Astrophys. J. Letters* **253**, L17.
Cavallo, G.: 1982, *Astron. Astrophys.* **111**, 368.
Cavallo, G., Palumbo, G. G. C., and Vettolani, G.: 1981a, *17th Int. Cosmic Ray Conf., Paris* **9**, OG 4–14, 230.
Cavallo, G., Caraveo, P. A., and Bignami, G. F.: 1981b, *17th Int. Cosmic Ray Conf., Paris* **9**, XG5, 1–6, 80.
Caswell, J. L. and Lerche, I.: 1979, *Monthly Notices Roy. Astron. Soc.* **187**, 201.
Chevalier, R. A.: 1981, *Fund. Cosmic Phys.* **7**, 1.
Colgate, S. A.: 1975, *Origin of Cosmic Rays*, D. Reidel Publ. Co., Dordrecht, Holland, pp. 425 and 447.
Craven, P. G., Cavallo, G., and Palumbo, G. G. C.: 1978, *Astron. Astrophys.* **64**, 87.
Elias, J. H., Frogel, J. A., Hackwell, J. A., and Persson, S. E.: 1981, *Astrophys. J. Letters* **251**, L13.
Fransson, C.: 1982, *Astron. Astrophys.* **111**, 140; and Paper II, in press.
Fransson, C., Benvenuti, P., Gordon, C., Hempe, K., Palumbo, G. G. C., Panagia, N., Reimers, D., and Wamsteker, W.: 1983, *Monthly Notices Roy. Astron. Soc.*, in press.
Ginzburg, V. L. and Syrovatskii, S. I.: 1964, *The Origin of Cosmic Rays*, Pergamon Press, New York.
Gottesman, S. T., Broderick, J. J., Brown, R. L., and Balick, B.: 1972, *Astrophys. J.* **174**, 383.
Kirshner, R. P.: 1981, *17th Int. Cosmic Ray Conf., Paris* **12**, 119.
Meyerott, R. E. and Gillespie, G. H. (ed.): 1980, *Supernova Spectra. Atomic and Spectroscopic Data Needs*, American Institute of Physics.
Milne, D. K.: 1979, *Australian J. Phys.* **32**, 83.
Palumbo, G. G. C.: 1983, *Adv. Space Res.* **2**, No. 9, p. 51.

Palumbo, G. G. C. and Cavallo, G.: in G. Setti, G. Spada, and A. Wolfendale (eds.), 'Origin of Cosmic Rays', *IAU Symp.* **94**, 51.

Palumbo, G. G. C., Maccacaro, T., Panagia, N., Vettolani, G., and Zamorani, G.: 1981, *Astrophys. J.* **247**, 484.

Panagia, N.: 1980, *Proc. Second European IUE Conf.*, ESA SP-157, p. XXVII.

Panagia, N.: 1982, *Proc. Third European IUE Conf.*, ESA SP-176, p. 31.

Panagia, N., Vettolani, G., Boksenberg, A., Ciatti, F., Ortolani, S., Rafanelli, P., Rosino, L., Gordon, C., Reimers, D., Hempe, K., Benvenuti, P., Clavel, J., Heck, A., Penston, M. V., Macchetto, F., Stickland, D. J., Bergeron, J., Tarenghi, M., Marano, B., Palumbo, G. G. C., Parmar, A. N., Pollard, G. S. W., Sanford, P. W., Sargent, W. L. W., Sramek, R. A., Weiler, K. W., and Matzik, P.: 1980, *Monthly Notices Roy. Astron. Soc.* **192**, 861.

Tammann, G. A.: 1978, *Mem. Soc. Astron. Italiana* **49**, 315.

Wefel, J. P.: 1981, in G. Setti, G. Spada, and A. Wolfendale (eds.), 'Origin of Cosmic Rays', *IAU Symp.* **94**, 39.

Weiler, K. W., Van Der Hulst, J. M., Sramek, R. A., and Panagia, N.: 1981a, *Astrophys. J. Letters* **243**, L151.

Weiler, K. W., Van Der Hulst, J. M., Sramek, R. A., and Panagia, N.: 1981b, in M. J. Rees and R. J. Stoneham (eds.), *Supernovae: A Survey of Current Research*, NATO Workshop, D. Reidel Publ. Co., Dordrecht, Holland, p. 281.

Wheeler, J. C. (ed.): 1980, *Texas Workshop on Type I Supernova*, Austin.

Wheeler, J. C.: 1981, *Rep. Prog. Phys.* **44**, 85.

GAMMA-RAY LINE ASTRONOMY*

R. RAMATY

Laboratory for High Energy Astrophysics, NASA/Goddard Space Flight Center, Greenbelt, MD 20771, U.S.A.

and

R. E. LINGENFELTER

Center for Astrophysics and Space Sciences, University of California, San Diego, La Jolla, CA 92093, U.S.A.

Abstract. Recent gamma-ray line observations and their interpretations are reviewed and prospects for future line detections are discussed.

1. Introduction

Very important advances have taken place in gamma-ray line astronomy since our previous review (Lingenfelter and Ramaty, 1980). These include new results on solar flares, gamma-ray bursts and the Galactic Center. Solar gamma-ray lines, observed (Chupp, 1982) from many flares by the spectrometer on SMM, imply nuclear interactions in thick targets by protons and nuclei confined to closed loops with little escape into interplanetary space. The timing of the gamma-ray line fluxes indicate that the acceleration of the particles is very impulsive.

Lines have been seen from gamma-ray bursts (Mazets *et al.*, 1981). Electron-positron pair production and annihilation are probably responsible for the emission line seen just below 0.5 MeV from several bursts. The shift in energy is due to either a gravitational redshift or grasar action (Ramaty *et al.*, 1982a).

The 0.511 MeV line from the Galactic Center has been shown to be time variable (Riegler *et al.*, 1981). The theoretical analysis of this observation, together with observations at other wavelengths, strongly suggests (Lingenfelter and Ramaty, 1982) that the positrons are produced by photon-photon collisions in the vicinity of a massive black hole.

We discuss these topics below. We also review the prospects for future observations of other gamma-ray lines, especially lines from processes of nucleosynthesis.

2. Solar Flares

The interactions of solar flare accelerated particles with the ambient solar atmosphere are a source of gamma rays, both lines and continuum. The first detailed calculation (Lingenfelter and Ramaty, 1967) of the expected energetic particle interaction rates in flares predicted observable gamma-ray line fluxes at Earth.

---

* Proceedings of the XVIII General Assembly of the IAU: *Galactic Astrophysics and Gamma-Ray Astronomy*, held at Patras, Greece, 19 August 1982.

Space Science Reviews **36** (1983) 305–317.   0038–6308/83/0363–0305$01.95.

Gamma-ray lines from solar flares were first observed (Chupp *et al.*, 1973) with a Na I spectrometer flown on board the OSO-7 satellite. The lines were observed at 0.511 MeV from positron annihilation, at 2.223 MeV from neutron capture on ^{1}H, and at 4.438 and 6.129 MeV from deexcitations of nuclear levels in ^{12}C and ^{16}O, respectively. These lines, as well as other nuclear deexcitation lines, have been seen from a number of subsequent flares by detectors on the HEAO-1 (Hudson *et al.*, 1980), HEAO-3 (Prince *et al.*, 1982), and SMM (Chupp, 1982; Chupp and Forrest, 1981; Chupp *et al.*, 1981) satellites.

Gamma-ray continuum from solar flares below an MeV is electron bremsstrahlung. But above an MeV, Doppler broadened unresolved nuclear lines make a significant contribution to the continuum, and in the energy range from 4 to 7 MeV nuclear radiation from C, N, and O constitutes the dominant radiation mechanism (Ibraginov and Kocharov, 1977; Ramaty *et al.*, 1977). Continuum emission at energies greater than 10 MeV has been observed (Chupp *et al.*, 1982). This emission probably is bremsstrahlung of relativistic electrons (Ramaty *et al.*, 1983).

The strongest predicted and observed line from solar flares is at 2.223 MeV from neutron capture on hydrogen, ^{1}H(n, γ)^{2}H. Studies of neutron production in flares (Lingenfelter *et al.*, 1965; Ramaty *et al.*, 1975) indicate that the bulk of the neutrons responsible for this line result from the breakup of helium by protons at energies greater than about 20 MeV/nucleon, ^{4}He(p, pn)^{3}He and ^{4}He(p, 2pn)^{2}H, with lesser contributions from spallation of heavier nuclei and from π^+ production, ^{1}H(p, nπ^+)^{1}H. The neutron production may take place above the photosphere, but the 2.223 MeV line emission comes from captures in the photosphere where the density is high enough ($> 10^{16}$ H/cm^{-3}) for the bulk of neutrons to be slowed down and captured before they decay. Calculations (Wang and Ramaty, 1974) of neutron slowing down and capture in the solar atmosphere show that the principal capture reactions are ^{1}H(n, γ)^{2}H and ^{3}He(n, p)^{3}H.

Comparisons of the observed (Chupp *et al.*, 1981) time dependence of the intensity of prompt nuclear deexcitation lines to that of the 2.223 MeV line show delays of $\sim 10^2$ s which are due to the mean thermal neutron capture time. The time required for the neutrons to slow down is much less than that required for their capture. A capture time of $\sim 10^2$ s implies (Wang and Ramaty, 1974) that the mean density of the gas where the neutrons are captured is $\sim 10^{17}$ H cm^{-3}, a density corresponding to a depth of ~ 300 km into the photosphere. Independent evidence for neutron capture in the photosphere comes from the relative attenuation, or limb darkening, of the neutron capture line from solar flares occurring close to the visible limb of Sun. Comparisons (Chupp, 1982) of the neutron capture line fluence to that of nuclear deexcitation lines show that the capture line, while essentially unattenuated for disk flares, is attenuated by at least a factor of 10 for limb flares. This attenuation results from Compton scattering in the photosphere (Wang and Ramaty, 1974). The width of the 2.223 MeV line, determined by the photospheric temperature, is expected to be very narrow (~ 100 eV), a result consistent with the high resolution HEAO-3 observations (Prince *et al.*, 1982) which have set an upper limit of several keV on the width of this line.

A significant fraction of the fastest ($\gtrsim 100$ MeV) neutrons can travel as far as the

Earth before they decay, resulting (Lingenfelter et al., 1965) in detectable neutron fluxes at the Earth following large flares. High energy solar neutrons were observed from a large flare in 1980 (Chupp et al., 1982).

The next most intense solar flare line is that at 0.511 MeV from the annihilation of positrons. There are many astrophysically important positron production mechanisms, but in solar flares the 0.511 MeV line results (Ramaty et al., 1975) from nuclear interactions producing short-lived radionuclei (e.g. ^{11}C, ^{13}N, ^{15}O, ^{17}F) and π^+ mesons which decay by positron emission, as well as excited ^{16}O in the 6.052 MeV level which decays by electron-positron pair emission. The initial energies of the positrons range from several hundred keV to tens of MeV, but only a few annihilate at these high energies. The bulk of the positrons slow down to energies comparable with those of the ambient electrons, where annihilation takes place either directly or via positronium (e.g. Crannell et al., 1976).

Positronium in astrophysical sites is formed by radiative combination with free electrons and by charge exchange with neutral hydrogen (Crannell et al., 1976); 25% of the positronium atoms decay from the singlet state and 75% from the triplet state. Singlet positronium annihilation and direct annihilation produce a line at 0.511 MeV, while triplet positronium annihilates into three photons which form a continuum below 0.511 MeV. But if the ambient density is $\gtrsim 10^{15}$ H cm^{-3}, as may be the case for solar flare positrons, then much of the positronium is broken up by collisions before it can decay (Crannell et al., 1976). The width of the 0.511 MeV line from solar flares depends on the temperature of the annihilation region, and could range from a few keV to tens of keV, depending on whether the annihilation takes place predominantly in the cool photosphere or the hot flare plasma. Measurements of the positronium continuum and the width of the 0.511 MeV line could thus provide important information on the positron annihilation site, but such observations are not yet available.

A variety of gamma-ray lines are produced by the deexcitation of nuclear levels. In solar flares these levels are populated by inelastic collisions (e.g. $^{12}C(p, p')^{12}C^{*4.44}$), spallation reactions (e.g. $^{20}Ne(p, p\alpha)^{16}O^{*6.13}$), nonthermal fusion reactions (e.g. $^{4}He(\alpha, p)^{7}Li^{*0.478}$) and the decay of radionuclei produced by spallation reactions (e.g. $^{16}O(p, p2n)^{14}O(e^+)^{14}N^{*2.31}$. Using laboratory measurements (e.g. Dryer et al., 1981) of the excitation functions of a great number of such reactions, calculations have been made (Ramaty et al., 1979) of gamma-ray spectra produced by the interaction of energetic particles in cooler ambient matter.

In the solar atmosphere these gamma-ray spectra have two components: a narrow-line component resulting from the deexcitation of ambient nuclei excited by interactions with energetic protons and α particles, and a broad-line component from the deexcitation of energetic heavy nuclei excited by interactions with ambient hydrogen and helium. The relative widths of the narrow lines, broadened by the recoil velocities of the heavy target nuclei, are on the order 1% to 2%, while those of the broad lines, reflecting the velocities of the projectiles themselves, are about an order of magnitude larger. If the elemental and isotopic compositions of both the energetic particles and the ambient medium resemble that of the solar photosphere, the strongest narrow lines are at

6.129 MeV from ^{16}O, 4.438 MeV from ^{12}C, 2.313 MeV from ^{14}N, 1.779 MeV from ^{28}Si, 1.634 MeV from ^{20}Ne, 1.369 MeV from ^{24}Mg, 1.238 MeV and 0.847 MeV from ^{56}Fe, all produced primarily by direct excitation of these nuclei, and at two lines, 0.478 MeV from ^{7}Li and 0.431 MeV from ^{7}Be, which result from the reactions ^{4}He$(\alpha, p)^7$Li* and ^{4}He$(\alpha, n)^7$Be*. As already mentioned, the broad lines, together with many unresolved narrow lines, contribute significantly to the gamma-ray continuum, in particular in the 4 to 7 MeV range.

The implications of the gamma-ray observations of solar flares concern the timing of the acceleration, the confinement of particles at the Sun, the fraction of the total flare energy that resides in energetic nucleons, chemical and isotopic abundances and the possible beaming of the energetic particles. In particular the gamma-ray observations show (Ramaty et al., 1982b; Von Rosenvinge et al., 1981) that as much as a few percent of the total flare energy resides in protons and nuclei which are accelerated to tens of MeV per nucleon on time scales of a few seconds in closed magnetic loops with little escape into the interplanetary medium. Further analysis of data should provide important and potentially unique information on abundances and on geometric effects such as beaming. The latter would follow from shifts in the peak line energies (Ramaty and Crannell, 1976) and modifications in the line widths (Kozlovsky and Ramaty, 1977).

3. Rapid Gamma-Ray Transients

Temporal variability is a common property of a large fraction of the astronomical sources of high-energy radiation. In fact, many gamma-ray sources have so far been observed only by their intense transient emission. The gamma-ray bursts are the most common class of these transients. We first consider these bursts, including the possibly unique March 5, 1979 burst. We then briefly review the properties of two very unusual transient that last for tens of minutes and have only been seen in line emission. The observations of emission lines and absorption features in the energy spectra of gamma-ray bursts has added a new dimension to the study of these transients (see Teegarden, 1982, for review). The absorption features, observed (Dennis et al., 1982; Mazets et al., 1981) at energies below about 100 keV, are probably due to cyclotron absorption in intense magnetic fields of the order 10^{12} G which are expected around neutron stars.

The most commonly observed emission line falls in the energy range from 0.40 to 0.46 MeV, as seen (Mazets et al., 1981) by low resolution NaI detectors in the spectra of a third of the most intense gamma-ray bursts. In the spectrum of the November 19, 1978 burst, a Ge detector has resolved (Teegarden and Cline, 1980) two emission lines at ~ 0.42 MeV and ~ 0.74 MeV, which the NaI detectors saw as one broad feature from 0.3 to 0.8 MeV. Line emission in the range of 0.4 to 0.46 MeV is probably optically thin $e^+ - e^-$ annihilation radiation redshifted by the strong gravitational field of a neutron star. In an optically thick region, however, stimulated annihilation radiation (Ramaty et al., 1982a) could produce a line at ~ 0.43 MeV without a gravitational redshift. The line at 0.74 MeV could be either collisionally excited and gravitationally redshifted 0.847 MeV emission from ^{56}Fe (Teegarden and Cline, 1981), or gravitationally redshifted single photon $e^+ - e^-$ annihilation (Daugherty and Bussard, 1980; Katz, 1982) radiation at 1.022 MeV in a very strong ($\gtrsim 10^{13}$ G) magnetic field. In all cases, the implied redshifts of 0.1 to 0.3 are consistent with those expected from neutron stars.

The ~ 0.43 MeV $e^+ - e^-$ annihilation line was also seen (Mazets *et al.*, 1979) from the March 5, 1979 burst suggesting that the source of this burst was also a neutron star. But other characteristics of this burst seem to place it in a different class from that of the typical galactic bursts (Cline, 1980, 1982).

Current theoretical ideas on gamma-ray bursts generally involve strongly magnetized neutron stars. These ideas have developed, in part, as a result of the detailed March 5 observations, even though it is quite likely that the underlying energy source of this burst is not typical of all gamma-ray bursts.

The most probable energy source of gamma-ray bursts is either gravitational or nuclear. Gravitational energy can be released impulsively from a neutron star when a large amount of solid matter such as an asteroid or comet is accreted onto its surface (Colgate and Petcheck, 1981; Harwit and Salpeter, 1973). Such accretion releases about 100 MeV/nucleon, the potential energy at the neutron star surface. Gravitational energy could also be released in a corequake of a neutron star (Ramaty *et al.*, 1980; Tsygan, 1975). Such quakes can set up neutron star vibrations which dissipate mainly by gravitational radiation. A fraction of the vibrational energy, however, can be converted into magnetoacoustic waves which dissipate by accelerating particles in the magnetosphere. Radiation from these particles is then responsible for the observed gamma-ray emission.

Alternatively, impulsive energy release from neutron stars could result from a nuclear detonation of degenerate matter accumulated over a relatively long period of time by accretion of gas (Woosley, 1982; Woosley and Taam, 1976). Such detonations release several MeV per nucleon from the burning of helium to the iron peak nuclei.

All three of these processes, solid body accretion, a corequake, or a nuclear detonation, appear to be quite capable of providing the 10^{37} to 10^{40} ergs required for typical galactic gamma-ray bursts. But to account for the $\sim 10^{44}$ ergs of the March 5, 1979 burst, very large amounts of accreted matter must be involved and this probably rules out solid body accretion and nuclear detonation for this burst. Corequakes, however, which could in principle release energies up to a fraction of the gravitational binding energy of a neutron star ($\sim 10^{53}$ erg), appear to be adequate for the March 5 burst (Ramaty *et al.*, 1980). But no detailed calculations on these possibilities have yet been published.

Electron-positron pairs probably play an important role in producing radiation from gamma-ray bursts. As already mentioned, pair annihilation is responsible for the observed emission line between 0.40 and 0.46 MeV. Since these lines have relatively narrow widths requiring a narrow and well defined range of gravitational redshift, the emitting material must be confined to a thin region close to the neutron star surface. This confinement could be achieved by the strong magnetic field ($\sim 10^{12}$ G) of a neutron star

(Ramaty *et al.*, 1980). Magnetic confinement is necessary, especially for the March 5 burst where the inferred radiation pressure greatly exceeds the gravitational pull of the neutron star. Magnetic fields similarly play an important role in nuclear detonation models of galactic bursts (Woosley, 1982) where magnetic confinement of the nuclear burning products, or lack of it, may constitute the difference between a gamma-ray burst and an X-ray buster. Lastly, if the absorption features, observed below 100 keV in gamma-ray bursts, are due to cyclotron absorption, then they provide direct observational evidence for $\gtrsim 10^{12}$ G magnetic fields in the burst sources.

The principal continuum emission processes suggested for gamma-ray burst sources are bremsstrahlung (Gilman *et al.*, 1980), Comptonization (Bussard and Lamb, 1982; Fennimore *et al.*, 1982; Liang, 1981) and synchrotron radiation (Liang, 1982; Ramaty *et al.*, 1981). In the March 5 burst the continuum below about 300 keV could be synchrotron emission from electron-positron pairs (Ramaty *et al.*, 1981), while the continuum at higher energies could be due to Compton scattering of the synchrotron photons by the same $e^+ - e^-$ pairs that produce the synchrotron radiation (Liang, 1981).

An important property of gamma-ray burst spectra is that they appear to be optically thin (Gilman *et al.*, 1980), especially at the higher energies ($\gtrsim 100$ keV). An optically thin emission region is also required (Ramaty *et al.*, 1981) to produce the ~ 0.43 MeV emission line, except in the case where grasar action is important (Ramaty *et al.*, 1982a). An optically thin source requires a sufficiently small ratio of source depth to source area, so that the small opacity can be consistent with the high observed luminosity. The gamma-ray emission should therefore be produced in a thin layer containing a high density of radiating matter. The most extreme conditions are found in the March 5 event, where in the model of Ramaty *et al.* (1981) the oserved radiation comes from a magnetically confined thin layer (~ 0.1 mm) of dense ($\sim 10^{26}$ cm^{-3}) $e^+ - e^-$ pairs covering the surface of a neutron star. The instantaneous energy content of this layer is orders of magnitude smaller than the total energy of the burst, so that energy must be supplied continuously to the layer. This is achieved by the neutron star vibrations discussed above. An attractive consequence of the continuous energization by vibrations is that the duration of the burst is determined by the damping time of the vibrations. Indeed, the neutron star mass-to-radius ratio, deduced from the observed gravitational redshift, implies a vibrational damping time which is almost exactly the same as the duration of the main emission spike of the burst (Ramaty *et al.*, 1980).

In addition to the gamma-ray bursts, there are apparently two other types of gamma-ray transients in which all of the radiation observed so far is in emission lines. One such gamma-ray line transient was discovered (Jacobson *et al.*, 1978; Ling *et al.*, 1982) with a high resolution Ge detector on June 10, 1974 from a unknown source. This event, lasting about twenty minutes, was characterized by strong emission in four relatively narrow energy bands at 0.40–0.42 MeV, 1.74–1.86 MeV, 2.18–2.26 MeV, and 5.94–5.96 MeV with no detectable continuum. Subsequent searches for similar line transients (Heslin, 1981), however, failed to observe such transients and therefore imply that their frequency is less than 30 per year.

It has been suggested (Lingenfelter *et al.*, 1978) that the June 10, 1974, gamma-ray line transient could result from episodic accretion onto a neutron star from a binary companion leading to redshifted lines from the neutron star surface and unshifted lines from the atmosphere of the companion star and that the lines are due to neutron capture and positron annihilation. Specifically, positron annihilation and neutron capture on hydrogen and iron at and near the surface of the neutron star with a surface redshift of ~ 0.28 would produce the observed redshifted line emission at about 0.41, 1.79, and 5.95 MeV, respectively. The same processes in the atmosphere of the companion star would produce unshifted lines, of which only the 2.223 MeV line from neutron capture an hydrogen was observed. The unshifted 0.511 MeV positron annihilation line could not have been seen because of the large atmospheric and detector background at this energy, while the line emission from neutron capture on iron should be significant only from the iron rich surface of the neutron star but not from the companion star.

The other type of transient line emission is observed in the pulsed spectrum of the Crab pulsar. This very narrow (FWMM < 4.9 keV) emission line, which may vary slightly in energy from 73 to 77 keV was first observed (Ling *et al.*, 1979) from the Crab nebula. The line was subsequently shown (Strickman *et al.*, 1982) to be pulsed with the Crab pulsar period of 0.033 s and to persist only for about 20 min and then turn off. The most likely source of this line is cyclotron emission in an intense ($\sim 8 \times 10^{12}$ G) magnetic field at the polar cap of a neutron star. In addition, a very narrow 0.4 MeV line was observed (Leventhal *et al.*, 1977) from a broad field of view that included both the Crab nebula and the source direction of the June 10, 1974 transient.

4. Galactic Gamma-Ray Line Emissions

Intense positron annihilation radiation at 0.511 MeV has been observed from the Galactic Center, and gamma-ray line emission at this and other energies is expected from a variety of discrete and diffuse sites in the Galaxy.

Annihilation radiation from the Galactic Center was first seen in a series of balloon observations with low-resolution Na I detectors, starting in 1970 (Haymes *et al.*, 1975; Johnson and Haymes, 1973; Johnson *et al.*, 1972). But it was not until 1977 that the annihilation line energy of 0.511 MeV was clearly identified with high-resolution Ge detectors (Leventhal *et al.*, 1978). The latter observation also revealed that the line is very narrow (FWHM $\lesssim$ 3.2 keV) and that it shows evidence for three-photon positronium continuum emission below 0.511 MeV, implying that $\sim 90\%$ of the positrons annihilate via positronium. Thus, the observed intensity of $\sim 2 \times 10^{-3}$ photons cm^{-2} s^{-1} implies an annihilation rate of $\sim 4 \times 10^{43}$ positrons s^{-1} or an annihilation radiation luminosity of $\sim 6 \times 10^{37}$ ergs s^{-1} at the 10 kpc distance of the Galactic Center.

Recent Ge detector observations (Riegler *et al.*, 1981) on HEAO-3 confirmed the narrowness (FWHM < 2.5 keV) of the line and provided more precise information on the location of the source and strong constraints on the size of the emission region. These measurements showed that the direction of the source is coincident with that of the Galactic Center (within the $\pm 4°$ observational uncertainty) and that the line intensity varies with time, decreasing by a factor of three in six months from the fall of 1979 to the spring of 1980. This six month variability implies that the sizes of both the annihilation region and the positron source are less than the light-travel distance of 10^{18} cm.

The nature of the positron annihilation region is further constrained by the observed line width and intensity variations. The line width (FWHM < 2.5 keV) requires (Bussard et al., 1979) a gas temperature in the annihilation region less than 5×10^4 K and an ionization fraction greater than 10%. If the gas were neutral, the line width would be larger than observed, because it would be Doppler broadened, not by the thermal motion of the gas, but by the velocity of energetic positrons forming positronium in flight by charge exchange with neutral hydrogen. In a partially ionized gas, however, the positrons lose energy to the plasma fast enough that they thermalize before they annihilate or form positronium. The line width thus reflects the temperature of the medium, requiring it to be $\leq 5 \times 10^4$ K. The intensity variation not only constrains the size of the annihilation region to be $< 10^{18}$ cm, but it requires that the density of gas in it be high enough that the positrons can slow down and annihilate in less than half a year. If the positrons are produced with kinetic energies on the order of their rest mass, then the time it takes for them to slow down by Coulomb collisions is longer than the time it takes for them to form positronium in such a gas once they have slowed down. Both times are inversely proportional to the gas density. A slowing down time of $\lesssim 1.5 \times 10^7$ s requires a density of $\gtrsim 10^5$ H cm^{-3}. Such regions appear to exist in both the peculiar warm clouds (Lacy et al., 1980) and the compact non-thermal source (Kellermann et al., 1977) within the central parsec of the Galaxy.

The nature of the positron source is also strongly constrained by the observed variation of the 0.511 MeV intensity and by observations at other wavelengths. The decrease of a factor of three in the line intensity in six months clearly excludes any of the multiple, extended sources, such as cosmic rays, pulsars (Sturrock and Baker, 1979), supernovae (Ramaty and Lingenfelter, 1979), or primordial black holes (Okeke and Rees, 1980), previously proposed. Instead, it essentially requires (Lingenfelter and Ramaty, 1982) a single, compact ($< 10^{18}$ cm) source which is apparently located either at or close to the Galactic Center and which is inherently variable on time scales of six months or less. With a luminosity of at least 6×10^{37} ergs s^{-1}, this source is the most luminous gamma-ray source in the Galaxy.

The various possible positron production processes and the observational constraints on them have been reviewed recently (Lingenfelter and Ramaty, 1982). It has been found that the observational (Matteson, 1982; Matteson et al., 1979) upper limits on accompanying continuum emission at energies $> m_e c^2$ appear to set the strongest constraints on the positron production process, requiring high efficiency such that more than 30% of the total radiated energy $> m_e c^2$ goes into electron-positron pairs. Under the conditions of positron production on time scales comparable to that of the observed variation and in an optically thin, isotropically emitting region, only photon-photon pair production among $\sim$ MeV photons can provide the required high efficiency.

Moreover, the absolute luminosity of the annihilation line requires that the photon-

photon collisions take place in a very compact source ($< 5 \times 10^8$ cm). Pair production in an intense radiation field around an accreting black hole of $< 10^3 \, M_\odot$ appears to be a possible source. Other mechanisms (Lingenfelter and Ramaty, 1982), such as pair production in an electromagnetic cascade in a strong electric field of an accreting and rotating black hole, would be possible if the above constraints are relaxed.

Turning now to the other sources of galactic line emission, thermonuclear burning in supernovae and novae (e.g. Clayton, 1982; Woosley *et al.*, 1981) and nuclear interactions of low-energy cosmic rays with interstellar gas (Ramaty *et al.*, 1979) are all expected (Ramaty and Lingenfelter, 1981) to produce throughout the Galaxy a variety of nuclear deexcitation lines, as well as additional positron-annihilation line emission. Observations (Gardner *et al.*, 1982) of galactic 0.511 MeV emission with wide ($\sim 100°$) field-of-view detectors have found considerably higher line intensities than would be expected from the Galactic Center source alone, suggesting that there may be a spatially diffuse source of 0.511 MeV line emission in the Galaxy.

The most abundant radionuclide expected from explosive nucleosynthesis in supernovae is ^{56}Ni (Clayton *et al.*, 1969) which decays with a 6.1 day half-life to ^{56}Co, which, in turn, decays with a half-life of 78.8 days to ^{56}Fe; 20% of the ^{56}Co decays are via positron emission. Nucleosynthesis of ^{56}Ni in supernovae is thought to be the primary source of galactic ^{56}Fe (e.g. Woosley *et al.*, 1981). The bulk of the gamma rays (Colgate and McKee, 1969) and positrons (Arnett, 1979) from the ^{56}Ni decay chain, however, are absorbed in the expanding nebula and their energy emerges only as lower energy radiation. The characteristic light curves of type I supernovae, in fact, appear to follow the ^{56}Ni and ^{56}Co decay (Colgate and McKee, 1969) and optical lines from both ^{56}Co and the resulting ^{56}Fe have recently been detected (Axelrod, 1980) in the spectrum of an extragalactic supernova, SN 1972e. Any direct gamma-ray line emission from the decay which could escape from the nebula would be detectable for only of few years after the supernova explosion. But a fraction of the positrons from ^{56}Co decay could escape into the interstellar medium. Since in the tenuous interstellar gas the positron lifetime against annihilation is quite long (10^5 yr in a density of $1 \, \text{cm}^{-3}$), positrons should accumulate from several thousand supernovae, assuming that galactic supernovae occur about once every 30 yr. Their annihilation should thus produce diffuse galactic gamma-ray line emission at 511 keV (Ramaty and Lingenfelter, 1979, 1981). Conclusive measurements of such diffuse line emission can put constraints on the fraction of positrons that escape from supernovae and on the average rate of galactic nucleosynthesis during the last 10^5 yr.

Similarly, the long-lived radionuclei ^{60}Fe (half life $\sim 3 \times 10^5$ yr) and ^{26}Al (half life $\sim 7.2 \times 10^5$ yr), which are also expected from explosive nucleosynthesis, should accumulate from $\sim 10^4$ or more supernovae and be well distributed through the interstellar medium before they decay. Diffuse galactic line emission is thus expected at 1.809 MeV from ^{26}Al decay to ^{26}Mg (Arnett, 1977; Ramaty and Lingenfelter, 1977) and at 1.332 MeV, 1.173 MeV and 0.059 MeV from ^{60}Fe decay to ^{60}Co and its subsequent decay to ^{60}Ni (Clayton, 1971). The line at 1.809 MeV from ^{26}Al decay has recently been deserved (Mahoney *et al.*, 1982).

Another important radionuclide from explosive nucleosynthesis in supernovae is ^{44}Ti (Clayton *et al.*, 1969). This isotope decays with a half-life of 47 yr into ^{44}Sc, producing lines at 0.078 and 0.068 MeV. ^{44}Sc subsequently decays into ^{44}Ca with line emission at 1.156 MeV. The ^{44}Ti half life is comparable to the average time between galactic supernova explosions and therefore gamma-ray lines from this decay chain could be observed from the few youngest galactic supernova remnants.

Explosive nucleosynthesis in novae is expected to produce ^{22}Na (Clayton and Hoyle, 1974) and ^{26}Al (Woosley and Weaver, 1980). Since about 40 novae occur in the Galaxy every year, the 1.275 MeV line emission from ^{22}Na with a 2.6 yr half life should be observable from $> 10^2$ novae at any particular time. Thus, both ^{22}Na and ^{26}Al from novae can also provide diffuse galactic line emission, and observational limits on their intensity can constrain nuclesynthetic models of novae.

The most intense deexcitation lines resulting from low-energy (< 100 MeV nucl^{-1}) cosmic-ray interactions are expected at 6.129 MeV from ^{16}O*, at 4.438 MeV from ^{12}C* and at 0.847 MeV from ^{56}Fe*. Of special interest are the very narrow lines (FWHM ~ 5 keV), such as that at 6.129 MeV from ^{16}O, resulting from deexcitation of nuclei in interstellar grains (Lingenfelter and Ramaty, 1977). The line broadening, which in gases in caused by the recoil velocities of the excited nuclei, is greatly reduced in solids where these nuclei or their radioactive progenitors can come to rest before deexcitation. The detection of gamma-ray lines from low-energy cosmic-ray interactions in the interstellar medium would measure the unknown interstellar density of these cosmic rays, and provide information on the distribution, motion, composition and size of interstellar dust grains.

5. Extragalactic Gamma-Ray Line Emission

Extragalactic gamma-ray line emission has so far been reported (Hall *et al.*, 1976) only from the radiogalaxy Centaurus A. The observed lines at 4.4 and 1.6 MeV could be produced (Lingenfelter *et al.*, 1978) in nuclear reactions in the vicinity of a massive black hole, but the statistical significance of these observations is quite low.

As in the nucleus of our Galaxy, electron-positron pair production could play an important role in active galaxies as well. Gamma rays have been observed from some of the brightest active galaxies, from the radio galaxy Centaurus A (Baity *et al.*, 1981; Grindlay *et al.*, 1975; Hall *et al.*, 1976), from the Seyfert galaxy NGC 4151 (Perotti *et al.*, 1979, 1981) and the quasar 3C273 (Bignami *et al.*, 1981; Swanenburg *et al.*, 1978). The comparison of these observations with observations at lower energies shows (e.g. Ramaty and Lingenfelter, 1982) that the luminosities of active galaxies peak at gamma-ray energies somewhat above 0.1 MeV suggesting that observations in these energy regions can directly probe the central source of power of these objects. The fact that these energies are close to the electron or positron rest mass energy may be due to the onset of pair production which prevents the sources from emitting a large fraction of their luminosities at higher energies. The resultant pairs could produce an annihilation feature. But unlike the nucleus of our Galaxy where $e^+ - e^-$ pairs annihilate in relatively cool regions, in an active galaxy the annihilation region could be much hotter in which case the line would be both broadened and blueshifted (Ramaty and Mészáros, 1981). This would explain, for explain, for example, the absence of a narrow 0.511 MeV line from the spectrum of Centaurus A (Baity *et al.*, 1981; Hall *et al.*, 1976).

Observable extragalactic gamma-ray lines could also result from nucleosynthesis in supernovae. In particular, the gamma-ray lines from ^{56}Co decay, at 0.847 and 1.238 MeV, could be detected from type I supernovae at distance as large as that of the Virgo cluster (e.g. Ramaty and Lingenfelter, 1981). About one supernova per year is detected optically from this cluster (Tamman, 1974), but the actual rate could be larger if some of them are obscured by dust.

6. Summary and Conclusions

We have discussed in this paper the interpretations of new astrophysical gamma-ray line observations. Such lines have been seen from solar flares, gamma-ray transients and the Galactic Center. Gamma-ray lines from solar flares are excellent probes of acceleration mechanisms and interaction models of energetic protons and nuclei in the solar atmosphere. The continuing observations with the gamma-ray spectrometer on SMM during the current maximum of solar activity are providing much new insight into these aspects of solar physics.

Gamma-ray lines seen in the spectra of gamma-ray bursts suggest that neutron stars are the sources of many of these bursts. The most commonly observed emission line is in the range from 400 to 460 keV, where it is likely to be positron-electron annihilation radiation, either redshifted by the gravitational field of a neutron star or produced at energies < 0.511 MeV by grasar action. Precise measurements of the energy and width of this line as well as the detection of other lines are very important objectives of future observations.

The 0.511 MeV line from the Galactic Center, first observed by balloon-borne detectors, has been confirmed by the HEAO-3 gamma-ray spectrometer. Moreover the HEAO-3 and subsequent balloon observations have shown that the line is time-variable. This result, together with gamma-ray continuum observations implies that the positrons are produced by photon-photon collisions, probably close to a massive black hole, and that they annihilate in a region no larger than a light year. Important objectives for future Galactic Center observations are the better determination of the position of this source, continued monitoring of the temporal variability of the 0.511 MeV line intensity, and the determination of whether this variability is correlated with other observations, especially X-rays.

No gamma-ray lines have yet been seen from processes of nucleosynthesis. Good prospects, however, exist for detecting a diffuse galactic 0.511 MeV line from ^{56}Co decay and deexcitation lines from ^{26}Al and ^{44}Ti, produced by galactic nucleosynthesis, and for seeing the lines of ^{56}Co from an extragalactic supernova.

Acknowledgments

We wish to acknowledge the support of NASA through Grant NSG 7541 and the Solar Terrestrial Theory Program.

The editors thank G. Manzo for assistance in evaluating this paper.

References

Arnett, W. D.: 1977, *Ann. N.Y. Acad. Sci.* **302**, 90.

Arnett, W. D.: 1979, *Astrophys. J. Letters* **230**, L32.

Axelrod, T. S.: 1980, Ph.D. Thesis, U.C. Santa Cruz.

Baity, W. A. *et al.*: 1981, *Astrophys. J.* **244**, 1981.

Bignami, G. F. *et al.*: 1981, *Astron. Astrophys.* **93**, 71.

Bussard, R. W. and Lamb, F. K.: 1982, in *Gamma-Ray Transients and Related Astrophysical Phenomena*, AIP, New York, p. 189.

Bussard, R. W., Ramaty, R., and Drachman, R. J.: 1979, *Astrophys. J.* **228**, 928.

Chupp, E. L.: 1982, in *Gamma-Ray Transients and Related Astrophysical Phenomenon*, AIP, New York, p. 363.

Chupp, E. L. *et al.*: 1979, *Nature* **241**, 333.

Chupp, E. L. *et al.*: 1981, *Astrophys. J. Letters* **244**, L171.

Chupp, E. L. *et al.*: 1982, *Astrophys. J. Letters* **263**, L95.

Clayton, D. D.: 1971, *Nature* **234**, 291.

Clayton, D. D.: 1982, in *Essays in Nuclear Astrophysics*, Univ. Press, Cambridge (in press).

Clayton, D. D. and Hoyle, F.: 1974, *Astrophys. J.* **187**, L101.

Clayton, D. D., Colgate, S. A., and Fishman, G. J.: 1969, *Astrophys. J.* **155**, 75.

Cline, T. L.: 1980, *Comm. Astrophys.* **9**, 13.

Cline, T. L.: 1982, in *Gamma-Ray Transients and Related Astrophysical Phenomena*, AIP, New York, p. 17.

Colgate, S. A. and McKee, C.: 1969, *Astrophys. J.* **157**, 623.

Colgate, S. A. and Petchek, A. G.: 1981, *Astrophys. J.* **248**, 771.

Crannell, C. J., Joyce, G., Ramaty, R., and Werntz, C.: 1976, *Astrophys. J.* **210**, 582.

Daugherty, J. K. and Bussard, R. W.: 1980, *Astrophys. J.* **238**, 296.

Dennis, B. R. *et al.*: 1982, in *Gamma-Ray Transients and Related Astrophysical Phenomena*, AIP, New York, p. 153.

Dyer, P., Bodansky, D., Seamster, A. G., Norman, E. B., and Maxson, D. R.: 1981, *Phys. Rev.* **C23**, 1268.

Fenimore, E. E., Klebesardel, R. W., Laros, J. G., Stockdale, R. E., and Kane, S. R.: 1982, *Nature* **297**, 665.

Gardner, B. M., Forrest, D. J., Dunphy, P. P., and Chupp, E. L.: 1982, in *The Galactic Center*, AIP, New York, p. 144.

Gilman, D., Metzger, A. E., Parker, R. H., Evans, L. G., and Trombka, J. I.: 1980, *Astrophys. J.* **236**, 951.

Grindlay, J. E., Helmken, H. F., Hanbury-Brown, R., Davis, J., and Allen, L. R.: 1975, *Astrophys. J. Letters* **197**, L9.

Hall, R. D., Meegan, C. A., Walraven, G. D., Djuth, F. T., and Haymes, R. C.: 1976, *Astrophys. J.* **210**, 631.

Harwit, M. and Salpeter, E. E.: 1973, *Astrophys. J. Letters* **187**, L97.

Haymes, R. C. *et al.*: 1975, *Astrophys. J.* **201**, 593.

Heslin, J. P.: 1981, *Bull. Am. Astron. Soc.* **13**, 901.

Hudson, H. S. *et al.*: 1980, *Astrophys. J. Letters* **236**, L91.

Ibragimov, I. A., and Kocharov, G. E.: 1977, *Soviet Astron. Letters* **3**, (5), 221.

Johnson, W. N. and Haymes, R. C.: 1973, *Astrophys. J.* **184**, 103.

Johnson, W. N., Harnden, F. R., and Haymes, R. C.: 1972, *Astrophys. J.* **172**, L1.

Jacobson, A. S., Ling, J. C., Mahoney, W. A., and Willett, J. B.: 1978, in *Gamma Ray Spectroscopy in Astrophysics*, NASA, Goddard, p. 228.

Katz, J.: 1982, *Astrophys. J.* **260**, 371.

Kellermann, K. I., Shaffer, D. B., Clark, B. G., and Geldzahler, B. J.: 1977, *Astrophys. J.* **214**, L61.

Kozlovsky, B. and Ramaty, R.: 1977, *Astrophys. Letters* **19**, 19.

Lacy, J. H., Townes, C. H., Geballe, T. R., and Hollenbach, D. J.: 1980, *Astrophys. J.* **241**, 132.

Leventhal, M., MacCallum, C. J., and Watts, A. C.: 1977, *Astrophys. J.* **216**, 491.

Leventhal, M., MacCallum, C. J., Stang, P. D.: 1978, *Astrophys. J.* **225**, L11.

Liang, E. P. T.: 1981, *Nature* **292**, 319.

Liang, E. P. T.: 1982, *Nature* **299**, 321.

Ling, J. C., Mahoney, W. A., Willett, J. B., and Jacobson, A. S.: 1979, *Astrophys. J.* **231**, 896.

Ling, J. C., Mahoney, W. A., Willett, J. B., and Jacobson, A. S.: 1982, in *Gamma-Ray Transients and Related Astrophysical Phenomena*, AIP, New York, p. 143.

Lingenfelter, R. E. and Ramaty, R.: 1967, in *High Energy Nuclear Reactions in Astrophysics*, Benjamin, New York, p. 99.

Lingenfelter, R. E. and Ramaty, R.: 1977, *Astrophys. J.* **211**, L19.

Lingenfelter, R. E. and Ramaty, R.: 1980, in *Non-Solar Gamma Rays (COSPAR)*, Pergamon, Oxford, p. 103.

Lingenfelter, R. E. and Ramaty, R.: 1982, in *The Galactic Center*, AIP, New York, p. 148.

Lingenfelter, R. E., Flamm, E. J., Canfield, E. H., and Kellmann, S.: 1965, *J. Geophys. Res.* **70**, 4077 and 4087.

Lingenfelter, R. E., Higdon, J. C., and Ramaty, R.: 1978, in *Gamma-Ray Spectroscopy in Astrophysics*, NASA, Greenbelt, p. 252.

Mahoney, W. A., Ling, J. C., Jacobson, A. S., and Lingenfelter, R. E.: 1982, *Astrophys. J.* **262**, 742.

Matteson, J. L., Nolan, P. L., Peterson, L. E.: 1979, in *X-Ray Astronomy*, Pergamon Press, Oxford, p. 543.

Matteson, J. L.: 1982, in *The Galactic Center*, AIP, New York, p. 109.

Mazets, E. P., Golenetskii, S. V., Illinskii, V. N., Aptekar, R. L., and Guryan, Yu. A.: 1979, *Nature* **282**, 587.

Mazets, E. P., Golenetskii, S. V., Aptekar, R. L., Guryan, Yu. A., and Illinskii, V. N.: 1981, *Nature* **290**, 378.

Okeke, P. N. and Rees, M. J.: 1980, *Astron. Astrophys.* **81**, 263.

Perotti, F. *et al.*: 1981, *Astrophys. J. Letters* **247**, L63.

Perotti, F. *et al.*: 1979, *Nature* **282**, 484.

Prince, T. A., Ling, J. C., Mahoney, W. A., Riegler, G. R., and Jacobson, A. S.: 1982, *Astrophys. J. Letters* **255**, L81.

Ramaty, R. and Crannell, C. J.: 1976, *Astrophys. J.* **203**, 766.

Ramaty, R. and Lingenfelter, R. E.: 1977, *Astrophys. J. Letters* **213**, L5.

Ramaty, R. and Lingenfelter, R. E.: 1979, *Nature* **278**, 127.

Ramaty, R. and Lingenfelter, R. E.: 1981, *Phil. Trans. Roy. Soc. London* **A301**, 671.

Ramaty, R. and Lingenfelter, R. E.: 1982, *Ann. Rev. Nucl. Part. Sci.* **32**, 235.

Ramaty, R. and Mészáros, P.: 1981, *Astrophys. J.* **250**, 384.

Ramaty, R., Kozlovsky, B., and Lingenfelter, R. E.: 1975, *Space Sci. Rev.* **18**, 341.

Ramaty, R., Kozlovsky, B., and Suri, A. N.: 1977, *Astrophys. J.* **214**, 617.

Ramaty, R., Kozlovsky, B., and Lingenfelter, R. E.: 1979, *Astrophys. J. Suppl.* **40**, 487.

Ramaty, R. *et al.*: 1980, *Nature* **287**, 122.

Ramaty, R. *et al.*: 1983, *Solar Phys.* **86**, 395.

Ramaty, R., McKinley, J. M., Jones, F. C.: 1982a, *Astrophys. J.* **256**, 238.

Ramaty, R., Lingenfelter, R. E., and Kozlovsky, B.: 1982b, in *Gamma-Ray Transients and Related Astrophysical Phenomena*, AIP, New York, p. 211.

Riegler, G. R. *et al.*: 1981, *Astrophys. J. Letters* **248**, L13.

Sturrock, P. A. and Baker, K. B.: 1979, *Astrophys. J.* **234**, 612.

Strickman, M. S., Kurfess, J. D., and Johnson, W. N.: 1982, *Astrophys. J. Letters* **253**, L23.

Swanenburg, B. N. *et al.*: 1978, *Nature* **275**, 298.

Tammann, G. A.: 1974, in *Supernovae and Supernova Remnants*, D. Reidel Publ. Co., Dordrecht, Holland, p. 155.

Teegarden, B. J.: 1982, in *Gamma-Ray Transients and Related Astrophysical Phenomena*, AIP, New York, p. 153.

Teegarden, B. J. and Cline, T. L.: 1980, *Astrophys. J. Letters* **236**, L67.

Teegarden, B. J. and Cline, T. L.: 1981, *Astrophys. Space Sci.* **75**, 181.

Tsygan, A. I.: 1975, *Astron. Astrophys.* **44**, 21; **49**, 159.

Von Rosenvinge, T. T., Ramaty, R., and Reames, D. V.: 1981, *17th Int. Cosmic Ray Conf. Papers* **3**, 28.

Wang, H. T. and Ramaty, R.: 1974, *Solar Phys.* **36**, 129.

Woosley, S. E.: 1982, in *Gamma-Ray Transients and Related Astrophysical Phenomena*, AIP, New York, p. 273.

Woosley, S. E. and Taam, R. E.: 1976, *Nature* **261**, 101.

Woosley, S. E. and Weaver, T. A.: 1980, *Astrophys. J.* **238**, 1017.

Woosley, S. E., Axelrod, T. S., and Weaver, T. A.: 1981, *Comm. Nucl. Part. Phys.* **9**, 185.

Zdziarski, A. A.: 1980, *Acta Astron.* **30**, 371.

COSMIC GAMMA-RAY BURSTS*

G. VEDRENNE and G. CHAMBON

Centre d'Etude Spatiale des Rayonnements, B. P. 4346–31029 Toulouse, Cedex, France

Abstract. The observational characteristics of gamma-ray bursts are reviewed, concerning their spectra as well as their temporal structure and spatial distribution. From this data, it is suggested that the sources belong to a thick halo population (scale height > 3 kpc), and that the mean recurrence time for one source is greater than 5 yr. The implications of these results are discussed, concerning the future experimental perspectives of detection of gamma-ray bursts, and also the constraints on theoretical models.

1. Introduction

The origin of gamma-ray bursts has still not been established with certainty, despite the numerous results which are being accumulated. In this report, we will discuss essentially two important aspects which can advance our understanding of this phenomenon: the energy spectra of bursts and the $\log N - \log S$ distribution associated with their spatial distribution. We will conclude with some remarks about the optical identifications. For a complete and recent review of the observational characteristics of gamma-ray bursts, see the review papers (Hurley, 1982a, b; Vedrenne, 1981); for a summary of the very interesting results of the Leningrad group, see the recent paper of Mazets (Mazets *et al.*, 1981a). For theoretical models, a complete review by Ruderman has appeared (Ruderman, 1975); since this review, the basic phenomena have not really changed, even though the details of some models have been refined, especially to take into account the new experimental results but also theoretical constraints.

After examining the general characteristics of gamma-ray bursts, with special emphasis on their spectra, the $\log N - \log S$ distribution, and the spatial distribution of bursts localized by Mazets (Mazets *et al.*, 1980) and by the international network, we will conclude by examining the implications of the most recent results.

1.1. GENERAL CHARACTERISTICS OF BURSTS

A. *Temporal Characteristics*

It is well known that gamma-ray bursts are in general short and intense emissions of gamma-ray photons, having a time structure which goes down to the millisecond range (Vedrenne, 1981). The distribution of their durations may be represented by an exponential law with a time constant $T_0 = 24$ s (Chambon, 1982). However, it is difficult to give an upper limit to the duration of bursts due to the experimental method used to detect them: burst data are generally stored in memories of limited capacity aboard satellites. Besides, the detection systems have a sensitivity threshold which would allow long, weak bursts either to be masked by the detector background or, more frequently, not detected

---

* Proceedings of the XVIII General Assembly of the IAU: *Galactic Astrophysics and Gamma-Ray Astronomy*, held at Patras, Greece, 19 August 1982.

Space Science Reviews **36** (1983) 319–335.   0038–6308/83/0363–0319$02.55.

at all by the trigger system, which requires a count rate level several σ above the background level. Nevertheless several attempts have been made to classify gamma-ray bursts according to their time histories (Klebesadel, 1982a; Mazets and Golenetskii, 1981a; Barat *et al.*, 1981a b; Desai, 1981). Three general classes appear: short, double, and long. Can specific energy spectra be attributed to each class? Studies have generally given negative results (Klebesadel, 1982a; Mazets, 1981a), although short events tend to have exponential spectra (Hurley, 1982b; Mazets *et al.*, 1981b).

Similarly, attempts to find periodicities have not been very fruitful (Pizzichini, 1980); apart from the events of 5 March 1979 and 29 October 1977 which have periods around 8 and 4 s, respectively (Mazets *et al.*, 1979; Barat *et al.*, 1979; Terrell *et al.*, 1980; Wood *et al.*, 1981), little can be concluded. Chambon (1982) has recently pointed out the possibility of a 280 ms pseudo period in the 4 November 1978 event, as well as a pseudo period of 143 ms in the 23 October 1978 event. These pseudo periodicities were observed by the 3 experiments in the Signe program, built at the CESR in Toulouse, France, and launched aboard Soviet spacecraft for the Franco–Soviet collaboration. Other possible periodicities have been recently suggested by Loznikov and Kuznetsov (1982).

B. *Spectra of Gamma-Ray Bursts*

The first results gave the same spectral shape for all the events (Cline, 1973). Since then, the results of numerous experiments indicate that gamma ray bursts in fact have different spectra (Vedrenne, 1981; Kane and Anderson, 1976; Barat *et al.*, 1981b; Kane and Share, 1977). However, Mazets *et al.* (1981a, b) have shown that the spectra of numerous bursts are well fitted by an optically thin thermal bremsstrahlung law with kT which can reach $\simeq 1$ MeV. Recently, Fenimore *et al.* (1981) has demonstrated clearly that a thermal emission at high temperature with Comptonization can also give excellent agreement and does not need the high temperatures ($kT \sim 200-300$ keV) required for an optically thin thermal bremsstrahlung fit; but in this model the presence of a strong magnetic field is not taken into account. Moreover, both Fenimore *et al.* (1982a) and Chambon (1982) have shown that the optically thin thermal bremsstrahlung hypothesis is difficult to apply to all gamma-ray busts, since it imposes source distances of less than several tens of parsecs.

Another characteristic of burst spectra is their extreme variability as a function of time (Barat *et al.*, 1981b). A good example of variability on a 250 ms time scale is found in the 19 November 1978 event (Figure 1). It therefore seems difficult, with this variability, to ascribe a single temperature thermal bremsstrahlung law on a 4 s time scale, to a phenomenon which varies significantly over several hundred milliseconds. This point has already been noticed (Lamb, 1982; Bussard and Lamb, 1982).

Note also that the detector dimensions (thicknesses greater than 1–2 cm) and the hardness of some burst spectra (E^{-1}) conspire to provide a large Compton contribution in the 100 keV range due to photons of higher energies; in contrast to the case of hard X- and gamma-ray sources, this contribution is not at all negligible and in fact it may even be essential. For example, for the first 4 s of the 19 November 1978 event, more than 80% of the count rate around 60 keV is due to the Compton contribution of

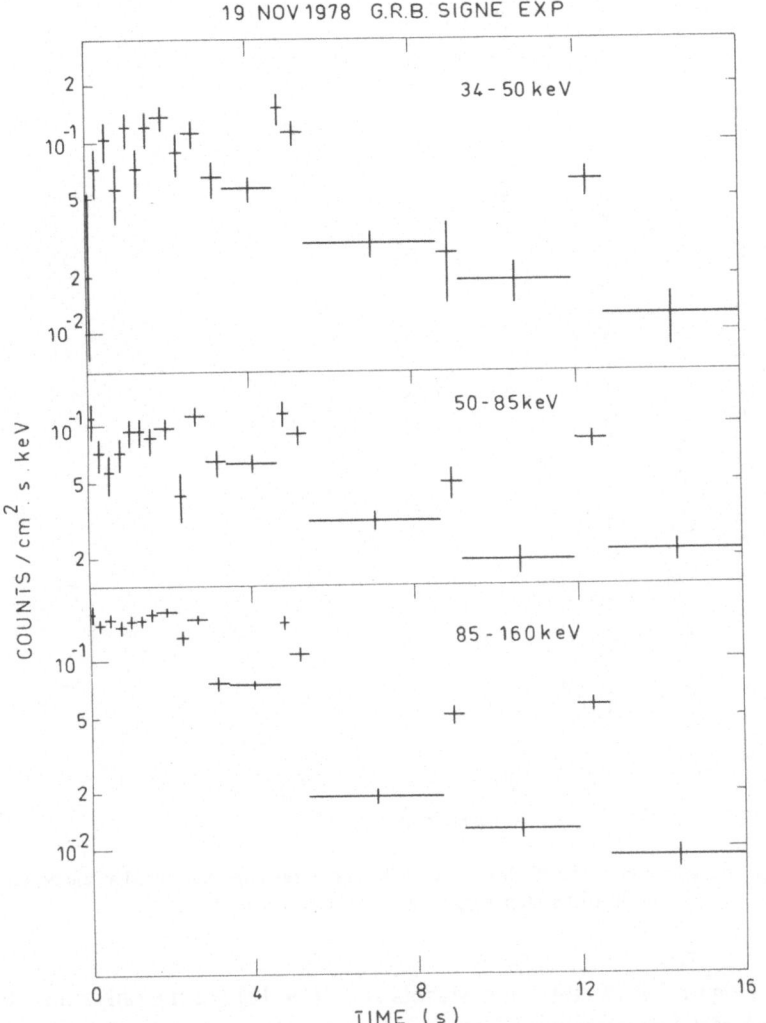

Fig. 1. The spectral variability on a 250 ms time scale for the 19 November 1978 gamma-ray burst. This variability is measured in 3 energy channels.

photons of higher energy (Barat *et al.*, 1982b). A change in the spectrum at high energy can therefore change the low energy part considerably, and even create dips in the spectrum. This naturally brings us to the very important results on the spectral features observed at 400 keV and 60 keV by Mazets *et al.* (1981a) in 10 and 30 spectra of 150 events, respectively. As far as the line at 400 keV is concerned, it was also detected in the 19 November event by Teegarden and Cline (Teegarden and Cline, 1981). This feature is also apparent in our results but on a short time interval (Figure 2) (Barat, 1982b). The same kind of structure in the spectrum can be also seen in the spectrum of the 4 November 1978 (Figure 3a) but if the spectrum is integrated on 3.8 s it

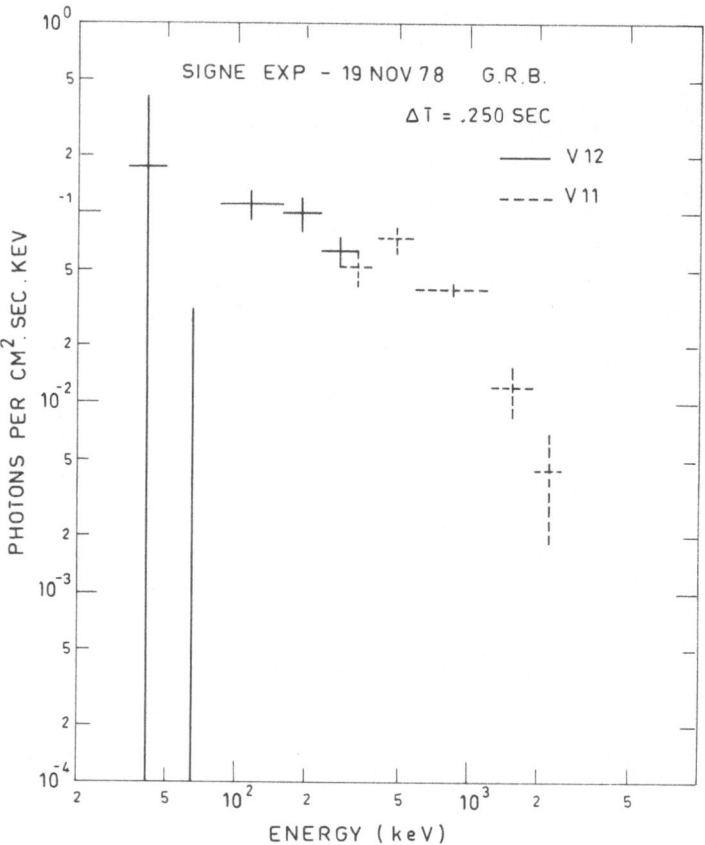

Fig. 2. The photon spectrum of the 19 November 1978 gamma-ray burst measured at the beginning of the event (0.3 s after triggering) and integrated over 0.25 s.

disappears (Figure 3b). At this time Mazets *et al.* (1981c) are the only ones* to report the presence of a feature around 60 keV (the energy is not always the same, and the feature can be either a peak or a valley). The SMM results, which seem to confirm this feature for the event of 19 April 1980 (Dennis *et al.*, 1982), in fact show a break in the spectrum below 150 keV which continues to low energies (30 keV); this can also be interpreted as an absorption without necessarily invoking a cyclotron process. Moreover, the spectral shape is well fitted by an exponential law. The results which we have obtained for the 19 November 1978 even show a similar phenomenon on a short time scale, (Figure 2) Fenimore *et al.* have shown (Fenimore *et al.*, 1982b), that a small change in the gain, and thus in the detection threshold, can introduce a very significant modification in the photon spectrum particularly it can introduce a bump around 400 keV. Such a possibility cannot be excluded; to be convinced, one need only to

---

* On 25 March 1978, Hueter and Gruber (1982) reported recently on the presence of an apparent absorption feature at 55 ± 5 keV.

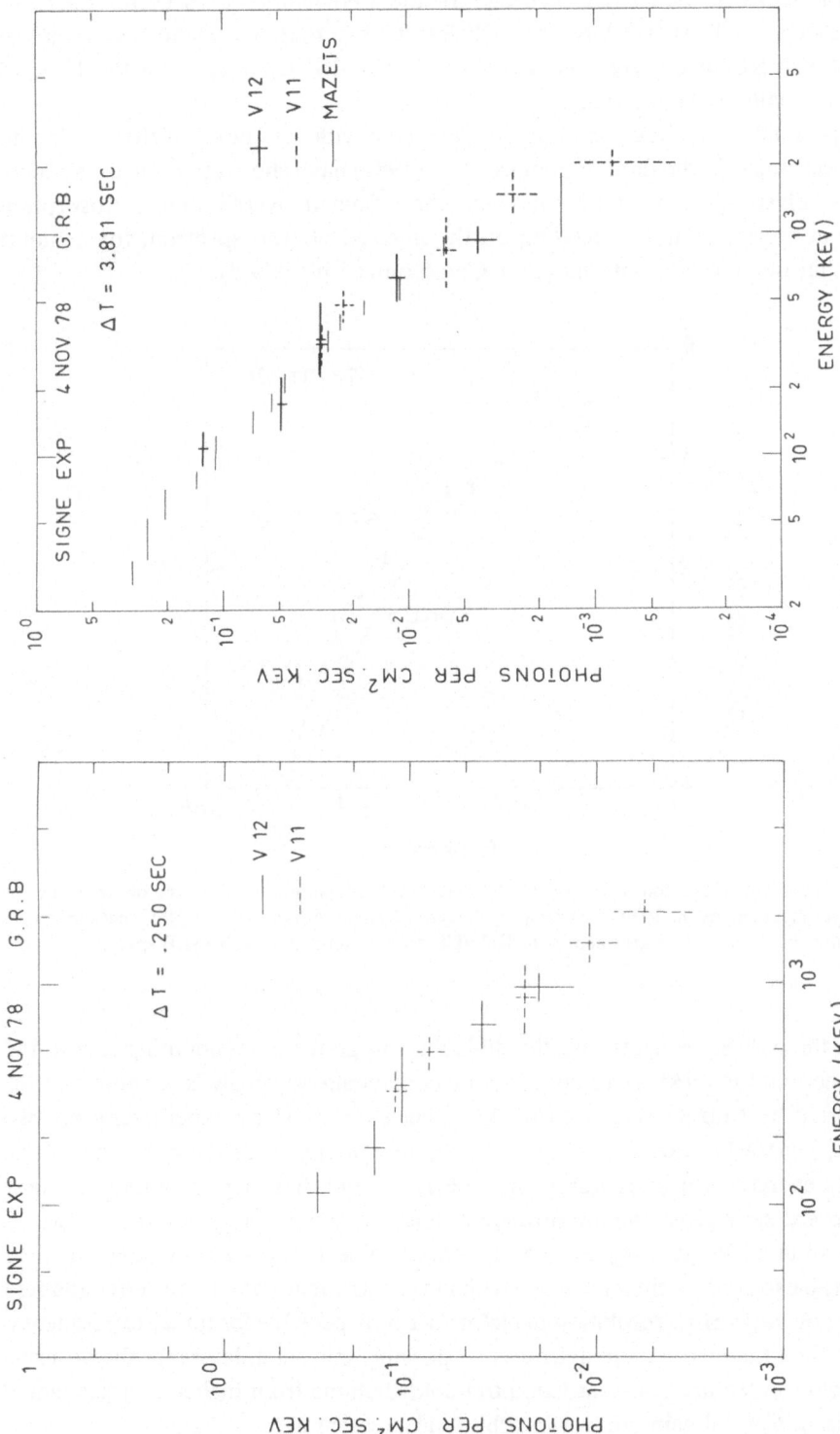

Fig. 3b.   The integrated spectrum during the first 3.8 s of this event does not
show a similar feature.

Fig. 3a.   The observed spectrum of the 4 November 1978 gamma-ray burst
integrated over 0.25 s during one of the peak of the event. Note the bump
at about 500 keV (C. Barat, 1983).

examine the KONUS and ISEE-3 results for the 4 November 1978 event (Figure 4) (Fenimore *et al.*, 1982a) (we have nevertheless to say that our Signe results are in agreement with Konus one for this event) and SMM–ISEE-3 results for the 19 April 1980 burst (Fenimore *et al.*, 1982b).

The appearance of features can be also connected with the choice of the continuum photon spectrum to fit the data. Fenimore *et al.* (1981) have shown that the data points in a deconvolved spectrum may vary with the assumed overall shape. Absorption features can appear or not depending on the injected photon spectrum for instance optically thin thermal bremsstrahlung or Comptonized blackbody.

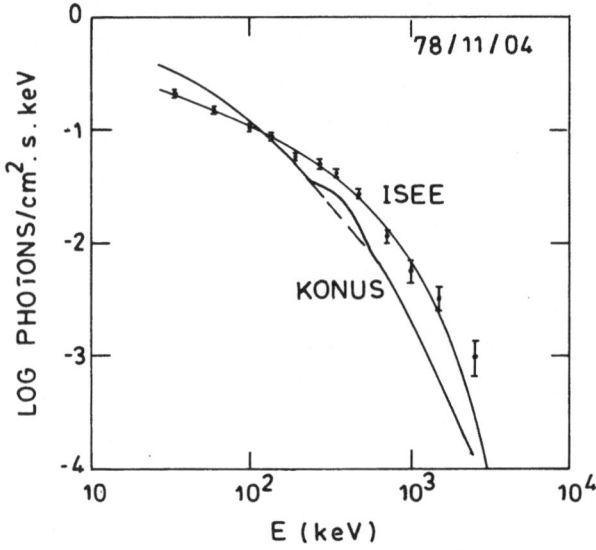

Fig. 4. A comparison of the continuum of the 4 November 1978 gamma-ray burst as observed by the KONUS experiment and by the ISEE-3 gamma-ray detector (Fenimore *et al.*, 1982a). Note that our results (Figure 3b) show better agreement with KONUS observations than with ISEE spectrum.

Despite these disagreements and the difficulties to go from the counting rate to the photon spectrum the presence of lines is quite convincing especially is we note that the lines reported by Mazets *et al.* (1981c) have been observed by experiments on two spacecraft, Venera 11 and 12, that they almost always appear at the beginning of the events with energies which are not always the same, and that they are absent in many bursts. Nevertheless given the importance of lines in gamma-ray burst spectra which constitute an unmistakable signature of a compact object with a strong magnetic field – it is desirable to confirm their presence with new experiments which are better adapted to this type of work. High resolution detectors are indispensible for future experiments: these could be either germanium detectors, or possibly thin scintillators for the observation of cyclotron features, so that Compton contributions from higher energies would be minimised. Careful gain control in orbit is also essential.

C. *Distribution of Bursts*

The distribution of gamma-ray bursts is important for testing their spatial origin, i.e., for determining whether they are galactic or extragalactic. In the absence of a significant number of localized gamma-ray bursts, the first method used was to establish the $\log N - \log S$ curve; but this classic method must be used with caution for gamma-ray bursts, which are highly variable emissions. It is difficult to know whether the total energy received from a burst should be used, or the average power, as suggested by Mazets after verifying that durations and total energies were related (Mazets *et al.*, 1981b).

In any case, the $\log N - \log S$ curve, in its classic form, if often used to attempt to determine the origin of bursts (Fishman, 1979; Jennings and White, 1980; Jennings,

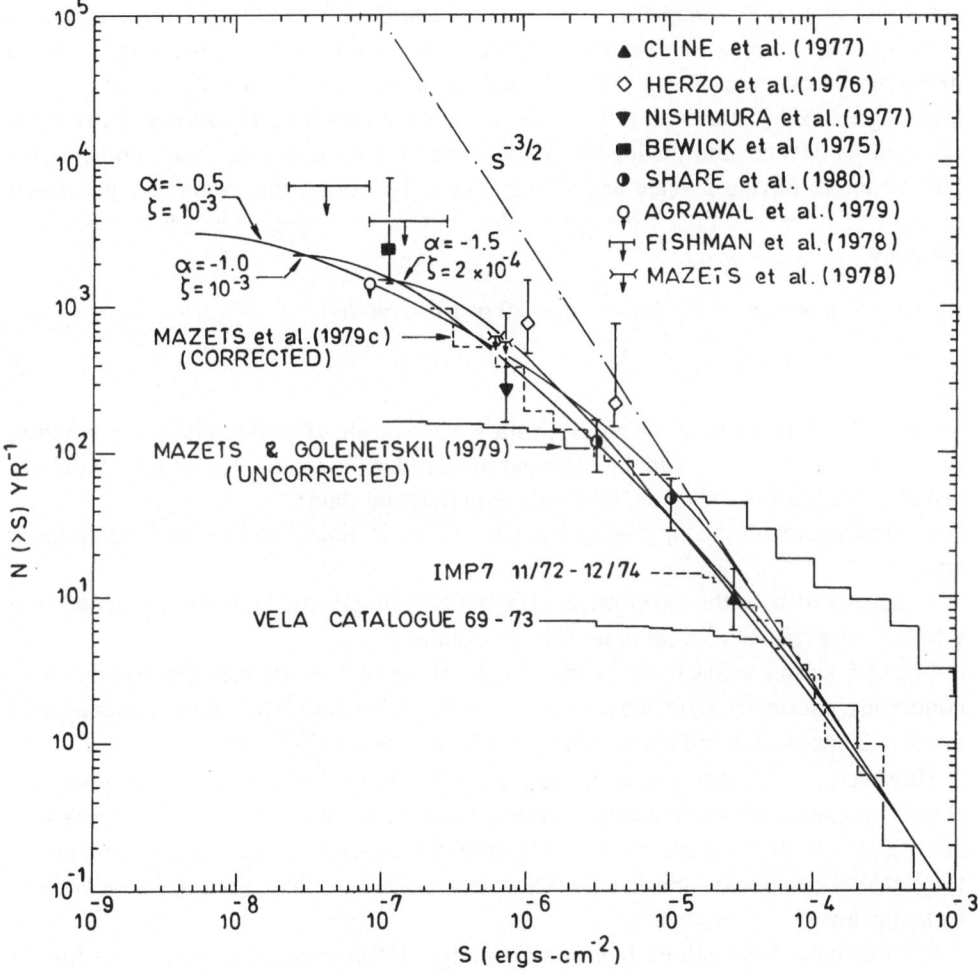

Fig. 5.   The $\log N - \log S$ curve for gamma-ray bursts (Jennings, 1982). Theoretical results for the two halo models are also shown. Note that the Mazets results are reported but not used for fitting the models.

1982; Mazets *et al.*, 1981a; and Yoshimori, 1978). The most recent calculations, based on both satellite observations and balloon results using large area detectors for weak bursts, have been carried out by Jennings and White (1980) (Figure 5); they study both disc and halo models. An extragalactic origin can be excluded by the fact that such a distribution should give a marked concentration of bursts from the direction of the Virgo Cluster, which represents the largest nearby extragalactic mass.

The effect of a variable luminosity from burst sources was controversial (Fishman, 1979; Yoshimori, 1978). However, Jennings (1982) recently demonstrated that it plays an important role; in particular, distributions which are moderately peaked toward low luminosity and with a dynamic range 10^2 have a strong effect on the $\log N - \log S$ curve. These results therefore modify the conclusions concerning the galactic monoluminosity gamma-burst model of Jennings and White (1980) because they permit spherical halo models, which had been rejected in the monoluminosity case. Presently the author concludes there is no compelling observational evidence for a burst distance much less than 1–2 kpc. Limits are set for a source distance scale $r < 1.5$–2.5 kpc, with a perceived burst rate $R < 2$–$10 \, \mathrm{yr}^{-1}$ and a space density $n > 2$–$9 \times 10^{-11} \, \mathrm{pc}^{-3}$ (Jennings, 1982). Chambon, (1982), taking into account both the source distribution reported by Mazets and the $\log N - \log S$ curve, arrives at similar conclusions, with a disc structure having a scale height $H \gtrsim 3$ kpc. To obtain this result, a 5 parameter model was used to simulate the spatial distribution of sources. The source density is given by

$$n = n_0 \, e^{-|z - z_0|/H}, \qquad z_0 = 20 \text{ pc (disc model)}$$

$$n = n_0 \text{ (halo)},$$

limited by the structure of the galaxy, with a scale height H and a luminosity function $\psi(L) \propto L^{-\alpha}$ between L_1 and L_2. The parameters $(H, \alpha, L_1, L_2, n_0)$ of the model are tested for conformity with the following experimental data:

– agreement with the $\log N - \log S$ curve: a $\chi^2 < 20$ must be obtained for 13 points; and

– agreement with the experimental distribution in galactic latitude b, grouped into $6 \times 30°$ intervals: a $\chi^2 < 20$ must also be obtained.

Figure 6 shows projections in the (L_1, α) plane of the volumes defined by these conditions, parametrized by the ratio L_2/L_1, with H constant. The volumes overlap only for $H > 3$ kpc, which indicates either a thick disc or a halo structure.

However, a thick disc model appears more probable than a halo model, since:

(a) A departure from isotropy has been noticed for weak bursts. Selecting bursts with intensities $S < 10^{-5} \, \mathrm{erg \, cm}^{-2}$ (36 events) from the list of 70 bursts localized up to now, the distribution appears anisotropic. These events, being weak, are the ones most likely to be far away.

(b) For halo distributions Jennings (Jennings, 1982) imposes a small value for the average recurrence time for bursts $(T < 5 \text{ yr})$. Using localized bursts, we have tried to study the possible cases of recurrence, realizing that the results could be subject to some

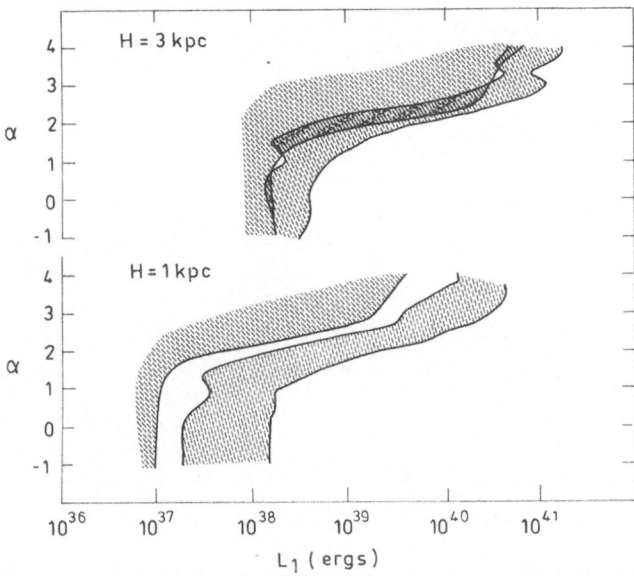

Fig. 6. Projected on the $(\alpha, L1)$ plane, the values allowed by the χ^2 test on the b distribution of localized gamma-ray bursts are denoted by the upper thatched region. The lower one denotes the values allowed by the χ^2 test on $\log N - \log S$ curve fit. The models are tested for two values of the scale height ($A : H = 3$ kpc, $B : H = 1$ kpc).

uncertainty: particularly because errors can exist for some localizations (Hurley, 1982b; Klebesadel, 1982b). Error boxes were looked for which overlapped once or twice; the total number of overlapping positions was compared with the number expected on a random basis. For 385 d of measurements, we found 5 ± 3 cases which may be due to real recurrence (excluding the events of 5 March 1979 and 25 March 1979) (Figure 7). From this result, it is possible to obtain a lower limit for T, the average recurrence time between bursts; we find $T < 3$ yr with a probability of 5‰ and $T < 5$ yr with a probability of 15%. This seems to exclude a halo model, as it was presented by Jennings (1982b).

Concerning the scale height H, note that the value of H is strongly dependant on the reality of the γ burst spatial distribution which as we will see has been questionned and also on the $\log N - \log S$ curve. This curve has been reconsidered taking into account all the available results. The disagreement between the results of Mazets and those of IMP-7 (Cline *et al.*, 1976) and VELA, PVO, ISEE-3 (Strong *et al.*, 1974) which reached a factor of 10, can be eliminated if the measurement of the burst energy above a common threshold is properly taken into account (Chambon, 1982; Barat *et al.*, 1982a) (Figure 8). Jennings' calculations (Figure 4) (Jennings, 1982) excluded the Mazets results. Introducing them extends the S^{-1} region of the $\log N - \log S$ curve to higher energy bursts, and in our opinion the curve used by Jennings to evaluate the halo model should be reconsidered.

Let's go now to the distribution of bursts: a strange characteristic of this distribution which has already been noted (Vedrenne, 1981) is the anisotropy between Northern and

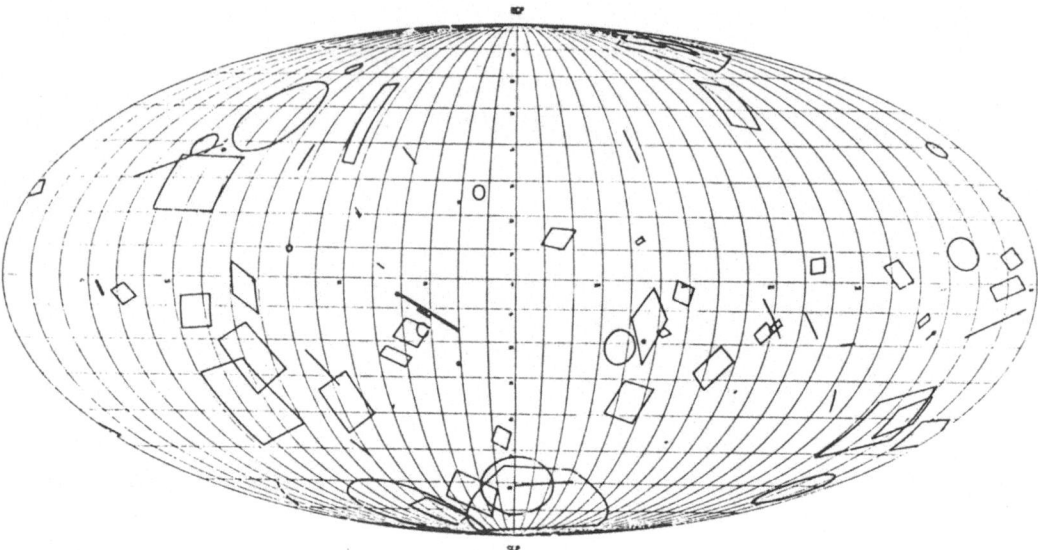

Fig. 7. The spatial distribution of all localized gamma-ray bursts in galactic coordinates with their error boxes. Data are from (Klebesadel *et al.*, 1982b; Mazets *et al.*, 1980).

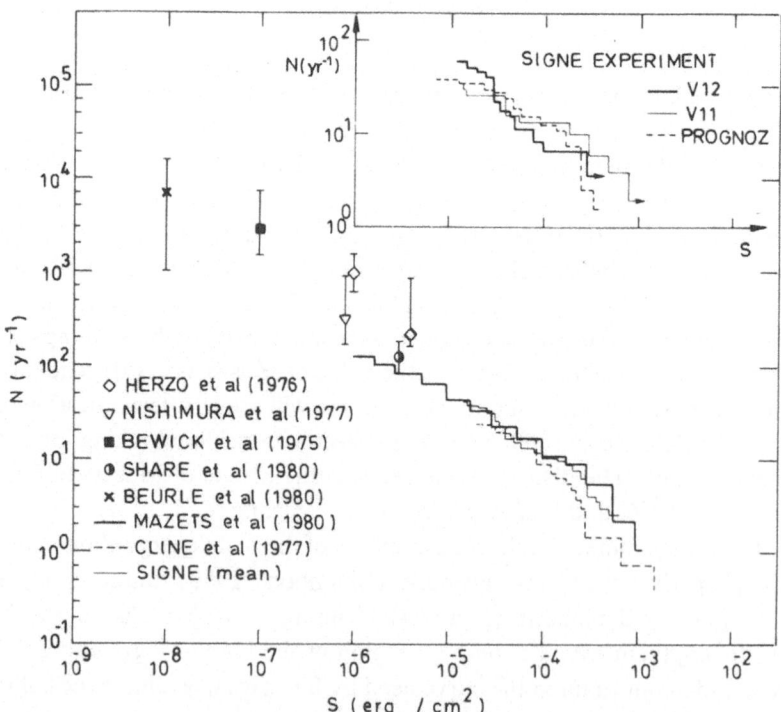

Fig. 8. $\log N - \log S$ curves of various experiments after renormalization (Barat *et al.*, 1982a). The disagreement noticed in Figure 5 has partly disappeared by using a common threshold for the measurement of the burst fluxes.

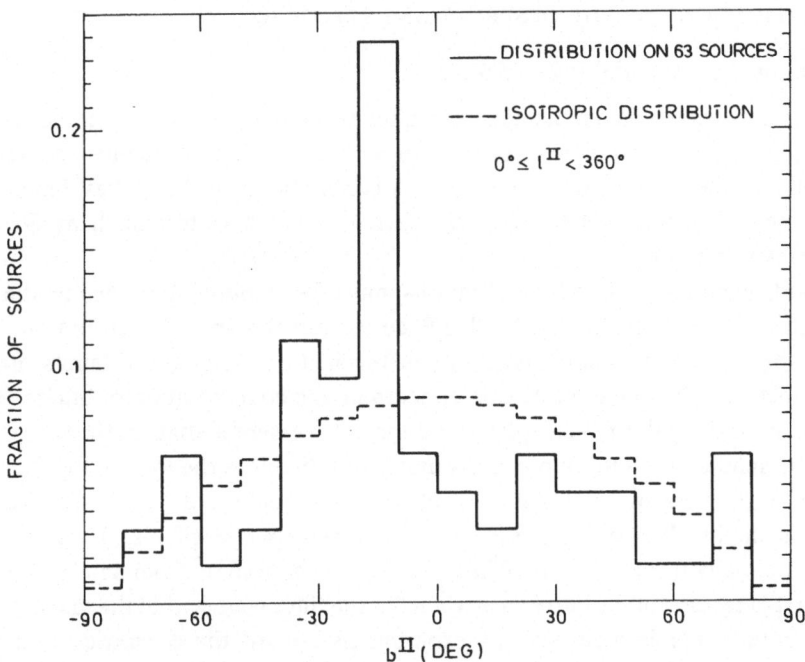

Fig. 9.   The galactic latitude distribution of bursts compared with the expected result for an isotropic distribution (dashed line) (Vedrenne, 1981).

Southern galactic Hemisphere (Figure 9), which cannot be explained by any class of objects. It was recently discussed by Laros *et al.* (1982) who showed that for the gamma ray bursts localized by Mazets, a relation exists between the right ascension of the events and the Earth to Venera vector. Taking into account this relation they tried to explain the anisotropy. But this geometrical effect has been explained recently by Mazets *et al.* (Mazets and Golenetskii, 1982) and their conclusion is that the sky coverage is truly uniform after 1 year measurement. It should be added, that using the events localized only by the international network (Klebesadel, 1982b) a difference appears between $0° < b < 30°$ and $0° > b > -30°$ for sources whose positions are determined using a method independent of that of Mazets. 6 of these sources appear between $0°$ and $30°$, while 11 are between $0°$ and $-30°$. These values have to be compared with Mazets' distribution: 22 sources between $0°$ and $-30°$, and 8 between $0°$ and $30°$. These results are not inconsistent but if we compare the sources localized in the northern hemisphere and the southern hemisphere, we have: for the international network, 14 sources in the northern hemisphere, and 17 in the southern hemisphere, while for the 50 sources localized by Mazets, 36 are in the southern hemisphere and 14 in the northern hemisphere. These results, and their differences, are still based on too few localizations to be completely significant. We will have to wait for the Venera 13 and 14 results, and those of the new international network (French Venera 13 and 14 experiments, PVO, SMM, ISEE, and possibly Vela) to re-evaluate these anomalies, and in any case to define the spatial distribution of bursts.

1.2. IMPLICATIONS OF THE OBSERVATIONS REPORTED

A. *From an Experimental Point of View*

The first conclusion which must be drawn after examining the main experimental results is that they are still insufficient, both from a point of view of quality and reliability, especially for the spectra at low energies. A confirmation of the 60 keV features must be obtained, as mentioned before, using better suited detectors such as Ge or high pressure Xe detectors.

Second, gamma-ray burst localizations must be continued to define the spatial distribution, since both the $\log N - \log S$ curve and the spatial distribution must be studied together before a conclusion can be reached on the origin of bursts. Presently, no argument can be made for certain concerning the concentration of bursts, even the weakest ones, along the galactic plane. It is therefore essential that sensitive experiments define the spatial distribution of weak bursts, and therefore the $\log N - \log S$ curve, for energies $< 10^{-7}$ erg cm^{-2}. Balloon results on non-confirmed gamma-ray bursts are always in doubt, due to the possibility of magnetospheric effects. These have been observed already from balloons (Cline *et al.*, 1977) as well as from satellites equipped with burst detectors, in the latter case when the satellites were within the magnetospheric boundary (e.g., the French SIGNE experiments aboard the eccentric earth orbiting Prognoz spacecraft, and the low altitude orbiting Signe III spacecraft). The northern–southern galactic hemisphere anisotropy must also be studied.

Finally, precise localization of gamma-ray bursts must be continued, as begun by the international network. This is the surest method of identifying bursts with objects at other wavelengths, and also of finding recurrent sources, and thus determining the recurrence rate. The present situation regarding optical identifications were treated by K. Hurley and T. Cline in recent review papers (Hurley, 1982b, c; Cline, 1982). It is clear that precisely localized bursts do not have positions consistent with cataloged objects. Nevertheless, from the results of deep optical searches it may be said that no gamma-ray burst error box is truly empty. Candidate optical objects are, however, very faint, leaving only low luminosity galactic objects as possibilities, if extragalactic objects are excluded. For example, for the 6 April 1979 event, the objects found in the error box are beyond 23rd magnitude (Chevalier, 1981), and could either be very faint stars or distant galaxies; a similar situation exists in other cases. Schaefer (1981) has recently reported a possible optical flash associated with the 19 November 1978 error box, found on a 1928 archival plate. The optical burst appears to have reached $m_B \simeq 3$ on about 1 s time scale. This result indicates that simultaneous optical and gamma-ray observations might be fruitful. A wide field camera network is now being planned to observe the sky simultaneously with the gamma-ray observations.

X-ray observations are also crucial, but the results are meager. Only for the 19 November 1978 has a marginal detection of a point X-ray source been reported, using the Einstein Observatory (G. Pizzichini *et al.*, 1981). Negative results were obtained for the 6 April 1979 and the 4 November 1978 burst.

The launch of EXOSAT will once more allow observations of precisely localized sources, to determine at least whether the emitting object has a steady-state emission associated with it, possibly due to continuous accretion of matter. This point is a very important one, especially for thermonuclear flash models, which require sufficient accretion rates (Ventura, 1982; Bonazzola et al., 1982). For three well localised bursts (4 November and 19 November 1978, and 6 April 1979), measures from the Einstein Observatory have put constraints on the temperature of the neutron star and on the accretion rate (Pizzichini et al., 1982).

Note also that the simultaneous observation of X-ray and optical emission at the time of the burst or very shortly after would be very important for determining the nature of the source. The confrontation of the X-ray and optical luminosities has been used by Grindlay et al. (1982) to explain the 19 November 1978 gamma-ray burst.

It would also be very useful to observe gamma-ray bursts at lower energies (1–10 keV) to study the possible relation between gamma- and X-ray bursts. A recent result concerning the association of a long, recurrent (on a 100 s time scale) X-ray burst with a gamma-ray burst might indicate a relation between these two phenomena (Terrell et al., 1982). Even if only one gamma-ray burst has been associated with a steady X-ray source, it is not impossible that X-ray sources can give rise to gamma-bursts although it is obvious that GRB sources are not strong X-ray sources. The average distance to X-ray sources (several kpc) might explain the apparent absence of gamma-ray bursts from these sources: present instrumentation might not be able to detect gamma-ray bursts at these distances. Nevertheless this argument would not hold if the scale height for the gamma ray bursts detected were several kpc, as we have proposed.

B. *From a Theoretical Point of View*

Here, there is general agreement, based on all the results known up to now, that gamma-ray bursts are certainly associated with compact objects such as neutron stars (Mazets, 1981a; Colgate and Petschek, 1981; Lamb et al., 1973; Hoshi, 1976; Takahara and Hoshi, 1978; Harwit and Salpeter, 1973; Woosley and Taam, 1976; Woosley and Wallace, 1981; Fabian et al., 1976; Bisnovatyi-Kogan and Chechetkin, 1980; Tsygan, 1975; Pacini and Ruderman, 1974; Brecher, 1982).

These neutron stars, with the proposed scale height of several kpc, would be old objects ($> 10^8$ yr). The large amount of energy liberated must be conserved as radiation and not dissipated in other forms by the expansion of the hot gas, so a large magnetic field is needed. Have these old objects conserved a strong magnetic field, as observations of cyclotron lines, seem to indicate? The question of the decay of the magnetic field has been extensively debated (Lamb, 1980) and a conclusion given by the author is: the accreting neutron stars, including Her X-1, which may be as old as 10^8 yr, seem to have large dipole magnetic moments, and this argues against the universality of rapid magnetic field decay.

Besides processes of accretion of matter (from a companion, interstellar medium or comet) the liberation of a large quantity of energy (10^{37-38} ergs for an object at 100 pc) could be the result of a change of structure of a neutron star [(starquakes: Fabian et al.,

1976; Bisnovaty-Kogan and Chechetkin, 1980; Tsygan, 1975); (vibrations: Ramaty *et al.*, 1981); (glitches: Brecher, 1982; Tsygan, 1975)].

In the case of slow accretion processes (low luminosity companion, interstellar medium (ISM), supernova remnant) nuclear energy would be liberated through a thermonuclear C or He flash, after non-explosive combustion of H (Woosley and Taam, 1976; Woosley and Wallace, 1982). Accretion from the ISM is compatible with the accretion rate given by Woosley and Wallace (1982) $10^{-13} M_{\odot}$ yr^{-1} if the medium is very dense ($n \simeq 10^3$ cm^{-3}). However no association has been found between γ bursts and dense clouds.

Another possibility is to consider the accretion on a neutron star in the interstellar medium but considering a slow relative speed for the neutron star; in this case accretion can be obtained at a good rate even without going through a dense cloud region. The critical accretion rate has been studied by Hameury *et al.* (1982) and by Bonazzola *et al.* (1982).

Nevertheless this attractive possibility is not without difficulties. The speed of rotation of the neutron star is a critical problem. Due to the centrifugal forces at the Alfvén surface the magnetosphere can be as a propeller expelling the matter which is approaching (Iliarionov and Sunayev, 1975). So the accretion can be strongly inhibited. In any case accretion in the interstellar medium is completely incompatible with the scale height we have tentatively determined.

Among other candidates one is particularly attractive: a low mass X-ray binary containing an accreting neutron star with a low mass optical companion: a dwarf star. The difficulty of this model is connected with the quite low luminosity of the optical candidates found in the error boxes of γ ray bursts, particularly if the distance of the burst is only 300 pc. One solution to this problem which does not involve increasing the distance to the source is to consider a companion with a very low mass ($< 0.1 M_{\odot}$) star; the behaviour of this system has been studied by Ventura *et al.* (1982) and Ventura (1982b).

Finally it should be noted in these models with an accreting compact object (Woosley and Wallace, 1982; Hameury *et al.*, 1981; Ventura, 1982a) that accretion of matter above a minimum rate is necessary, and that this imposes a permanent, low luminosity for the objects: so the absence of steady X-ray sources established by the Einstein Observatory for some of the precisely localized error boxes of the international network, especially 6 April 1979, and 4 November 1978, can be used to exclude some of these models (Ventura, 1982a) if the distance of the source is less than 500 pc. But another possibility which cannot be ruled out is a highly variable accretion rate. In this case optical and X-ray observations of the same region must be repeated.

The key to these problems is therefore to be found in the accumulation of more results, especially on the fine spectrometry of bursts, and the continuation of precise localization by the triangulation method, which have proven to be very fruitful. Correlated observations in the X- and optical domains must be done, with a delay as short as possible after the gamma burst has been localized.

Acknowledgement

The editors thank G. Pizzichini for assistance in evaluating this paper.

References

Barat, C.: 1983, *Workshop on e⁺ e⁻ Pairs in Astrophysics*, January 1983, GSFC.

Barat, C., Chambon, G., Hurley, K., Niel, M., Vedrenne, G., Estulin, I. V., Kurt, V. G., and Zenchenko, V. M.: 1979, *Astron. Astrophys.* **79**, L24.

Barat, C., Chambon, G., Hurley, K., Niel, M., Vedrenne, G., Estulin, I. V., Kuznetsov, A. V., and Zenchenko, V. M.: 1981a, *Space Sci. Instr.* **5**, 229.

Barat, C., Chambon, G., Hurley, K., Niel, M., Vedrenne, G., Estulin, I. V., Kuznetsov, A. V., and Zenchenko, V. M.: 1981b, *Astrophys. Space Sci.* **75**, 83.

Barat, C., Chambon, G., Hurley, K., Niel, M., and Vedrenne, G.: 1982a, *Astron. Astrophys.* **109**, L9.

Barat, C.: 1982b, in *Symposium on Gamma Ray Astronomy in Perspective of Future Space Experiments*, Ottawa, May 1982, 2–4–4.

Bisnovatyi-Kogan, G. S. and Chechetkin, W. M.: 1980, preprint Inst. for Space Research Acad. Nauk SSSR 561.

Bonazzola, S., Hameury, J. M., Heyvaerts, J., and Ventura, J.: 1982, in *Proc. Workshop Accreting Neutron Stars MPI*, Garching, July 1982, p. 241.

Brecher, K.: 1982, in *Gamma Ray Transients and Related Astrophysical Phenomena*, AIP Conference Proceedings No. 77, AIP, New York.

Bussard, R. W. and Lamb, F. K.: 1982, in *Gamma Ray Transients and Related Astrophysical Phenomena*, AIP Conference Proceedings No. 77, New York, p. 189.

Chambon, G.: 1982, 'Etude des sursauts gamma dans le cadre du programme SIGNE', Doctorat thesis No. 1045, UPS Toulouse, France.

Chevalier, C., Ilovaisky, S., Motch, Ch., Barat, C., Hurley, K., Niel, M., Vedrenne, G., Laros, J., Evans, W., Fenimore, E., Klebesadel, R., Estulin, I. V., and Zenchenko, V.: 1981, *Astron. Astrophys. Letters* **100**, L1.

Cline, T. L.: 1982, in *Symposium on Gamma Ray Astronomy in Perspective of Future Space Experiments at the XXIV COSPAR*, Ottawa 2–4–2.

Colgate, S. A. and Petschek, A. G.: 1981, *Astrophys. J.* **248**, 771.

Cline, T. L. and Desai, U. D.: 1976, *Astrophys. Space Sci.* **42**, 17.

Cline, T. L., Desai, U. D., Klebesadel, R. W., and Strong, I. B.: 1973, *Astrophys. J. Letters* **185**, L1.

Cline, T. L., Desai, U. D., Schmidt, W. K. H., and Teegarden, B. J.: 1977, *Nature* **266**, 694.

Dennis, B. R., Frost, K. J., Kiplinger, A. L., Orwig, Desai, U., and Cline, T. L.: 1982, in *Gamma Ray Transients and Related Astrophysical Phenomena*, AIP Conference Proceedings No. 77, New York, p. 153.

Desai, U. D.: 1981, *Astrophys. Space Sci.* **75**, 15.

Fabian, A. C., Icke, V., and Pringle, J. E.: 1976, *Astrophys. Space Sci.* **42**, 77.

Fenimore, E., Klebesadel, R. W., Laros, J. G., Stockdale, R. E., and Kane, S. R.: 1981, *Nature* **297**, 665.

Fenimore, E. E., Laros, J. G., Klebesadel, R. W., Stockdale, R. E., and Kane, S. R.: 1982a, in *Gamma Ray Transients and Related Astrophysical Phenomena*, AIP Conference Proceedings No. 77, AIP, New York, p. 201.

Fenimore, E., Klebesadel, R. W., and Laros, J. G.: 1982b, in *Symposium on Gamma Ray Astronomy in Perspective of Future Space Experiments*, XXIV COSPAR, Ottawa, 2–4–6.

Fishman, G. J.: 1979, *Astrophys. J.* **238**, 851.

Grindlay, J. E., Cline, T., Desai, U. D., Teegarden, B. J., Pizzichini, G., Evans, W. D., Klebesadel, R. W., Laros, J. G., Hurley, K., Niel, M., and Vedrenne, G.: 1983, submitted to *Nature*.

Hameury, J. M., Bonazzola, S., Heyvaerts, J., and Ventura, J.: 1982, *Astron. Astrophys.* **111**, 242.

Harwit, M. and Salpeter, E. E.: 1973, *Astron. J. Letters* **187**, L97.

Hoshi, R.: 1976, *Prog. Theoret. Phys.* **56**, 542.

Hueter, G. J. and Gruber, D. E.: 1982, in *Proc. Workshop on Accreting Neutron Stars MPI*, Garching July 1982, p. 213.

Hurley, K.: 1982a, *Workshop on Gamma Ray Transients and Related Astrophysical Phenomena*, La Jolla, August, 1981.

Hurley, K.: 1982b, in *Symposium on Gamma Ray Astronomy in Perspective of Future Space Experiments*, XXIV COSPAR, Ottawa, Pergamon Press.

Hurley, K.: 1982c, in *Gamma Ray Transients and Related Astrophysical Phenomena*, AIP Conf. Proc. No. 77, AIP, New York, p. 85.

Illarionov, A. F. and Sunyaev, R. A.: 1975, *Astron. Astrophys.* **39**, 185.

Imhof, W. L., Nakano, G. H., Johnson, R. G., Kilner, J. R., Reagan, J. B., Klebesadel, R. W., and Strong, I. B.: 1975, *Astrophys. J.* **198**, L7.

Jennings, M. C.: 1982, *Astrophys. J.* **258**, 110; also in *Gamma Ray Transients and Related Astrophysical Phenomena*, La Jolla, 1981, p. 107.

Jennings, M. C. and White, S.: 1980, *Astrophys. J.* **238**, 110.

Kane, S. R. and Anderson, K. A.: 1976, *Astrophys. J.* **210**, 875.

Kane, S. R. and Share, G. H.: 1977, *Astrophys. J.* **217**, 549.

Klebesadel, R. W., Fenimore, E. E., Laros, J. G., and Terrell, J.: 1982a, in *Gamma Ray Transients and Related Astrophysical Phenomena*, AIP Conference Proceedings No. 77, New York, U.S.A.

Klebesadel, R. W., Evans, D., Laros, J., Cline, T., Desai, U., Barat, C., Hurley, K., Niel, M., Vedrenne, G., Estulin, I. V., Kuznetsov, A. V., and Zenchenko, V. M.: 1982b, *Astrophys. J. Letters* **259**, L51.

Lamb, D. Q.: 1982, in *Gamma Ray Transients and Related Astrophysical Phenomena*, AIP Conference Proceedings No. 77, AIP, New York, p. 249.

Lamb, D. Q., Lamb, F. K., and Pines, D.: 1973, *Nat. Phys. Sci.* **246**, 52.

Lamb, F. K.: 1980, *Compact Galactic X-Ray Sources*, Lamb and Pines Ed.

Lamb, F. K., Pethick, C. J., and Pines, D.: 1973, *Astrophys. J.* **184**, 271.

Laros, J. G., Evans, W. D., Fenimore, E. E., and Klebesadel, R. W.: 1982, *Astrophys. Space Sci.* **88**, 243.

Loznikov, V. and Kuznetsov, A. V.: 1982, Acad. Sci. USSR, Preprint No. 743.

Mazets, E. P. and Golenetskii, S. V.: 1981, *Astrophys. Space Sci.* **75**, 47.

Mazets, E. P. and Golenetskii, S. V.: 1982, *Astrophys. Space Sci.* **88**, 247.

Mazets, E. P., Golenetskii, S. V., Ilinsky, V. N., Panov, V. N., Aptekar, R. L., Guryan, Yu. A., Proskura, M. P., Sokolov, I. A., Sokolova, Z. Ya., Kharitronova, T. V.: 1979, *Nature* **282**, 587.

Mazets, E. P., Golenetskii, S. V., Ilinsky, V. N., Panov, V. N., Aptekar, R. L., Guryan, Yu. A., Proskura, M. P., Sokolov, I. A., Sokolova, Z. Ya., and Kharitronova, T. V.: 1980a, Dokl. Acad. Sci. USSR No. 686.

Mazets, E. P., Golenetskii, S. V., Ilinsky, V. N., Panov, V. N., Aptekar, R. L., Guryan, Yu. A., Proskura, M. P., Sokolova, Z. Ya., Kharitronova, T. V., Diatchkov, A. V., and Khavenson, N. G.: 1980b, *Astrophys. Space Sci.* **80**.

Mazets, E. P. and Golenetskii, S. V.: 1981a, *Astrophys. Space Sci. Rev.* **1**, Sunyaev (ed.).

Mazets, E. P., Golenetskii, S. V., Aptekar, R. L., Guryan, Yu. A., and Illinsky, V. N.: 1981b, *Nature* **290**, 378.

Pacini, F. and Ruderman, M.: 1974, *Nature* **251**, 399.

Pizzichini, G.: 1980, *Bull. Am. Phys. Soc.* **12**(2), 448; also in *Gamma Ray Transients and Related Astrophysical Phenomena*, AIP Conf. Proceedings No. 77, New York.

Pizzichini, G.: 1982, in *International Symposium and Workshop on Accreting Neutron Stars*, Max Planck Inst., Garching, June 1982, p. 237.

Pizzichini, G., Danziger, J., Grosbol, P., Tarenghi, M., Cline, T. L., Desai, U. D., Mushotsky, R., Teegarden, B. J., Evans, W. D., Klebesadel, R. W., Laros, J. G., Barat, C., Hurley, K., Niel, M., Vedrenne, G., Esulin, I. V., Mersov, G., Zenchenko, V. M., and Kurt, V.: 1981, *Space Sci. Rev.* **30**, 467.

Ramaty, R., Lingenfelter, R. E., and Bussard, R. W.: 1981, *Astrophys. Space Sci.* **75**, 193.

Ruderman, M.: 1975, *Ann. New York Acad. Sci.* **262**, 164.

Schaefer, B. E.: 1981, *Nature* **294**, 722.

Share, G. H., Stricman, M. S., Kinzer, R. L., Chupp, E. L., Forrest, D. J., Ryan, J. M., Rieger, E., Reppin, C., and Kanbach, G.: 1981, *Proceedings 17th ICRC*, Paper XG-2-1-4, Paris.

Share, G. H., Kurfess, J. D., Dee, S., Chupp, E. L., Ryan, J. M., Forrest, D. J., Lanigan, J., Rieger, E., Kanbach, G., and Reppin, C.: 1982, in *Gamma Ray Transients and Related Astrophysical Phenomena*, AIP Conf. Proceedings No. 77, AIP, New York, p. 45.

Strong, I. B., Klebesadel, R. W., and Olson, R. A.: 1974, *Astrophys. J. Letters* **188**, L1.

Takahara, F. and Hoshi, R.: 1978, *Prog. Theor. Phys.* **59**, 425.

Teegarden, B. J. and Cline, T. L.: 1981, *Astrophys. Space Sci.* **75**, 181.

Terrell, J., Evans, W. D., Klebesadel, R. W., and Laros, J. G.: 1980, *Nature* **285**, 383.

Terrell, J., Fenimore, E., Klebesadel, R. W., and Desai, U. D.: 1982, *Astrophys. J.* **254**, 279.

Tsygan, A. I.: 1975, *Astrophys. Space Sci.* **44**, 21.

Vedrenne, G.: 1981, *Phil. Trans. Roy. Soc. London* **A301**, 645.

Ventura, J.: 1982a, in *Symposium on Gamma Ray Astronomy in Perspective of Future Space Experiments*, XXIV COSPAR, Ottawa 2–4–3, Pergamon Press.

Ventura, J.: 1982b, in *Proc. Workshop on Accreting Neutron Stars*, MPI Garching, July 1982, p. 250.

Ventura, J., Bonazzola, S., Hameury, J. M., and Heyvaerts, J.: 1982, *Nature*, preprint.

Wood, K. S., Byram, E. T., Chubb, T. A., Friedmann, H., Meekins, J. F., Share, G. H., and Yentis, D. J.: 1981, *Astrophys. J.* **247**, 632.

Woosley, S. E., and Livermore, L.: 1982, in *Gamma Ray Transients and Related Astrophysical Phenomena*, AIP Conf. Proceedings No. 77, AIP, New York, p. 273.

Woosley, S. E. and Taam, R. E.: 1976, *Nature* **263**, 101.

Woosley, S. E. and Wallace, R. K.: 1982, *Astrophys. J.* **258**, 696.

Yoshimori, M.: 1978, *Australian J. Phys.* **31**, 189.